Saving Time

Saving Time

Discovering a Life
Beyond the Clock

Jenny Odell

THE BODLEY HEAD
LONDON

1 3 5 7 9 10 8 6 4 2

The Bodley Head, an imprint of Vintage, is part of the Penguin Random House group
of companies whose addresses can be found at global.penguinrandomhouse.com.

Penguin
Random House
UK

First published in the US by Random House in 2023
First published in the UK by The Bodley Head in 2023

www.penguin.co.uk/vintage

A CIP catalogue record for this book is available from the British Library

Hardback ISBN 9781847926845
Trade paperback ISBN 9781847926852

Printed and bound in Great Britain by Clays Ltd, Elcograf S.p.A.

The authorised representative in the EEA is Penguin Random House Ireland,
Morrison Chambers, 32 Nassau Street, Dublin, D02 YH68

Penguin Random House is committed to a sustainable future
for our business, our readers and our planet. This book is made
from Forest Stewardship Council® certified paper.

Design by Fritz Metsch

To my family,
broadly defined

I wish the idea of time would drain out of my cells and leave me quiet even on this shore.

—AGNES MARTIN, *Writings*

Contents

Introduction

A Message for the Meantime

At some point in the spring of 2019, I noticed that my apartment had received some unexpected visitors. They likely arrived not through the door but through the window, and for a long time, I didn't know they were there. It was not until I happened to see the fronds of a moss growing in a pig-shaped ceramic planter next to the window that I had any idea we'd been invaded.

The moss spores took up residence around a tiny bunny ear cactus a friend had given me for my birthday years earlier. I'd always hated how cold and damp the area next to my kitchen window was—it never gets any sun—and the cactus probably did, too, but the moss found it hospitable. It set about dividing, differentiating, grabbing hold of the potting soil with hairlike rhizoids and growing tiny green leaves. Then it sent up long, slender sporophytes, ready to continue the same process as its ancestors outside the apartment. Soon, a miniature forest was fairly bursting from the top of the pig.

Compared to vascular plants, mosses have a relatively unmediated relationship to water and air. In her book *Gathering Moss,* Robin Wall Kimmerer notes that the one-cell-thick "leaves" of mosses are like the alveoli of human lungs, given that they require moisture and come into direct contact with air. On Antarctica, which has no trees, scientists have used mosses in a similar manner to tree rings. The mosses are capable of "laying down a record" each summer because of the way their leaves take in chemicals from the surroundings, growing from their tips. Sitting in my kitchen, I obviously could not read the record this errant moss was writing, but it did at least tell me, *I'm alive.* And the next day: *Still alive.*

I reread *Gathering Moss* during the early lockdowns of the Covid-19 pandemic. Time felt frozen then, but the moss kept growing, both inside and outside the apartment—and the pandemic had shrunk the scale of my attention. I walked around Oakland like a pedestrian conspiracy theorist, peering at things from odd angles. Moss loved the interstices, which meant it was often exactly where I wouldn't think to look: between cracks in the walkway outside my apartment, between the asphalt of the road and the manhole cover, between the grocery store wall and the sidewalk, between bricks. I came to understand moss as the signature of water, both in placement and appearance, as it grew wherever water had collected in the past, but it also answered the rain in real time, expanding and turning greener within minutes of a light shower.

The moss made me consider both very short timescales—like the minute-to-minute changes in moisture, or the moment of a spore growing in my planter—and very long evolutionary timescales, as mosses were some of the very first plants to live on land. Yet both ends of the temporal spectrum were also reminders of how impossible it is to pinpoint a moment (a very human thing to want to do). On one end, for example, I found disagreement about when a moss spore has officially germinated. Is it when it has become swollen with water to a certain degree or when the germ tube forms and the cell wall ruptures? On the other end, the earliest mosses evolved from aquatic algae at some point hundreds of millions of years ago, but it would be absurd to try to identify the exact "moment" of this innovation or even the speciation of the guest on my windowsill.

This slipperiness easily spilled over into other questions. Could a moss be meaningfully separated from its surroundings? Was a moss spore considered alive? What about a moss that was frozen, like the one from Antarctica that was brought back to life after fifteen hundred years? Even outside extreme conditions, mosses complicate the idea of uniform time with the ability of some species to go dormant for more than a decade without water; under the right circumstances, they will revive. It was this very quality, as Kimmerer mentioned in a 2020 interview in *The Believer*, that made moss especially worth paying attention to during the Covid-19 pandemic. Noting that her students took inspiration from mosses' rootedness and dormancy, Kimmerer suggested that these plants could teach humans how to inhabit this historical moment.

The moss arrived in my apartment around the time I started thinking about this book. It was still growing when I finished. True, it will probably not live in that spot for five thousand years, as has one moss shelf on Elephant Island, in Antarctica. But in the long meantime, it has taken in three years of sunlight, breathed three years of air, and witnessed three years of me at the kitchen table. Like an emissary from somewhere outside clock time, it has populated my mind with questions of porousness and response, of inside and outside, of potentiality and imminence. Most of all, it has been a reminder of time: not the monolithic, empty substance imagined to wash over each of us alone, but the kind that starts and stops, bubbles up, collects in the cracks, and folds into mountains. It is the kind that waits for the right conditions, that holds always the ability to begin something new.

IMAGINE YOU'RE AT a bookstore. In one section are time management books that give advice for adapting to a general sense of time scarcity and a world always speeding up: either counting and measuring your bits of time more effectively or buying time from other people. In a different section, you find cultural histories of how we came to see time the way we do and philosophical inquiries into what time even is. If you're scrabbling for time and feeling burned out, which section would you turn to? It would seem to make sense to look in the first section, which is more directly concerned with everyday life and practical reality. Ironically, there never seems to be enough time to do something as idle as contemplate the very nature of time. But what I want to suggest is that some of the answers we might seek in the first section live in the second. That's because, without exploring the social and material roots of the idea that "time is money," we risk entrenching a language about time that is itself part of the problem.

Consider the difference between work-life balance and the notion of leisure outlined by the German-Catholic philosopher Josef Pieper in his 1948 book, *Leisure, the Basis of Culture*. In work, he writes, time is horizontal, a pattern of forward-leaning labor time punctuated by little gaps of rest that simply refresh us for more work. For Pieper, those little gaps are not leisure. True leisure, instead, exists on a "vertical" axis of time, one whose totality cuts through or negates the entire dimension of workaday time, "run[ning] at right angles to work." If such moments happen to refresh us for work, that is merely secondary. "Leisure does not exist for the sake of work," Pieper wrote, "however much strength it may give a man to work; the point of leisure is not to be a restorative, a pick-me-up, whether mental or physical; and though it give new strength, mentally and physically, and spiritually too, that is not the point." Pieper's distinction strikes an intuitive chord for me, as it probably does for anyone else who suspects that productivity is not the ultimate measure of the meaning or value of time. To imagine a different "point" means also imagining a life, identity, and source of meaning outside the world of work and profit.

I think the reason most people see time as money is not that they want to, but that they have to. This modern view of time can't be extricated from the wage relationship, the necessity of selling your time, which, as common and unquestionable as it seems now, is as historically specific as any other method of valuing work and existence. The wage relationship, in turn, reflects those same patterns of empowerment and disempowerment that touch everything else in our lives: Who buys whose time? Whose time is worth how much? Whose schedule is expected to conform to whose, and whose time is considered disposable? These are not individual questions, but cultural, historical ones, and there are few ways to liberate your time or anyone else's without considering them.

One of the lessons in a popular 2004 book called *In Praise of Slowness* is that employer and employee could both benefit from work-life balance, as "studies show that people who feel in control of their time are more relaxed, creative and productive." While I'm sure that everyone would enjoy some extra hours in the day, the reasoning here matters. As long as slowness is invoked merely to make the machine of capitalism run faster, it risks being a cosmetic fix, another little gap on the horizontal plane of work time. I'm reminded of an episode of *The Simpsons* where Marge gets a job at the nuclear plant and notices low employee morale. In a conversation with Mr. Burns, she points to one employee sobbing, another walleyed and pouring a drink, and a third polishing a gun while saying, "I am the angel of death. The time of purification is at hand." In an attempt to be helpful, Marge gamely suggests "Funny Hat Day" and piping in some Tom Jones. We then see the same three employees again: sobbing (while wearing a sombrero), drinking (in a moose hat), and walking offscreen while cocking the gun (in a propeller hat) as "What's New Pussycat?" plays in the background. "It's working!" says Mr. Burns (in a Viking-style hat with horns).

Just as I suspect we all want something more than funny hats, I doubt burnout has ever been solely about not having enough hours in the day. What first appears to be a wish for more time may turn out to be just one part of a simple, yet vast, desire for autonomy, meaning, and purpose. Even when external circumstance or internal compulsion forces you to live entirely on Pieper's horizontal axis—work and refreshment-for-more-work—it remains possible to harbor desire for the vertical realm, that place for the parts of our selves and our lives that are not for sale.

Nor has the clock, even if it runs our days and lifetimes, ever completely conquered our psyches. Under the grid of the timetable, we each know many other varieties of time: the stretchy quality of waiting and desire, the way the present may suddenly feel marbled with childhood memory, the slow but sure procession of a pregnancy, or the time it takes to heal from injuries, physical or emotional. As planet-bound animals, we live inside shortening and lengthening days; inside the weather, where certain flowers and scents come back, at least for now, to visit a year-older self. Sometimes time is not money but these things instead.

Indeed, it's this very awareness of overlapping temporalities that invites a deep suspicion that we are living on the wrong clock. Nothing in the horizontal realm can answer that more spiritual form of burnout: the simultaneous experience of time pressure and a growing awareness of just how out of joint the climate is. Even for a very privileged person who is isolated from the effects of climate change, toggling between a Slack window and headlines about a soon-to-be-uninhabitable earth produces, at the very least, a sense of dissonance and, at the very worst, a kind of spiritual nausea and nihilism. There is a lonely absurdity in the idea of racing against the clock at the end of time, as evidenced in a headline by the parody site Reductress: "Woman Waiting for Evidence That World Will Still Exist in 2050 Before She Starts Working Toward Goals."

At least in part, this absurdity stems from how hopelessly unrelated the two timescales seem. From where we stand, the processes of the planet seem to take place somewhere on the periphery of the clock and the calendar, outside human social, cultural, and economic time. Thus, as researcher Dr. Michelle Bastian puts it, "the clock can tell me whether I am late for work, [but] it cannot tell me whether it is too late to mitigate runaway climate change." Yet these two seemingly unbridgeable realms of experience—individual time pressure and climate dread—share a set of deep roots, and they have more in common than just fear. It was European commercial activity and colonialism that occasioned our current system for measuring and keeping time and, with it, the valuing of time as interchangeable "stuff" that can be stacked up, traded, and moved around. As I will expand on in chapter 1, the origins of the clock, calendar, and spreadsheet are inseparable from the history of extraction, whether of resources from the earth or of labor time from people.

In other words, someone who today struggles to reconcile time pressure with climate dread is dealing on both ends with an outcome of a distinct worldview, one that occasioned both the measurement of work time and ecological destruction for profit. In the body, chronic pain can result from an imbalance in a different place from where you feel it. While you can massage a painful spot and feel better for a few days, if the cause is repetitive stress, the real fix is usually to change what you're doing. In a similar manner, time pressure and climate dread, experienced as distinct forms of pain, stem from the same set of relationships in a larger "body," one distorted into an unsustainable posture after centuries of an extractive mindset. For this reason, being able to connect one's own personal experience of time to the experience of a collapsing climate clock is no mere mental exercise, but a matter of urgency for everyone involved. The only way to address the pain is to fundamentally change what we are doing. The earth, too, needs more than funny hats.

Part of that fundamental change has to do with the way we speak and think about time. While it's true that the clock doesn't determine the entirety of our psychological experience, the quantitative view of time that arose with industrialism and colonialism remains the lingua franca for time in much of the world. This creates challenges for trying to speak a different language, but it also shows how meaningful the effort can be. At an online event called "Is There Time for Self-Care in a Climate Emergency?"—a title that implies no small amount of confusion and shame—I witnessed an example of this challenge. Minna Salami, the author of *Sensuous Knowledge: A Black Feminist Approach for Everyone*, was ultimately able to answer the titular question only by refusing its premise. Obviously self-care was necessary, but the way the question was posed was part of the problem, since it upheld the idea that everyday cultural time and ecological time are unrelated. If we merely saw self-care as "stealing little moments in which we can prioritize the self," imagining that self-care and climate justice would vie for our hours and days in a zero-sum game, we'd be furthering the problem by speaking that old lingua franca. For Salami, it couldn't be a question of either-or. Instead, learning to speak a different language about time would bring climate justice and self-care together into the same effort.

In Ancient Greek, there are two different words for time, *chronos* and *kairos*. *Chronos*, which appears as part of words like *chronology*, is the realm of linear time, a steady, plodding march of events into the future. *Kairos* means something more like "crisis," but it is also related to what many of us might think of as opportune timing or "seizing the time." At the climate event, Salami described *kairos* as qualitative rather than quantitative time, given that, in *kairos*, all moments are different and that "the right thing happens at the right point." Because of what it suggests about action and possibility, I too have found the distinction between *chronos* and *kairos* to be crucial when it comes to thinking about the future.

On the surface, it might seem that stable *chronos* is the realm of comfort and unstable *kairos* is the place for anxiety. But what comfort can *chronos* give when we are, in the words of the 1990s antiwork magazine *Processed World,* "marching in lockstep toward the abyss"?* What I find in *chronos* is not comfort but dread and nihilism, a form of time that bears down on me, on others, relentlessly. Here, my actions don't matter. The world worsens as assuredly as my hair is graying, and the future is something to get over with. In contrast, what I find in *kairos* is a lifeline, a sliver of the audacity to imagine something different. Hope and desire, after all, can exist only on the differential between today and an undetermined tomorrow. It is *kairos* more than *chronos* that can admit the unpredictability of action, in the sense that Hannah Arendt describes it: "The smallest act in the most limited circumstances bears the seed of the same boundlessness, because one deed, and sometimes one word, suffices to change every constellation." In that sense, the issue of time is also inextricable from the issue of free will.

This book grew out of my feeling that a significant part of climate nihilism and other painful experiences of time comes from an inability to recognize or access that fundamental uncertainty that lives at the heart of every single moment, where our agency also lives. In the context of climate, this is not to say that we can undo the damage that we have already ensured. But a foregone conclusion is self-fulfilling: In any situation, if we believe the battle is over, then it is. The difference between perceiving *chronos* and perceiving *kairos* may begin in the conceptual realm, but it doesn't end there: It directly affects what seems possible in every moment of your life.

It also affects whether we see the world and its inhabitants as living or dead-alive. This is perhaps the most far-reaching consequence of the idea that (European) Man is the sole mover and shaker in a natural world that lives by predictable, mechanistic laws. When it emerged, this distinction relegated colonized people to a kind of permanent stasis within *chronos,* the same agency-less category as their lands and all the other life within them. This conception not only justified colonists' exploitation of these "resources" but also set the stage for both the climate crisis and the racial injustices of today. To (re)learn to see action and decision outside

---

*I will return to *Processed World* in chapter 6.

such a narrow realm—to admit that everything and everyone previously left out of the picture is equally real, in *kairos* together—is to see time not as *happening to* objects in the world, but as being *co-created with* the actors of the world. To me, this is an issue both of justice and of practicality, as I read the climate crisis as the expression of beings (human and nonhuman) that need not be so much "saved" as heard.

Initially, I set out to try to find a conception of time that wasn't painful—something other than time as money, climate dread, or fear of dying. It was more of a personal question than an academic one. In my search, I found something I didn't expect: that while one temporal sense can make you feel dead before your time, a different one can make you feel undeniably alive. During the Covid-19 pandemic, I witnessed several processes of shedding: on a local webcam, where baby falcons, still mostly gray and fluffy, were growing individual feathers on the very tips of their wings, like fingers; on an Oakland hillside, where I found a snakeskin, its owner having disappeared into the brambles; on my desk in my apartment, where the leading edge of a plant stem peeled itself back to allow a new section to stretch toward the window. There was something difficult and self-defying about these processes of shedding, I thought. And this quality was also mine to claim, given that I, too, had desires to follow, a will to express, and a container to supersede. Tomorrow was growing raw out of the husk of today, and in it, I'd be different. All of us would.

IN 2021, TROPICFEEL, a Barcelona-based shoe company, asked British travel influencer Jack Morris to "Say Yes" and go on a last-minute adventure somewhere in Indonesia. The result is an eight-minute video called "Say Yes to Climbing an Active Volcano!" Morris chose to watch the sunrise from a volcano called Ijen in East Java, documenting an adventure that is also product placement for a company whose 2018 Kickstarter produced "the most funded shoe ever."

Through nostalgic color filters and yellow blips mimicking Super 8 film, we see Morris leaving Bali and taking a boat and a car to a resort in East Java. He strides confidently across a rice terrace in slow motion as the sun sets. The next day, in order to watch the sun rise from Ijen, he wakes up early and gets to a crowded, ramshackle coffee shop at the Ijen base camp. It's two A.M. After a woman in a head scarf, speaking in halting English, shows him a bread fryer, the camera pans to a man lying on a padded bench; he is wrapped in a scarf and a hoodie, his eyes closed.

"What about your friend? Is he sleeping?" Morris asks.

"Yeah, sleep, yeah," the woman says.

After an hour and a half spent hiking in the dark and then waiting, Morris does some more slo-mo striding at the top of the volcano. A drone-mounted camera takes it all in, exposing the vast, rocky landscape while mostly omitting Morris's assistant and the tourist crowd. It's just Morris, the mountains, and bright-white sneaker soles (Monsoon, in fresh black, $121) set off by layers of ancient rock. To emphasize the drama of the sunrise, this part of the video is set to a song I can only describe as epic and vaguely non-Western.

Once the sun is fully up, Morris heads down the mountain, where he runs across a group of Javanese sulfur miners. They are harvesting the yellow rocks from pipes that have been set into the volcano's seams, breaking the rocks up with hammers and transporting giant loads of them in wicker baskets attached to carrying poles. Chatting with the miners, Morris learns about the hundreds of pounds of sulfur they carry out of the crater every day—as much as they can, being paid by the pound. As another Super 8 effect flashes across footage of a man shouldering his baskets, Morris says, "It's so crazy. These guys are super strong." As his assistant loiters with a camera in the background, Morris observes that the miners seem proud of their job, and he leaves with "nothing but respect."

What Morris has seen is one of the last sulfur mines where mining is done by hand. It is one of the last, in part, because the sulfuric gases emitted at such a site are incredibly toxic, capable of dissolving teeth over time. With disfigured shoulders, respiratory conditions, and little to no protection, these miners carry out a harsh calculus: because the journey to the hospital is too long to be practical, they elect not to address ongoing injuries and, rather, work until they simply can't anymore. "They say working here can shorten your life," one miner told the BBC, and he was

right: By one journalist's account, the life expectancy of the sulfur miners is just fifty years. While many are working there in the hope that the
comparatively high wage can send their children to school and break the
cycle of poverty, this shortened life expectancy means their sons must
sometimes take up their jobs. In the meantime, the work renders their
faces "both young and old at the same time, worn out to the point of
total age ambiguity."

Inside this strange encounter among the travel influencer, the café,
the mountain, the miners, and the sun is a dense intersection of different
lenses on time. Several things are being extracted at Ijen—a marketable
picture of Nature, an experience of leisure, a bunch of sulfuric rocks—
and one of these things is labor time. Whether miners are paid by the
piece or by the hour, time for them is a wage, a means of survival, and
the most valuable thing they have to sell. The man trying to sleep in the
café might have been a miner, given that, like the hundreds of tourists
who climb the volcano each weekend during the high season, the miners
must also make predawn trips up the mountain from the base. They do
it out of necessity, to avoid the heat and the winds that might blow toxic
smoke toward them. While labor time is disembodied and uniform for
the buyer, who can always buy more, this is not the case for the laboring
person, who gets only one life and one body.

As the economic historian Caitlin Rosenthal has noted, the tools
that we would now call spreadsheets were used on colonial plantations
in America and the West Indies to measure and optimize productivity,
and they concerned work like sulfur mining—mindless, backbreak-
ing, repetitive. The labor hours recorded in these ledgers were as interchangeable as the pounds of tobacco or sugarcane being shipped away.
As it so happens, sulfur and sugar are linked at Ijen. Most of the sulfur being hauled out by the miners there is processed and sent directly
to local factories, where it's used to bleach and refine cane juice into
whitened granules of sugar—that commodity so intertwined with the
history of colonialism and European wealth. Ultimately, what describes
rock-as-commodity and sugar-as-commodity also describes labor-time-
as-commodity: In one sense, all are standardized, free-floating, infinitely

divisible. In another sense, they are indelibly linked to both human and ecological exhaustion.

Meanwhile, awake at night and running a business that caters to tourists, the woman at the café adjusts her sense of time to accommodate the temporal needs of people who arrive to consume the image of a sunrise. This phenomenon, in which one adapts her temporal rhythms to those of something or someone else, is called entrainment, and it often plays out on an uneven field of relationships that reflects hierarchies of gender, race, class, and ability. How much someone's time is valued is not measured simply by a wage, but by who does what kind of work and whose temporality has to line up with whose, whether that means rushing or waiting or both. Keeping this field in sight is all the more important amid exhortations to "slow down" for which one person's slowing down requires someone else to speed up.

"Slowness" is an ideal that often dovetails with leisure, and though he's technically working, Morris is performing leisure time in his video. Travel influencers are a keystone species in the experience economy, itself just one part of the elaborate relationship between leisure time and consumerism. When they coined the term *experience economy* in the 1990s, B. Joseph Pine II and James H. Gilmore were thinking of cut-and-dried examples like the Rainforest Cafe (a jungle-themed restaurant chain with animatronic crocodiles, fog machines, and simulated thunderstorms). Since then, Instagram has turned every corner of the world into a menu of backdrops and experiences. Now you can shop for life itself in a virtual mall where posts about self-care and retreat come across as ads for self-care and retreat. *Tap to add this to your life.* On the Tropicfeel website, you can find shoes, backpacks, and a sweatshirt similar to what Morris wears in the video. In this case, "shop the look" and "shop the experience" are even closer than usual.

In the experience economy, nature (and everything else) appears devoid of agency, a backdrop to be consumed. But the Ijen volcano sits uneasily in such a framing. It's alive. Its story began around 50 million years before Morris visited—when the Indo-Australian ocean plate collided with and then subducted beneath the Eurasian Plate. As the ocean plate melted, lava rose to the surface of the Eurasian Plate through a series of volcanoes that formed the islands of the Sunda Arc, of which Java is a part. A giant stratovolcano (now known as Old Ijen) formed, erupted, and collapsed, leaving an enormous caldera (depression) whose outline you can see on Google Earth. Inside that ancient caldera some smaller stratovolcanoes popped up, including modern-day Ijen. This, too, erupted and collapsed, creating a depression that filled with mete-oric water. When Ijen erupted in 1817, the depth of the caldera doubled, the soon-to-be-Instagrammable lake got bigger, and dead forests stood in twenty feet of ash. Meanwhile, sulfur that used to be part of the sub-ducted seafloor escaped—and is still escaping—through vents in the cra-ter and into the miners' pipes. At night, escaping sulfur gas reacts with the air and burns with a blue flame.

In 1989, Bill McKibben wrote, "I believe that we are at the end of nature." Then he clarified: "By this I do not mean the end of the world. The rain will still fall, and the sun will still shine. When I say 'nature,' I mean a certain set of human ideas about the world and our place in it." An active volcano provides as good an opportunity as any to consider "our place" and what it means to see "Nature" not as an object but as a subject, as something (someone) acting in time. The lava moves, and it's not because of us.

At the start of the Covid-19 pandemic, when the structure of my life was held constant, I began noticing changes that used to escape me: a hill slowly turning yellow; water carrying rocks down a hill; a buckeye branch budding, flowering, and dying. A red-breasted sapsucker regis-tered time each day by adding to a dense pattern of holes on the same tree, and the tree limb became like a calendar. The Mojave poet Natalie Diaz asks, "How can I translate—not in words but in belief—that a river is a body, as alive as you or I, that there can be no life without it?" What if these actions were not the mindless ticking of a clockwork universe, but the actions of a *who*? At the time, I was learning that whether you see an inert world or an agential one—whether something like Ijen is a pile of stuff or a subject deserving of regard—is an outgrowth of an age-old distinction about who gets to occupy time and who (and what) does not.

The second time I watched the Tropicfeel video, I used Shazam to look up the vaguely non-Western song that accompanied the sunrise. It turned out to be a track by Daniel Deuschle called "Rite of Passage," and it was also the number five suggestion in the Travel section of Musicbed, a website for licensable music. Deuschle's bio there reads, "Raised in Zimbabwe, Daniel Deuschle is a singer, songwriter, and producer. . . . He brings worlds together as he infuses the African sound into soaring melodies and heart-hitting progressions." I'm not suggesting that Morris (or whoever edited the video) consciously chose the song for its "African sound" or even listened to it that closely; they were just doing their job for Tropicfeel by speaking the dominant language and employing digestible clichés. Nonetheless, "Rite of Passage" suggests a certain exoticizing attitude toward the place that exists in tension with its reality. After Morris praises the miners, there's an uncomfortable moment where, seemingly unsure of how to move on from their plight, the video transitions from shots of the workers to slow-panning views of the hillsides. The miners dissolve into the landscape, timeless and inexplicable, like the sulfur itself.

But Morris, too, has to be marketable. When Instagram was in its early days, he was cleaning carpets in Manchester for minimum wage, and it was by reposting content from niche brands on an array of accounts that he made enough money to go backpacking. The personal account where he posted travel photos had been his "fun side project." By 2019, it was his job, an account with 2.7 million Instagram followers. He was dating another travel influencer; their popularity largely traded on their image as a carefree, jet-setting couple. But in 2021, a year after building a custom house in Bali, the two broke up. Once you know this, Morris seems sad or, at least, lacking in energy as he performs his volcano visit. "For more than a year I've had a creative mental block and didn't really have the motivation to pick up the camera," he wrote on Instagram as he began a solo trip through Egypt. "Creating wasn't fulfilling me the way it used to . . . prob because I was running around stressing over getting the perfect shot without ever actually experiencing the beauty right in front of me." A brand image can lead to its own kind of objectification, and he hoped things would be different in Egypt: "I really want to slow down to take in everything I see and do. Experience new things, learn, appreciate, and then take photos." It sounded like Morris had lost some of the "acquisitive mood" that Susan Sontag once associated with tourist photography. Instead, he was looking for encounter.

Morris's dutiful video made me think of the time I myself ascended a volcano to see the sunrise and why I didn't want to go. This was in 2014, and my family—whose origins are in another volcanic chain of islands, the Philippines—was in Hawaii, ostensibly for a wedding but also to check off a list of tourist activities. On Maui, one popular thing to do is get up early to see the sunrise from atop the island's volcano. Though I knew it would be beautiful, I felt the visit would mean little more than collecting a postcard. "Do we have to?" I whispered to my mom as we prepared to leave in the dead of night. The view from the car window was pitch black; I had no sense of where we were. When we arrived at a parking lot at the top of Haleakalā ("House of the Sun"), the restive crowd had already assembled, wrapped, inadequately, in towels and blankets that snapped in the cold, sharp wind.

Sometime around six A.M., the sun began to rise above an even, frothy layer of clouds surrounding the volcano. A fainter echo of that sunrise happened in front of me, as the glowing orange rectangles of camera screens went up and jostled for position, selfie sticks being used to photograph over other people's heads. My mom and I were sharing the same blanket, attempting to hold it closed against the wind. I felt an icy gust come in as my mom, somewhat sheepishly, raised her arm to take a photo.

The future is always over the horizon, and to be alive is to be in transit. For a few minutes, a sunrise collects all that ineffable bittersweetness into a single burning point. People (including my mom) can be forgiven for wanting to hold on to it in a photo. Outside the camera, though, sunrises escape. They show us that time is passing and the earth is turning, one of two times in the day at most latitudes when the light changes fast enough for us to perceive the shift. To watch it is to understand that while the sun rises every day, no individual sunrise will ever happen again. Each one gives us an image of renewal, return, creation, and a "new day," and it fleetingly repairs that Western rift between time and space—especially at Haleakalā, from which some people claim to be able to see the curvature of the earth.

My mother's photo

Had I tried to photograph the sunrise, though, it would not have captured what stands out to me most in that memory. More than the emergence of the blinding sphere, it was the sensation of my mother's small, warm body inside that blanket, how improbable I felt we were, how fragile—as if we could be blown away any moment. Haleakalā is one of two volcanoes that built the entire island of Maui, a series of kairotic events millennia ago that now gave us some *chronos* to stand on in the middle of a vast ocean. One hundred forty miles to the southwest of us, the seamount Kamaʻehuakanaloa was still forming, the latest creation of the Hawaii hotspot, the volcanic plume over which the Pacific Plate was passing.* I am not Hawaiian, and I have no claim to that place—to any

* Kamaʻehuakanaloa was formerly known as the Lōʻihi Seamount, a name applied in the 1950s based on its shape, as *lōʻihi* means "long" in Hawaiian. Since then, cultural practitioners and scholars have recovered traditional Hawaiian stories of Kamaʻehu, a reddish child of Kanaloa, the sea god, that may have referred to an undersea volcano. For example, one excerpt (*O ka manu ai aku laahia / Keiki ehu, kama ehu a Kanaloa / Loa ka imina a ke aloha*) was translated to "The elemental acrid aroma [of the volcano] / Is the predictor for an *ehu* child of Kanaloa / The wait to greet this new island is long." In 2021, the Hawaii Board on Geographic Names officially updated the name.

place, really. But there was something about my dual proximity to my mother and to this other, much larger body that reminded me of something: that it was not I who threw myself into time and not I who would accept myself when I ended. After the sunrise was "done" and everyone drove off the mountain, the earth would keep moving, Haleakalā eroding, Kamaʻehuakanaloa rising. Of all the senses of time I will describe in this book, this is the one I most want to "save": that restlessness and change that runs through all things, making them anew, rending the crust of the present like the molten edges of a lava flow.

THIS BOOK IS not a practical means for making more time in the immediate sense—not because I don't think that's a worthwhile topic, but because my background is in art, language, and ways of seeing. What you will find here are conceptual tools for thinking about what "your time" has to do with the time you live in. Rather than despairing at the increasing dissonance among clocks, between the personal and the seemingly abstract, between the everyday and the apocalyptic, I want to dwell in that dissonance for a moment. I started thinking about this book before the pandemic, only to watch those years render time strange for so many people by upending its usual social and economic contours. If anything good can come out of that experience, perhaps it is an expansion upon doubt. Simply as a gap in the known, doubt can be the emergency exit that leads somewhere else.

For all their variety, the lenses on time that I offer in this book cannot be effective in isolation. We also dwell in practical reality, and one of the challenges in thinking about any valuation of time as something other than money is that this thinking has to happen in the world as it currently appears. In turn, looking for *kairos* while living largely in *chronos* puts you in that difficult gray area between personal agency and structural limits, an area long explored by social theorists but also simply experienced by anyone negotiating life in a social world. Some of the most helpful articulations of this relationship I've come across are from Jessica Nordell's book, *The End of Bias: A Beginning.* Nordell writes that individual and institutional biases are inseparable because it's people who make the "processes, structures, and organizational culture" in which our decisions play out. At the same time, we are each in turn influenced by the culture in which we live. Nordell thus describes the effort to address bias without refashioning structures like policies, laws, and algorithms as "running up a down escalator." With something like racial and gender bias, the potential and responsibility for justice lie both inside and outside the individual.

In a similar manner, the personal and collective project of thinking about time differently has to go hand in hand with structural changes that would help to pry open space and time where now there are only cracks. That is why I consider this book only one portion of a conversation. My deepest hope is that it can combine with the work of activists and those who do write expressly about policy—like Annie Lowrey, who has written on topics like universal basic income and the "time tax" imposed on the poor; or Robert E. Goodin, Lina Eriksson, James Mahmud Rice, and Antti Parpo, whose detailed analyses of policies in different countries inform their concluding recommendations in *Discretionary Time: A New Measure of Freedom*.* As will be clear in chapter 5, on climate nihilism, I also want to be precise in my placement of responsibility on the fossil fuel industry for the running down of the climate clock. I have a hard time imagining that my attention to the flowering schedule of my local ecosystem would in any way move the needle on a company like ExxonMobil's desire to continue existing. For that reason, I see this book also in conversation with climate activists and writers on climate policy like Naomi Klein and Kate Aronoff.

Beyond that, there is an even more basic sense in which this book requires someone else. To speak another language about time, to eke out a space different from the dominant one, you need at least one other person. That speaking can invoke a world, perhaps one less characterized by a cruel, zero-sum game. Writers like Mia Birdsong have taught me the role of culture shift, something that exists on the everyday level of personal interactions and politics with a lowercase *p*. In *How We Show Up*, Birdsong writes that the American Dream exploits our fears, creat-

---

*In the book's final section, Goodin et al. emphasize the importance of work-time flexibility, equitable divorce rules, a culture of equality, and public transfers and subsidies. I will return to their concept of discretionary time in chapter 2.

ing real and imagined scarcity, and she calls for "accessible, celebrated models of what happiness, purpose, connection and love look like" that are different from what we are ordinarily taught.

You can see this work as emancipatory and utopian, or you can see it as simply filling in the gaps left by the erosion of services under neoliberalism. In fact, both can be true. The rise of mutual aid at the beginning of the Covid-19 pandemic in 2020 gives one example. All those Google Docs and spreadsheets were, on the one hand, a response to horrific gaps in the social safety net and, on the other, a concrete, living experiment in nondominant ideas of value, responsibility, kinship, and deservedness. Yes, it would be great if something like mutual aid were not needed in such a way. But it is, and besides the very real help it provides to people, it keeps alive and even advances these ideas in the broader culture. It is to that type of shift in what seems possible that I want this book to contribute. I offer these images, concepts, and places as provocations that can defamiliarize an old language of time while pointing in the direction of something else. For that reason, I hope they are in conversation with you as much as you might put them into conversation with others.

Sometimes the best muse is the thing you're so afraid of you almost cannot speak it. For me, that is nihilism. In *How to Do Nothing,* I quote the painter David Hockney on what he wanted to achieve in one of his many nonorthogonal, cubist-inspired collages: He called them "a panoramic assault on Renaissance one-point perspective." If I could borrow that phrase, this book is my panoramic assault on nihilism. I wrote it in an effort to be helpful, but toward the end, I felt I was writing it to save my life. As the largest gesture of hope I could muster, the following is intended as a future shelter for any reader who feels the same heartbreak as I.

In the conclusion of *This Changes Everything: Capitalism vs. the Climate,* Naomi Klein writes honestly about her own fear regarding the future and motions toward *kairos* as it relates to action. She recognizes "upwellings" and "effervescent moment[s]" in which "societies become consumed with the demand for transformational change." These moments often come as a surprise, even to longtime organizers—the surprise that "we are so much more than we have been told we are—that we long for more and in that longing have more company than we ever imagined."

She adds that "no one knows when the next such effervescent moment will open."

I reread these words in 2020, in the weeks following George Floyd's murder, which were full of such upwellings. For me, this time was an unforgettable illustration of the relationship among *kairos*, action, and surprise. Time took on new topographies, and the author Herman Gray contrasted "the slow time of COVID and the hot time of the streets." In a July 2021 podcast, Birdsong suggested that the pandemic had invited some amount of culture shift simply by exposing how connected people were to those they'd never thought about, like farmworkers and nurses. It changed how the world and the people in it looked, and it was in this opening that Floyd's death and the uprisings occurred. She suggested that here, in this specific moment, "there was a greater sense of connection . . . from people who hadn't previously felt any connection to black people being murdered." It was a reminder of what Rebecca Solnit repeats several times in *A Paradise Built in Hell: The Extraordinary Communities that Arise in Disaster:* "Beliefs matter."

In the midst of calls to "get back to normal," this book was written in *kairos* for *kairos*—for a vanishing window in which the time is ripe. In any moment, we can choose whom and what we perceive as existing in time, just as we can choose to believe that time is the site of unpredictability and potential rather than inevitability and helplessness. In that sense, changing how we think about time is more than a means for confronting personal despair in a catastrophic meantime. It can also be a call to action in a world whose current state can't be taken for granted any more than its actors can remain unnamed, exploited, or abandoned. I believe that a real meditation on the nature of time, unbound from its everyday capitalist incarnation, shows that neither our lives nor the life of the planet is a foregone conclusion. In that sense, the idea that we could "save" time—by recovering its fundamentally irreducible and inventive nature—could also mean that time saves us.

Saving Time

Chapter 1

Whose Time, Whose Money?

THE PORT OF OAKLAND

Time to me is about life-span and the ageing of individuals against the
background of the history of our world, the universe, eternity.
— DOMINIQUE, a schoolteacher interviewed in Barbara Adam, *Timewatch*

Moments are the elements of profit.
—A nineteenth-century British factory master, quoted in Karl Marx, *Capital*

*We've emerged westward from the Seventh Street tunnel into the Port of
Oakland, in a sun-blasted sedan I have had since high school. The clock
in this car went dark at some undetermined point long ago, but my phone tells
us it's seven A.M., eight minutes after the sunrise.*

*Ahead is a wide cement expanse punctuated by palm trees and pieces of
things: trucks without containers; containers without trucks; chassis, tires,
boxes, pallets. All of them lumped together, sometimes stacked, partitioned in
ways not immediately legible to us. A landscape of work. As the BART train
tracks and their chain-link fence disappear underground, soon to pass beneath
the San Francisco Bay, they reveal a different kind of train, double-stacked with
containers in serendipitous color combinations: white and gray, hot pink and
navy blue, bright red and dark, dusty red. There are a few indications of human
bodily concerns: a picnic table painted red, a portable toilet, an empty food
stand, and a vinyl ad for chiropractic services.*

*We pull in to Middle Harbor Shoreline Park, which is separated from the
SSA Marine terminal by a see-through fence. Just on the other side, the stacks are
six containers high, giving the impression of an endless city made of corrugated
metal. Farther ahead are the dinosaur-like figures: blue-green straddle carriers
and white shipping cranes, some of them sixteen stories tall. A massive ship sits*

underneath them, having arrived from Shenzhen. But, for now, the equipment is sleeping; the workers are just clocking in.

IN JULY 1998, the Italian National Institute for Nuclear Physics (INFN) decided to make its researchers start clocking in and out of the lab. They could not have known the backlash this would inspire, not only at the institute but also across the world. Hundreds of scientists wrote in support of the INFN physicists' complaints, saying that the move was need-lessly bureaucratic, insulting, and out of step with how the researchers actually worked. "Good science can't be measured by the clock," wrote the former director of the American Institute of Physics. A physics pro-fessor from Rochester University surmised that "the US garment indus-try must be advising the INFN on how to improve productivity." And the deputy director of the Lawrence Berkeley National Laboratory wrote in with cutting sarcasm: "Maybe they will chain you to your desks and benches next, so you do not go out after you come in, or better yet, install brain monitors to make sure you are thinking physics and not other thoughts while you are at your desks."

In a compilation of the letters written in response to the new policy, only a few express ambivalence over the scientists' protest. The most straightforward disagreement comes from Tommy Anderberg, a rare contributor with no listed professional affiliation. Instead, he identifies as a taxpayer and one who is angry about this kvetching by public employees:

> Your employers, in this case anyone paying taxes in Italy (the real thing, money derived from earnings realized in the private sector, not the piece of accounting fiction being applied to your own, tax-financed paycheck), have every right to demand that you be at your place of employment at the times stipulated by your contract.
>
> If you don't like your terms of employment, quit.
>
> In fact, I have a great suggestion if you want real freedom. Do what I did: start your own business. Then you'll be able to call your own shots and work when, where and with whatever you feel like.

At its heart, this disagreement—between the working scientists on the one hand and the INFN and Tommy Anderberg on the other—isn't just about what work is and how it should be measured. It's also about what an employer buys when they pay you money. For Anderberg, it's a package deal including not only work but also life minutes, bodily pres-ence, and humiliation.

As attested to by the scientists' wry jokes about factories and being "chained to a desk" (an image that comes up in several of the letters), the concept of clocking in and out comes from an industrial model of work. Probably one of the best illustrations of this model is the beginning of Charlie Chaplin's 1936 film, *Modern Times*. The very first image in the film is that of a clock—severe, rectangular, and filling the entire screen behind the title credits. Then a shot of sheep being herded fades into a view of workers exiting the subway and heading to work at "Electro Steel Corp.," where two very different kinds of time exist side by side.

The first is leisurely: The president of the company sits alone in a quiet office, halfheartedly working on a puzzle and glancing idly through a newspaper. After an assistant brings him water and a supplement, he pulls up closed-circuit camera views of various sections of the factory. We see his face appear on a screen in front of a worker in charge of the factory's pace. "Section Five!" he barks. "Speed her up, four one."

Chaplin's character, the Tramp, is now subjected to the second temporality—that of time as punishing and ever intensifying. On an assembly line, he frantically works to screw nuts into pieces of machinery, falling behind when he has to scratch an itch or is distracted by a bee hovering around his face. When his foreman tells him to take a break, he walks away jerkily, unable to stop performing the motions of his job. In the bathroom, the manic soundtrack briefly turns to reverie, and the Tramp calms down a bit, beginning to relish a cigarette. But all too soon, the face of the president appears on the bathroom wall: "Hey! Quit stalling! Get back to work!"

Meanwhile, the company tries out an inventor's time-saving device. It comes with its own recorded advertisement: "The Billows Feeding Machine, a practical device which automatically feeds your men while at work. Don't stop for lunch! Be ahead of your competitor. The Billows Feeding Machine will eliminate the lunch hour." On his break, the Tramp is picked out as management's guinea pig and strapped into what is essentially a full-body vise behind a rotating platter of foods. Things get out of hand when the machine malfunctions and the corn cob rotator starts going too fast, slamming the spinning cob into the Tramp's face over and over again.

I consider the corn cob malfunction one of the funniest movie moments I have ever seen. On the one hand, the scene is a joke about the capitalist's desire to scrimp and save on the labor time for which he has paid—to squeeze more work from the worker in the same amount of time. (If humans could just eat corn faster, the crazily spinning cob might not be a problem at all.) On the other hand, it's a joke about the human assimilated to a disciplinary pacing: Just as he must keep up with the assembly line and minimize bathroom breaks, he must also comply with the feeding machine's rate of food delivery. He must become an eating machine.

Time, in this world, is an input just like water, electricity, or corn cobs. A 1916 advertisement by the International Time Recording Company of New York in *Factory Magazine* addressed the head of the factory and made this connection explicit: "Time costs you money. You buy it just as you buy a raw material." In order to wring the most value from this time material, the employer resorts to surveillance and control. In a 1927 issue of *Industrial Management,* Calculagraph, another time recorder company, put it this way: "You pay them CASH! How Much TIME do They pay You?"

This final question makes sense only from the point of view of the factory owner, who is counting not just elapsed time, but time spent specifically producing value for him. The Tramp illustrates this distinction when he dutifully punches out in order to go to the bathroom and punches in again after the boss ends his break. Nor is this an exaggeration. In the history of work, things could get pretty granular: In the one hundred thousand words that make up the eighteenth-century rule book for the Crowley Iron Works, deductions from time paid included "being at taverns, alehouses, coffee houses, breakfast, dinner, playing, sleeping, smoking, singing, reading of news history, quarelling, contention, disputes, or anything forreign to my business, any way loytering [*sic*]." In other words, a more accurate ad for the Calculagraph might have asked, "How Much LABOR TIME do They pay You?"

This experience of time may sound antiquated, relegated to specific jobs in the industrial era. But time in low-wage workplaces still plays out in the dimension of intensity and control, now augmented by algorithmic sorting and faster processing. In her 2019 book, *On the Clock: What Low-Wage Work Did to Me and How It Drives America Insane,* Emily Guendelsberger describes this reality:

> Working in an Amazon warehouse outside Louisville, Kentucky, I walked up to sixteen miles a day to keep up with the rate at which I was supposed to pick orders. A GPS-enabled scanner tracked my movements and constantly informed me how many seconds I had left to complete my task.
>
> Working at a call center in western North Carolina, I was lectured about how using the bathroom too often is the same thing as stealing from the company and had the minutes I spent in the bathroom tracked in a daily report sent to my supervisor.
>
> Working at a McDonald's in downtown San Francisco, we were underscheduled to the point of a constant, never-ending line of customers—everyone worked at the frantic speed of those in-the-weeds waitresses of my youth all shift, nearly every shift.

A century after the Calculagraph exhorted factory owners to "be certain—know with precision the exact time on every job for every man—to the last minute," Guendelsberger's Amazon scanner gun dutifully performs this function down to the last second. Describing the meticulously oppressive design of the Amazon workplace, she refers to Frederick Winslow Taylor, the mechanical engineer who, in the early twentieth century, spurred the mania for breaking down industrial tasks into minutely timed segments: "My scanner gun is [Taylor's] vision incarnate—my own personal stopwatch and pitiless robo-manager rolled into one. . . . Would Taylor be horrified that his fears about the abuse of his ideas had come true? Or would he jizz in his pants?"

Meanwhile, a "robo-manager" of sorts has spread outside the workplace. Installed on workers' computers, employee tracking systems like Time Doctor, Teramind, and Hubstaff saw huge increases during the Covid-19 pandemic as more people began to work from home. While some systems use self-reporting, others monitor employees with keystroke logging, screenshots, continuous video recording, and OCR (optical character recognition) that allows employers to search for words within the employee's chats and emails. "Get the Most out of Your Employees' Time," reads the site for Insightful (formerly Workpuls), which offers employee tracking systems. "Time is money. Discover exactly what your employees are up to every minute of the day with all-seeing employee monitoring and complete behavior analytics." In a *Vox* article on remote work, a contractor for a translation agency based in Australia complained, "My manager knows every single damn thing I do. . . . I barely get to stand up and stretch, as opposed to when I am physically in the office." This uneasy awareness of management exhibits the dual function of workplace surveillance as both a spur and a disciplinary mechanism.

A 2020 *PCMag* review of employee tracking systems claimed that the systems' features promote productivity rather than policing. But the same review mentions that these systems set automated alerts and "compile employee transgressions into reports that can later be used to build disciplinary cases against employees." Perhaps this confusion arises because productivity and policing are two sides of the same coin. "Just because the employees know about the computer monitoring software, they'll be more focused," writes Insightful, "and you can rest assured that their attention is where it needs to be." The aptly named software StaffCop shows the employer a spreadsheet of minutes worked, with those minutes graded into five categories: Premium, Productive, Neutral, Unproductive, and Incident. While some surveillance is meant to prevent data leaks, the entire structure seems implicitly designed to render more of the time paid for as "Premium." StaffCop's website, for example, includes both "Productivity Optimization" and "Insider Threat Detection" in the same tagline.

When Microsoft rolled out individual-level productivity data for Office 365 in 2020,* the critic and novelist Cory Doctorow was quick to identify this as the "The Shitty Tech Adoption Curve," in which oppressive technology makes its way up the "privilege gradient": "Asylum seekers, prisoners and overseas sweatshop workers get the first version. Its roughest edges are sanded off against their tenderest places, and once it's been normalized a little, we inflict it on students, mental patients, and blue collar workers." Doctorow writes that remote work surveillance had already been used on at-home call center employees, who tended to be poor Black women. During the pandemic, this type of surveillance spread more widely to university students participating in remote learning and, finally, to white-collar employees working from home.

It's entirely possible that you work somewhere that affords you more trust and temporal latitude than what I'm describing. Even if that's the case, this standardized and often punitive form of time reckoning is relevant to you for several reasons. First, it characterizes the experience of many workers' time "on the clock" in the present, across many sectors of work, including those that undergird the everyday lives of others. But, more generally, it embodies aspects of standardization, intensification, and discipline that affect how many of us think of productivity and even the "stuff" of time itself.

A BLACK PHOEBE lands on the chain-link fence and looks back at us, flicking its tail. Behind it, the containers are emblazoned with names in different fonts: Matson, APC, Maersk, CCA CGM, Hamburg Süd, Wan Hai, Cosco, Seaco, Cronos. With the exception of some exactly half-size ones, the containers are all the same size and shape—one that became the standard in the 1970s because it made transport between land and sea easier and faster. Their sameness and

---

* After unveiling the Productivity Score in the fall of 2020, Microsoft faced significant pushback from critics concerned about user privacy. Their subsequent version of the Productivity Score no longer featured the ability to associate data with end-user names.

opacity take an unthinkable jumble—things like frozen chicken tenders, wax, peaches, yarn, microfiber towels, leggings, pumpkinseeds, and plastic forks— and make it uniform and legible. To this day, containers are made to specifications by the International Organization for Standardization.

TIME AS MONEY (in the most literal sense) represents what Allen C. Bluedorn calls fungible time, meaning that, like currency, it is consistent and can be endlessly subdivided. Measuring fungible time is like envisioning standardized containers that can potentially be filled with work; in fact, there is a strong incentive to fill these units of time with as much work as possible. As opposed to the duration of life or even the processes of the human body, one hour is meant to be indistinguishable from another— decontextualized, depersonalized, and infinitely divisible. In its most dehumanizing form, this view sees individual people as interchangeable, separate repositories of this usable time stuff: as Marx put it, "nothing more than personified labour-time."

The idea of fungible time as money is so familiar that it's easy to take for granted. But it combines two things that are not as natural as they've come to seem: (1) the measurement of abstract and equal amounts of time like hours and minutes, and (2) the idea of productivity that divides up work into equal intervals. Any system of time reckoning and any measurement of value reflects the needs of its society. In our system of standard time units, grids, and zones, for instance, one can still read the marks of the Christian, capitalist, and imperialist crucibles in which it was formed. Understanding the invention of modern mechanical clocks, the historian David Landes writes, means first asking who needed them.

The ancient world was full of constructed apparatuses for sensing time within a day: sundials, which used the movement of the sun; clepsydrae, which used the flow of water; and fire clocks, which used the burning of incense. Yet, for most of human history, there has been no need to divide the day into equal numerical units, much less to know the hour at any particular moment. For example, when in the sixteenth century an Italian Jesuit brought mechanical clocks to China—which had a long tradition of astronomical clocks driven by water, but did not organize life or work around anything more numerically specific than calendar dates— they were not embraced. Even in the eighteenth century, a Chinese reference book called Western clocks "simply intricate oddities, destined for the pleasure of the senses," objects that "fulfil[led] no basic needs."

The actual story of how measurable, countable equal hours came into existence is not a straightforward one. Landes suggests that a crucial deviation happened with the development of Christian canonical hours, particularly under the sixth-century Rule of Saint Benedict. The Rule, which subsequently spread to other orders, specified seven times during the day when Benedictine monks should pray, as well as an eighth in the middle of the night. Determining that "idleness is the enemy of the soul," the Rule also described punishments for monks who failed to hurry sufficiently upon the signal for work or prayer.* Five centuries later, Cistercian monks, for whom spiritual enterprise also meant economic enterprise, would intensify this temporal discipline. With their bell towers and smaller bells throughout the monastery, the monks' "temporal sensibilities" emphasized punctuality, efficiency, and the abil-

---

*In the section "On Those Who Come Late to the Work of God or to Table," punishments include being made to stand "in a place set aside by the Abbot for such negligent ones in order that they may be seen by him and by all," eating alone, and having a portion of one's wine taken away.

ity to "profit from this precious gift of time by ordering it and using it." At the time, monks regularly hired labor and ran the most efficient farms, mines, and factory-like enterprises in Europe.

Canonical hours are not equal hours, and the monks' bells were more of an alarm system than a clock. But some of them did use escapement-like designs—pendulum mechanisms rather than the passage of water. Landes calls it "an unintended consequence" that this technology, having developed in the monastery, caught on in a new context: the public and private clocks that spread as European towns centralized power and commerce. Once again, the bells were tools of coordination, but this time it was a bourgeois class that needed them. The clocks not only helped them conduct trade, but also marked the outer bounds of the day's worth of labor bought from workers with nothing but labor time to sell. Unlike the canonical hours of the Catholic Church, the hours marked by the new mechanical turret clocks were equal, countable, and easy to calculate. While capitalism did not itself create standard time units, they proved useful for imposing uniformity on workers, seasonal activities, and latitudes.

The separation of time from its physical context is preserved in our everyday speech. As John Durham Peters points out in *The Marvelous Clouds*, "o'clock" means "of the clock," as opposed to less artificial standards (e.g., the light at one's particular location). Observing clock time signaled a supposed domination over the natural world that was similar to other rationalist ideals, like the imposition of an abstract grid onto a decidedly diverse landscape. A clock hour was meant to be an hour, no matter where or what the season, just as a man-hour would be expected to be an hour, no matter who the man. This was as useful for regulating labor as it was for conquering land. Water clocks could freeze, and sundials could be unreadable on cloudy days, but a clock with an escapement would keep marking its intervals—and it could be miniaturized. It's no accident that the marine chronometer, a clock that could keep time at sea, came about in eighteenth-century Britain, a colonial power rising to international dominance. As we will see shortly, this technology allowed not only for navigation but for clocks and clock time to be exported overseas.

Because this kind of time has become so common, it's all too easy to think countries like Britain were the first to grasp a "more accurate" or "real" sense of time. Here again, I would stress that each development responded to some culturally specific "basic need." Just as there had been no need to know the hour of the day, there was no need to coordinate temporally across great distances until the advent of the British mail coach service and, later, the British railway. Starting in the 1850s, "master clocks" in Greenwich, England, exported Greenwich Mean Time (GMT) via electrical pulses to "slave clocks" across the country so that all the trains would run on the same schedule. The United States and Canada, by contrast, had railways but no time zones until 1883, and the two countries' railway systems suffered for it. Thus, an 1868 railway guide gives an exasperated acknowledgment before a "comparative time-table" that compares noon in ninety cities to noon at the center of power, Washington, D.C.:

> There is no "Standard Railroad Time" in the United States or Canada; but each railroad company adopts independently the time of its own locality, or of that place at which its principal office is situated. The inconvenience of such a system, if system it can be called, must be apparent to all. . . . From this cause many miscalculations and misconnections have arisen, which not infrequently have been of serious consequences to individuals, and have, as a matter of course, brought into disrepute all Railroad Guides, which of necessity give the local times.

Perhaps unsurprisingly, the person who proposed international time zones in 1879 was an engineer who had become a standard-time enthusiast while helping design the Canadian railway network. In his 1886 treatise, "Time-Reckoning for the Twentieth Century," Sandford Fleming imagined the exact opposite of local time: Namely, everyone on earth would observe a "Cosmic Day" within one of twenty-four time zones starting in Greenwich, England, where the prime meridian had been established a few years earlier. "The Cosmic Day is a new measure of time entirely non-local," he wrote. For Fleming, a "necessary connection between the numbers of the hours and the position of the sun in each local firmament" was something inconvenient and outdated.

Fleming also advocated for a twenty-four-hour clock, somewhat similar to what we now call "military time." So keen was he on this standardized time reckoning that he wanted everyone to attach a paper "supplementary dial" to their watches showing hours thirteen through twenty-four. "The committee is aware that these seem trifling matters," he wrote, ". . . but questions of great moment not seldom hinge on small details." While neither the twenty-four-hour clock nor Fleming's specific proposal for time zones was adopted at the International Meridian Conference in 1884, twenty-four international time zones were eventually established with Greenwich, England, at their center. In the current Coordinated Universal Time (UTC), Greenwich, England, is still at the center (UTC+0).

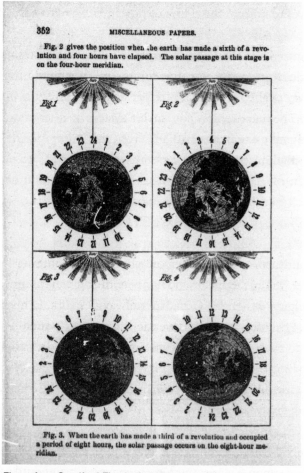

Figure from Sandford Fleming's 1886 report, "Time-Reckoning for the Twentieth Century"

All these strands came together in the nineteenth-century colonies, where a standardized approach to time and labor often accompanied colonists wherever they went. Historian Giordano Nanni writes that "the project to incorporate the globe within a matrix of hours, minutes, and seconds demands recognition as one of the most significant manifestations of Europe's universalizing will." Clocks arrived as tools of domination. Nanni quotes an 1861 letter by Emily Moffat, the daughter-in-law of Robert Moffat, British missionary to what is now known as South Africa: "You must know that today we have unpacked our clock and we seem a little more civilized. For some months we have lived without a timepiece. John's chronometer and my watch have failed, and we have left time and been launched onto eternity. However, it is very pleasing to hear 'tic tic tic' and 'ding ding.'"

The phrase "launched onto eternity" is indicative of most colonists' views of the time reckoning they found among native inhabitants. In short, the colonists were not able to perceive it at all, because the native sense of time and space did not exhibit the same abstraction and independence from natural cues as their own. On a larger scale, they graded native populations as being more or less "progressed" into modernity based on how removed their systems of time seemed from nature—a topic I will return to in the next chapter.

But Moffat's letter also suggests a fragile island of Western clock time amid something completely different. In some southern African towns, for example, the seven-day week, including the Sabbath, extended only as far as the audible range of the mission station bell tower. One reverend at a mission in South Africa carefully counted the number of inhabitants "living within the sound of the station-bell," while another was dismayed to find populations outside the mission's influence willfully ignorant of the Sabbath. Likewise, in the Philippines and Mexico, Spanish colonists would convert natives into Spanish subjects by placing them *bajo las campanas* ("under the bells").

The boundary of this audible range was not one between time and timelessness, but between two fully formed understandings of time, ritual observances, and age. Nanni quotes a faltering exchange in Coranderrk, Australia, between a colonial commissioner and an Aboriginal man who was not accustomed to measuring age numerically. Eventually, they must settle on the "lingua franca" of biological time:

> How old were you when you came [to Coranderrk]?—I have no
>   idea.
> Do you know how old you are now?—I am supposed to be about
>   twenty-two.
> Then you must have been about ten when you came in?—I was a
>   boy then, but I have no idea how old I was then.
> You had no whiskers then?—No, I had no whiskers then.

Bound up in this misunderstanding was much more than a system of measuring time; it was an entire way of thinking about what time is. Nanni notes that colonial missions tried "to induce people [not simply] to work . . . but to work in a regular and uniform manner, for a specific period of time per day." This view of abstract labor hours could not have been more alien to task-oriented communities who organized their activities based on different ecological and cultural cues—such as the flowering or fruiting of a certain plant—and where things took however much time they took. These communities, for whom work was not profit but part of a social economy, did not make the same distinctions between what was called "work time" and "nonwork time."

And just as colonists saw their abstract time reckoning as more evolved than that of their subjects, their attempts at "civilizing" meant inculcating in those subjects the perception of time as money. As E. P. Thompson observed, Puritanism entered a "marriage of convenience" with capitalism in the eighteenth and nineteenth centuries, becoming "the agent which converted men to new valuations of time; which taught children even in their infancy to improve each shining hour; and which saturated men's minds with the equation, time is money." For colonizing countries, this was true both at home and abroad. Nanni cites a passage "decidedly lacking in subtlety" from *Lovedale News,* a publication of a South African mission in 1876:

HOW MUCH HAVE YOU IN THE BANK? Not the Savings Bank, though it would be a good thing for you to have a little there too. This Bank is a better one. Perhaps you have nothing to put in the savings bank, and think you have nothing to put in any other. You are wrong. You may be putting money in every day. Did you ever count up how much or how little you had got there—in the Bank of which God is the manager, and over whose counter pass the well used moments of each day and all the good things a man thinks or says or does. We speak of spending time. Time spent does not go into the Bank, any more than money spent. But every moment you use well, for God, you put into the Bank. . . . I would advise you all to put something in—to put in all you can. For the Bank gives good interest.

FROM THE FENCED-IN accumulation of containers, we turn toward the San Francisco Bay on a hard, sandy path. Fully embedded in the ground is an old train track, burnished to near oblivion by time and ignored by the Canada geese, who are more interested in the park grass. A placard informs us that this used to be the western terminus of a transcontinental railroad. Long before the container terminals were built, this place was already a node in the battle against space and time, the end of a line that would eventually reduce the trip from New York to San Francisco from months to about a week.

Behind us and beyond the stacks of container chassis, you can see the East Bay Hills. They look like a cutout in the morning haze, a single layer of eucalyptus studded with houses. But if we could ascend to the level of the cranes, we might be able to see how far back they go, and if we went higher, we'd see the Central Valley and the formidable peaks of the Sierra Nevada. In the 1860s, Chinese railroad workers labored to connect the place where we are to Omaha, Nebraska, blasting tunnels, hacking through forests, building walls and trestles, and laying track with no machine tools to speak of. They continued working through the brutal winter of 1866/67, which saw forty-four separate storms.

Railroad baron Leland Stanford—yes, the progenitor of my workplace for eight years—had initially wanted to keep Asians out of California. But he changed his tune when there was a labor shortage, pleased to find that the Chi-

nese were "quiet, peaceable, patient, industrious and economical." What was
especially economical was paying them 30 to 50 percent less than white workers
he employed and charging them for room and board. In June 1867, the Chinese
workers went on strike for shorter hours, improved working conditions, and
wage parity in what was, at the time, the largest worker action in U.S. history.
The railroad responded by cutting off their food supply, though it did later qui-
etly raise the wages of some workers. Meanwhile, out there on the slopes, hours
and conditions remained the same.

WHEN LOOKING AT the history of how productivity has been measured,
it is always illuminating to ask: Who is timing whom? The answer to
this question often identifies a person who has purchased someone else's
time or owns it outright—and who, in either case, wants to make the
most of it. It's not hard to imagine how those who held slaves or kept
servants might have had reason to see people as "personified labor time"
long before employers bought employees' work hours. Capitalist practices also had roots in the organization of ancient armies. In *Technics and
Civilization*, Lewis Mumford observes:

> Before inventors created engines to take the place of men, the
> leaders of men had drilled and regimented multitudes of human
> beings: they had discovered how to reduce men to machines. The
> slaves and peasants who hauled the stones for the pyramids, pull-
> ing in rhythm to the crack of the whip, the slaves working in the
> Roman galley, each man chained to his seat and unable to perform
> any other motion than the limited mechanical one, the order and
> march and system of attack of the Macedonian phalanx—these
> were all machine phenomena.

It was a short step from seeing people as the embodiment of work
to turning the units of time they spent working into money. While the
systematic management of other people's time is often associated with
Taylorism, the roots of modern management can readily be found on
West Indian and southern U.S. plantations in the eighteenth and nineteenth centuries. In *Accounting for Slavery: Masters and Management*, Cait-

lin Rosenthal surveys the bookkeeping practices of these plantations and finds an uncomfortable analogy with more contemporary business strategies: "Though modern practices are rarely compared to slaveholders' calculations, many planters in the American South and the West Indies shared our obsession with data. They sought to determine how much labor their slaves could perform in a given amount of time, and they pushed them to achieve that maximum." Plantation owners were some of the earliest users of what we would now call spreadsheets, producing preprinted work logs and conducting labor-timing experiments similar to the ones Taylor would become famous for many decades later.

In the work logs, the enslaved appear only as names and quantities of labor. Justin Roberts, in *Slavery and the Enlightenment in the British Atlantic, 1750–1807*, describes how the Barbadian Society for Plantership "conceptualized a total pool of 'labour days'" that an estate had at its disposal. Although they were far more subject to natural factors like the weather, plantation labor days were considered as fungible as the industrial man-hour. And as with the man-hour, their standardization obscured brutal circumstances.

In a 1789 letter to one of his overseers, General George Washington emphasized that slaves should "[do] as much in the 24 hours as their strength, without endangering their health, or constitution will allow of." Anything less would be bad business sense, amounting to "throw[ing] . . . labour away." Thomas Jefferson produced his own experiments, writing in a memorandum, "Four good fellows . . . in 8 1/2 hours dug in my cellar a mountain of clay a place 3 f. deep, 8 f. wide and 16 1/2 f. long. . . . I think a mid[d]ling hand in 12. hours (including his breakfast) could dig & haul away the earth of 4 cubical yards, in the same soil."

As would be the case in many different contexts going forward, the science of recording labor days was inextricable from the project of intensifying them. Plantation accounting systems were designed both to maximize the amount of work done in a day and to increase the number of these maximal days. In fact, by the last part of the eighteenth century, some West Indian sugar planters began pushing for slaves to work on Sundays, their only day off. When clocks arrived on the plantation, they merely aided in this process.

These calculations were possible because much of plantation work represented fungible labor: pounds, bushels, yards per day and per hour. Whether in the fields or on the estate grounds, enslaved people performed the same actions over and over again and were always pushed to do them faster. Planters did not see them as people but as labor embodied, and that labor could be optimized. Rosenthal writes that, unlike wage workers, "[enslaved people] could not quit, and planters blended information systems with violence—and threat of sale—to refine labor processes, building machines made out of men, women, and children." Readable between the ledger lines of the plantation book is the violence underlying the system's "standards."

A MORE FAMILIAR form of time as money is the wage. But, just like the "tic tic" and "ding ding" in the midst of eternity, the widespread phenomenon of selling one's time is historically specific and surprisingly recent. In early-nineteenth-century America, which was still largely rural, self-employed people outnumbered wage earners. Even after a dramatic rise in wage labor after the Civil War, it was compared to prostitution or slavery, sometimes by white workers wanting to maintain distance from sex workers and enslaved Black people. But Black freedpeople, too, noted the similarity of a hireling to a slave. Richard L. Davis, a Black miner, maintained that "none of us who toil for our daily bread are free. At one time . . . we were chattel slaves; today we are, one and all, white and black, wage slaves." In 1830, the *Mechanic's Free Press* asked, "In What does slavery consist?" and concluded, "In being compelled to work for others so that they may reap the advantage." Wage labor, or "the unfettered ability to sell the self," seemed undemocratic if "liberty was defined as complete ownership of one's own labor and, by extension, oneself."

The world of wage labor—of work hours and work minutes—required discipline to maintain. Foreshadowing the flags and alerts of StaffCop, the waged workplace contained a paralegal structure of rules and penalties, one where a transgression could mean losing pay or getting fired. Penalties were often time-based: One could be punished for showing up too early or too late, working too slowly, or doing anything unrelated to producing value for the employer (as described earlier, "stealing time"). Such were the terms of employment, and until workers began to organize, these terms were generally nonnegotiable. When workers did organize, many of them immigrants, cities like Boston and New York followed London's lead and created formal police forces to suppress the unrest. Leaders of commerce urged northern cities to build armories in urban industrial areas where strikes were imminent. Labor historian Philip Dray writes that although "Americans have come to think of these dour, substantial buildings as historic rallying places for troops in the case of foreign threats to U.S. soil . . . their original purpose was to allow the rapid deployment of the militia to keep workingmen in check."

In theory, if you don't enjoy your employer's policies, temporal or otherwise, you should be able to seek a different job. ("If you don't like your terms of employment, quit," I hear Tommy Anderberg saying.) But even before unions began forming in the United States, northern industrialists had already begun to act collectively, agreeing to institute certain policies or to blacklist employees across the board. This kind of behavior provoked one of the first documented factory strikes in the country—at a textile mill in Pawtucket, Rhode Island, in 1824. Mill owners had announced a one-hour increase in the workday, which would be unpaid and taken out of the workers' meal time. Because multiple mill owners had colluded, this policy affected every factory in the city. One hundred two young women walked off the job, and at the end of the weeklong strike, one of the mills was set on fire, prompting the owners to institute a night watch. According to newspapers, the mills began running again when owners and workers reached a "compromise."

Within its policed borders, the industrial workplace of the time resembled many other institutions upholding the philosophy of hours invested in God's bank. Whether in the factory, the school, the prison, or the orphanage, it wasn't just a matter of productivity, but one of training—learning to eat the corn on the spinning cob. In this context, the clock was an unforgiving foreman. In *The Lowell Offering*, an 1840s publication run by women workers at a textile mill in Lowell, Massachusetts, a worker wrote, "I object to the constant hurry of everything. . . .

Up before day, at the clang of the bell—and out of the mill by the clang of the bell—into the mill, and at work, in obedience to that ding-dong of the bell—just as though we were living machines."

The British journalist John Brown, taking down the story of a former child laborer in 1832, described the timed movements of workers in a Manchester cotton mill: "If [workers] arrived only two or three minutes after the clock had struck, they were locked out; and those, who were within, were all locked in, till dinner time, and not only were the outward doors, below, locked; but every room above, and there was a door-keeper kept, whose duty it was, a few minutes before the respective hours of departure, to unlock the doors, by whom they were again locked, as soon as the work-people arrived!"

Time discipline could become even more minute. E. P. Thompson quotes from the rule book of the Methodist Sunday schools in York in 1819, in which a simple action like beginning a lesson was broken down into segments that both echo militarism and foreshadow Taylorism in the factory: "The Superintendent shall again ring,—when, on the motion of his hand, the whole School rise at once from their seats;—on a second motion, the Scholars turn;—on a third, slowly and silently move to the place appointed to repeat their lessons,—he then pronounces the word 'Begin.'"

This isn't detail for the sake of detail. Time discipline was and is a tool used both inside and outside the factory to render a more docile and productive workforce, whether by directing and intensifying work or generally instilling a pious "habit of industry" in would-be workers. (Whether it was ever fully internalized, however, is a question I will return to in chapter 6.) It is telling, for example, that owners of the mills in Lowell, Massachusetts, tried to argue that longer hours were actually good for the women. Without the "wholesome discipline of factory life," the women would be left to their own dangerous whims, "without a warrant that this time will be well employed." In a similar manner to the British colonists who sought to "save" native inhabitants, factory owners established Sunday schools where children were plied with lessons on the virtues of hard and unceasing work. One of the rules at Philadelphia's Eastern State Penitentiary in the 1840s could just as easily have applied to the school, the poorhouse, or the asylum: "5th. You must apply yourself industriously, at whatever employment is assigned you; and when your task is finished, it is recommended that your time be devoted to the proper improvement of your mind, either in reading the books provided for the purpose, or in case you cannot read, in learning to do so."

The idea of a life evenly filled with "well-employed" hours reaches a tragicomical extreme in Jeremy Bentham's designs for the panopticon. Bentham, an eighteenth-century British philosopher and social reformer, envisioned a new disciplinary architecture: a ring of cells arranged around a single tower where individuals would assume they were being watched at all times. Here, every moment would be accounted for and put to work—not as straightforward punishment, but as the rehabilitative "penitence" of the penitentiary. Accordingly, Bentham expected prisoners to work fourteen hours a day. But that was not all. Realizing the necessity of exercise for prisoners' health, Bentham imagined that prisoners could conveniently spend their recreation time walking in a giant wheel that would transport water to the top of the building. Not a drop of time would be wasted.

I'VE INVOKED THIS backstory of purchased and timed labor in order to defamiliarize, just for a moment, the concept of the wage. When the relationship of time to literal money is expressed as a natural fact, it obscures the political relationship between the seller of time and its buyer. This may seem obvious, but if time is money, it is so in a way that's different for a worker than for an employer. For the worker, time is a certain amount of money—the wage. But the buyer, or employer, hires a worker to create surplus value; this excess is what defines productivity under capitalism. From an employer's point of view, purchased time could always yield more money.

In the first volume of *Capital*, Marx describes the peculiar nature of labor time as a commodity in an industrial setting. Having finished off part 2 by describing the exchange of money for time between the worker and the employer—an exchange in which both act as equals—he ends on a creepy cliffhanger:

When we leave this sphere of simple circulation or the exchange of commodities . . . a certain change takes place, or so it appears, in the physiognomy of our dramatis personae. He who was previously the money-owner now strides out in front as a capitalist; the possessor of labour-power follows as his worker. The one smirks self-importantly and is intent on business; the other is timid and holds back, like someone who has brought his own hide to market and now has nothing else to expect but—a tanning.

Part 3 opens in the factory, where buyer and seller are far from equal. The employer is busy trying to get more labor out of the worker, while the worker is trying to keep herself from being worked to death. In the drive to make time equal more money, the employer has two strategies to pursue: extension (increasing the amount of time that money buys) or intensification (demanding more work in the same amount of time).

In the chapter "The Working Day," Marx gives an example of the extension approach by recounting a harrowing struggle between nineteenth-century British factory owners and workers over the length of the workday. It was only through protracted efforts by workers and British lawmakers that the workday was limited at all. Even then, management was quick to find ways around its limits, in particular by encroaching on break time or, as factory inspectors called it, the "'petty pilfering of minutes,' 'snatching a few minutes,' or, in the technical language of workers, 'nibbling and cribbling at meal times.'" Sometimes employers practiced outright deception, setting the clocks forward in the morning and back at night.

As industrialists came up against natural or regulatory limits on hours, they took the other route to increasing profit: intensifying the hours they had. Making them denser with value involved a "closer filling-up of the pores of the working day." In nineteenth-century textile mills in the United States, such innovations could include the "stretch-out" (making a worker responsible for more machines), the "speedup" (having foremen speed up the process, resulting in the Tramp's misery in *Modern Times*), and the "premium system" (which offered cash rewards to supervisors with the most productive workers).

At first glance, there seems to be a paradox here: While industrial capitalism spawned many machines that saved time and labor, it seemed only to take up more and more of workers' time. But unlike the Ancient Greeks, who imagined that, someday, machines might replace slave labor so that everyone might enjoy some free time, capital only "frees time in order to appropriate it for itself." In other words, the goal of capitalism is not free time but economic growth; any time freed up goes right back into the machine to increase profits. Thus the paradox: The factory is efficient, but it also produces "the drive toward the consumption of the person's time up to its outermost, physical limit." Or, as the workplace adage would have it, "The only reward for working faster is more work."

THE WEBSITE FOR SSA Marine says, "Accelerating the Pace of Business." Its terminal is now giving off a deafening whir: engine sounds, horns, beeps, and the echoes of workers shouting. The giant cranes lift containers off the ship, sliding them inward fast enough that they swing a little bit in midair. Currently, the bay is full of the haze-lightened silhouettes of container ships, players in that sprawling, fractal network whose workings have recently come to the fore in headlines about the supply chain.

In the restored marsh along the park, clusters of migrating shorebirds are keeping their own schedule. It's currently three hours from high tide, and on the shrinking islands, tiny sandpipers sit together so densely that they look like a tessellated pattern. Stalking around them are a variety of spidery birds, including long-billed curlews, which have surreal curved beaks more than half the length of their entire bodies. They are back for the time being, having traveled northeast to breed—possibly as far as Idaho—and in the meantime, they adjust their activities to the tides.

On the one hand, it is true that you can see multiple forms of time here. The containers pile up; the shorebirds probe the mud; the phoebe chases its flies; a small, brown mushroom pushes up from the grass; and the tide continues to rise. Your stomach rumbles. But one of these clocks is not like the others. In order to maintain its equilibrium, it has to run ahead faster and faster.

IT'S WORTH NOTING here that a scrupulous accounting of time is not in itself unique to capitalism. As I've mentioned, pre-industrial or pre-colonial societies were and still are imagined to have been inherently leisurely, or even "without time," in part because they were task oriented—a way of working that follows the contours of different tasks rather than a rigid, abstract schedule. But as the sociologist Michael O'Malley has pointed out, such societies exhibited their own "fierce attention to saving time." Besides the precision required for agricultural timing, every society makes social determinations about what is worth spending time on, as well as how much.

It can also be tempting to view the capitalist temporality as being especially associated with the clock. But while it certainly has played a crucial role in time discipline, the clock is just one tool among many for reckoning time, and its full meaning emerges only when it's joined with a particular goal or cosmology. O'Malley observes the "ambiguous position" of American clocks in the nineteenth century: "They could stand for industry and business, for the perfectibility of machinery, for linear time and progress into the future. But they could also stand for stasis, for the cycles of the seasons. The hands, after all, endlessly repeat their circuit around the dial, instead of moving into the future."

Nor was mechanical efficiency the sole domain of industrial capitalists. For one thing, it depends on how you define *mechanical,* as humans have studied their environments and designed labor-saving systems for millennia, systematizing them across generations. And even if you're looking for something very traditionally mechanical, you can find timesaving systems in Catharine Beecher's 1841 housekeeping bible, *A Treatise on Domestic Economy,* well predating Frederick Winslow Taylor's *Principles of Scientific Management.* Beecher's book was largely responsible for the rise of the fitted kitchen, envisioning both a living space and a mode of work that would be designed to reduce the time and effort women spent on housework. Yet the aims of this efficiency were clear: Beecher wasn't seeking profit but, rather, "economy of labor, economy of money, economy of health, economy of comfort, and good taste."

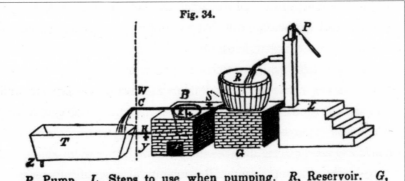

Fig. 34.

P, Pump. *L,* Steps to use when pumping. *R,* Reservoir. *G,* Brickwork to raise the Reservoir. *B,* A large Boiler. *F,* Furnace, beneath the Boiler. *C,* Conductor of cold water. *H,* Conductor of hot water. *K,* Cock for letting cold water into the Boiler. *S,* Pipe to conduct cold water to a cock over the kitchen sink. *T,* Bathingtub, which receives cold water from the Conductor, *C,* and hot water from the Conductor, *H.* *W,* Partition separating the Bathing-room from the Wash-room. *Y,* Cock to draw off hot water. *Z,* Plug to let off the water from the Bathing-tub into a drain.

Laundry system from Catharine Beecher's *A Treatise on Domestic Economy*

The capitalist version of time, in contrast, is defined by the ends to which intensity and standardization were ultimately directed: more capital for the company. After all, it is neither the assembly line nor the feeding machine that directly assaults Chaplin's character in *Modern Times;* it

is the president who speeds up the assembly line and the management
that straps the Tramp into the machine. Humans make these decisions,
just like humans today design call center and delivery app interfaces. (In
turn, Marx might have pointed out, those people are simply acting at
the behest of capital.) In *Labor and Monopoly Capital,* Harry Braverman
gives an example of this distinction by quoting the vice president of a
1960s insurance company, who feigns aloofness about the frantic pace
of his keypunch operators. "[He] remarked: 'All they lack is a chain,'
and explained himself by adding that the machines kept the 'girls' at
their desks, punching monotonously and without cease." In a footnote,
Braverman calls the executive's bluff: "This vice-president gives us a
clear illustration of the fetishism which puts the blame for the situation
on the 'machines' rather than on the social relations within which they
are employed. He knew when he made this remark that it was not the
'machines' but he himself who chained the workers to their desks, for in
his next breath he pointed out that a count of production was kept for
the workers in that machine room."

This is the lens through which it is most helpful to view Taylorism, a
set of practices stemming from Frederick Winslow Taylor's streamlin-
ing of the industrial work process. In his 1911 book, *Principles of Scientific
Management,* Taylor outlines methods for breaking down actions into the
smallest measurable components and reconfiguring them in the most
mechanically efficient manner. Proponents of scientific management
produced incredibly detailed timetables as well as "motion studies," long-
exposure photographs where lights were attached to the hands of workers to break down and better understand their motions. One article in
Factory Magazine that detailed some of these methods made the equation
explicit: "The way to effect savings is by reduction in time. . . . The man
whose business it is to increase the output per unit of time is the time-
study man."

As the foremost time-study man, Taylor was known for being somewhat fanatical in his pursuit of efficiency. Braverman mentions that, as
a youth, Taylor Taylorized himself: He counted his steps, timed his own
activities, and analyzed his own motions. When he became the gang
boss at a technologically advanced steel works in the 1870s, he directed

Stopwatch and motion study by Frank Gilbreth. Gilbreth and his wife Lillian conducted studies of workers' movements for industrial managers in the 1910s.

his drive for efficiency toward the group of machinists he oversaw. Taylor observed among the workers the phenomenon of "systematic soldiering," where the workers would agree among themselves on a reasonable pace and quantity of output—an output that happened to be far below what Taylor thought to be their capacity.

In a testimony before the Special Committee of the U.S. House of Representatives, Taylor described his arduous efforts over several years to break the solidarity of the workers in order to get them to adopt his more intense methods. He kept showing workers how to do it, only to have them turn around and do it the old way, and when he hired unskilled workers to train in the new methods, they fell in with the others and refused to work faster. At one point, he told them, "I am going to cut your rate in two tomorrow and you are going to work for half price from now on. But all you will have to do is turn out a fair day's work and you can earn better wages than you have been earning." Echoing George Washington's "24 hours," Taylor's definition of a fair day's work was a maxing out—what Braverman calls a "crude physiological interpretation: all the work a worker can do without injury to his health."

Once again, we may ask: Who is timing whom? Scientific management was a matter not just of measuring work and increasing productivity but of discipline and control. As Taylor's years-long battled showed, workers would have some control over the pace of work as long as they held knowledge about the work process. As much as it was about intensification, Taylorism was also about breaking apart and codifying this process in a way that concentrated knowledge in the hands of employers rather than employees. "Under our system the workman is told minutely just what he is to do and how he is to do it," Taylor wrote, "and any improvement which he makes upon the orders is fatal to success." In this way, Taylorism rendered labor more abstract and fungible, hastening a process that has often been referred to as "de-skilling." Among other things, this deepened the divide between how different kinds of time are valued. As Braverman puts it, "Every step in the labor process is divorced, so far as possible, from special knowledge of training and reduced to simple labor. Meanwhile, the relatively few persons for whom special knowledge and training are reserved are freed so far as possible from the obligations of simple labor. In this way a structure is given to all labor processes that at its extremes polarizes those whose time is infinitely valuable and those whose time is worth almost nothing."

The "time-study man" was the forerunner to "high-level thinkers," consultants, and idea men, many of whom can name their hourly price because something about their work seems ineffable, not yet alienated from them as people. The business guru David Shing, otherwise known as Shingy, would be one ridiculous extreme of someone who possesses his own "means of production" (of ideas). The INFN physicists, expressing disdain for manufacturing jobs because their work has never been so alienated from them, are somewhere in between. In contrast, for those timed by the time-study man, work becomes more like the Tramp's work on the assembly line in *Modern Times:* consistent and eminently timeable, with less and less left to the discretion of the worker, who in turn becomes more easily replaceable. That this development merely extended the old relationship between standard time and control has been noted by Dan Thu Nguyen, who observes that "metric time first gave us the rule of the seas and oceans, then the colonization of the land; it taught us how to structure our bodies and movements in work and how to rest when the job is done."

The Taylorist divide between the timed and the timer is just one step

DIAMETER OF BOLT	⅝ AND ¾				⅞ AND 1						TAYLOR SYSTEM ELEMENTARY UNIT TIME
LENGTH OF BOLT	6	12	14	18	6	12	18	24	30	36	
TIME FOR CLAMPING IN MINUTES	0.42	0.45	0.45	0.51	0.48	0.53	0.57	0.69	0.78	0.92	**CLAMPING** BOLT TO BE PUT IN SLOT — THE TIME GIVEN IN THIS SECTION TO BE USED ONLY FOR FIRST PIECE IN LOT OR ON OTHER PIECES WHEN BOLT HAS TO BE PUT INTO SLOT.
LIFT BOLT, CLAMP AND BLOCK TO TABLE	0.07	0.08	0.08	0.09	0.09	0.10	0.10	0.13	0.14	0.17	
PUT BOLT IN SLOT	0.04	0.04	0.04	0.05	0.04	0.04	0.05	0.05	0.06	0.07	
SLIP CLAMP ON BOLT AND ON WORK	0.05	0.05	0.05	0.05	0.05	0.05	0.05	0.06	0.07	0.08	
PUT BLOCK UNDER CLAMP	0.04	0.05	0.05	0.05	0.04	0.05	0.05	0.07	0.08	0.10	
SCREW NUT DOWN WITH FINGERS	0.05	0.06	0.06	0.07	0.06	0.07	0.08	0.10	0.11	0.14	
TIGHTEN NUT LIGHTLY WITH WRENCH	0.08	0.08	0.08	0.10	0.11	0.12	0.13	0.15	0.17	0.19	
DRAW NUT DOWN TIGHT WITH WRENCH	0.09	0.09	0.09	0.10	0.09	0.10	0.11	0.13	0.15	0.17	
TIME FOR REMOVING IN MINUTES	0.16	0.17	0.17	0.18	0.22	0.22	0.24	0.26	0.29	0.33	**REMOVING** BOLT NOT TO BE TAKEN FROM SLOT — THE TIME GIVEN IN THIS SECTION TO BE USED FOR ALL PIECES IN LOT EXCEPTING THE LAST PIECE.
LOOSEN NUT WITH WRENCH	0.11	0.12	0.12	0.13	0.17	0.17	0.19	0.20	0.22	0.25	
REMOVE CLAMP FROM BOLT	0.05	0.05	0.05	0.05	0.05	0.05	0.05	0.06	0.07	0.08	
TIME FOR CLAMPING IN MINUTES	0.30	0.33	0.34	0.37	0.35	0.39	0.42	0.51	0.58	0.68	**CLAMPING** BOLT ALREADY IN SLOT — THE TIME GIVEN IN THIS SECTION TO BE USED FOR ALL PIECES IN THE LOT AFTER THE FIRST PIECE WHEN BOLT HAS NOT BEEN REMOVED.
SLIP CLAMP ON BOLT AND ON WORK	0.05	0.05	0.05	0.05	0.05	0.05	0.05	0.05	0.07	0.08	
PUT BLOCK UNDER CLAMP	0.04	0.05	0.05	0.05	0.04	0.05	0.05	0.07	0.08	0.10	
SCREW NUT DOWN WITH FINGERS	0.05	0.06	0.06	0.07	0.06	0.07	0.08	0.10	0.11	0.14	
TIGHTEN NUT LIGHTLY WITH WRENCH	0.08	0.08	0.08	0.10	0.11	0.12	0.13	0.15	0.17	0.19	
DRAW NUT DOWN TIGHT WITH WRENCH	0.08	0.09	0.10	0.10	0.09	0.10	0.11	0.13	0.15	0.17	
TIME FOR REMOVING IN MINUTES	0.24	0.25	0.25	0.27	0.30	0.30	0.33	0.37	0.41	0.47	**REMOVING** BOLT TO BE TAKEN FROM SLOT — THE TIME GIVEN IN THIS SECTION TO BE USED ONLY FOR LAST PIECE IN LOT OR ON OTHER PIECES WHEN BOLT HAS TO BE TAKEN FROM SLOT.
LOOSEN NUT WITH WRENCH	0.11	0.12	0.12	0.13	0.17	0.17	0.19	0.20	0.22	0.25	
REMOVE CLAMP FROM BOLT	0.05	0.05	0.05	0.05	0.05	0.05	0.05	0.05	0.06	0.07	
REMOVE BOLT FROM SLOT	0.03	0.03	0.03	0.04	0.03	0.03	0.04	0.04	0.05	0.05	
PUT CLAMP, BOLT AND BLOCK IN TOTE BOX ON FLOOR	0.05	0.05	0.05	0.05	0.05	0.05	0.05	0.08	0.09	0.10	

FIGURE II: THE DETAILED TIMING OF SETTING AND REMOVING WORK WHEN PERFORMED UNDER VARIOUS CONDITIONS IS TAKEN UP BY OPERATIONS ON THIS FORM

Taylorist chart from "The Stop Watch as Inventor," in the February 1916 issue of *Factory: The Magazine of Management*

in a division of labor that has long run along gendered and racialized lines. There is, first of all, the line between waged and unwaged work—a question of whose time and which labor is even money at all (an issue taken up by feminist thinkers I'll turn to in chapter 6). In the United States, when domestic work did become waged, it was often done by Black women, and it was (and is) devalued compared to work that directly produced a profit.*

---

* In *Women, Race & Class,* Angela Y. Davis observes, "Since housework does not generate profit, domestic labor was naturally defined as an inferior form of work as compared to capitalist wage labor." Sociologist Barbara Adam made a similar point in *Timewatch:* "Research on women's caring and emotional work demonstrates that times which are not convertible into currency have to remain outside the charmed circle. . . . That is to say, time-generating and time-giving activities have no place in the meaning cluster of quantity, measure, dates and deadlines, of calculability, abstract exchange value, efficiency and profit."

The de-skilling associated with Taylorism represented a division within "productive" waged labor. Well into the twentieth century, Black workers in U.S. factories were prevented from working machine jobs and were kept in positions involving menial labor. During World War II, women employed in military intelligence did the tedious and repetitive calculations, which resulted in the term *kilogirl*. (One kilogirl was "equivalent to roughly a thousand hours of computing labor.") It seems that the more temporal surveillance a job entails, the less likely it is to be done by someone white or male. In 2014, when Amazon released data showing its workforce to be surprisingly diverse, further news came out that most of the "diversity" stemmed from Black and Latino workers in their fulfillment centers. The situation was still far from ideal in 2021.

The more fragmented and minutely timeable work becomes, the more meaningless it becomes. Echoing Marx's description of an "automaton" where "workers themselves are cast merely as its conscious linkages," one former worker at a clothing manufacturer complained in 2020 that "your movements are controlled, like friggin' everything is controlled." Jessica Bruder, in a story about work as a UPS driver, describes a sensor-outfitted truck that nearly drives its driver: "[The sensors] reported when he opened the bulkhead door. When he backed up. When his foot was on the brake. When he was idling. When he buckled his safety belt." This data flowed back to UPS in a system whose name, "telematics," originated, appropriately enough, in a military context. There were even efficiency-seeking time-and-motion studies, similar to Taylor's, that told the UPS driver "how to handle his ignition key, which shirt pocket to use for his pen (right-handed people should use the left pocket, and vice versa), how to pick a 'walk path' from the truck, and how to occupy time while riding in an elevator."

In this case, the immediate reasons for such efficiency are relatively simple. "Time is money, and management knows exactly how much," Bruder writes, before quoting the company's senior director of process management: "Just one minute per driver over the course of a year adds up to $14.5 million." But this time is also money in a different way. Data collected through telematics systems like the one used by UPS is also used to prepare the grounds for driverless cars.

A 2019 episode of Channel 4's *Secrets of the Superfactories* featured an Amazon warehouse where much of the sorting of products had been offloaded to shelves that moved themselves around with the eerie smoothness of Roombas; there were still human employees in the warehouse, but far fewer than there had been. And in "lights-out manufacturing," humans number next to none. At FANUC (Fuji Automatic Numerical Control), a complex of twenty-two factories in Japan, robots replicate themselves twenty-four hours a day, seven days a week. The robots are great workers who don't even need heat or air conditioning. An article from the software design firm Autodesk, citing clients like Tesla and Apple, wrote that "job stability for FANUC's self-replicating robots is at an all-time high."

What is less headline grabbing, however, is an eternal meantime in which some humans are not replaced by robots but, instead, must act more like them. In *On the Clock,* Guendelsberger describes feeling this reality in her body, lamenting how humans "increasingly have to compete with computers, algorithms, and robots that never get tired, or sick, or depressed, or need a day off." When she eventually collapses from pain and exhaustion at Amazon, a veteran employee helps her buy some ibuprofen from the dispensers the company keeps on the warehouse floor, and adds, "Be careful about overusing those. . . . I have to take four to get the effect of two now." Given her experience, Guendelsberger is remarkably understanding of Taylor, who hoped the resulting increase in productivity would produce value shared by the worker. But then she points to the graph of growth versus wages in the United States, where the line for wages drops off precipitously from the line of productivity after the 1970s. Now, not only does increased productivity not lead to free time, but it doesn't lead to money for American workers. Their time is more money, but for someone else.

Given all this, there is an understandable incentive to stay on the upper side of the division of labor—or what some have called "above the API"*—where designing a Taylorized interface means not having

---

* Peter Reinhardt, CEO of a carbon capture company, may have coined the expression "above the API" in a 2015 blog post called "Replacing Middle Management with APIs." Describing automated processes at Uber (freelance drivers) and 99designs (freelance designers), he gives examples where a line of code is executed by humans: "The Uber

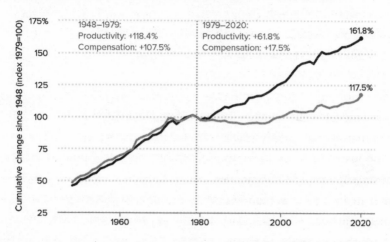

The gap between productivity and a typical worker's compensation has increased dramatically since 1979

Productivity growth and hourly compensation growth, 1948–2020

1948–1979:
Productivity: +118.4%
Compensation: +107.5%

1979–2020:
Productivity: +61.8%
Compensation: +17.5%

161.8%

117.5%

Cumulative change since 1948 (index 1979=100)

175% 150 125 100 75 50 25

1960 1980 2000 2020

Notes: Data are for compensation (wages and benefits) of production/nonsupervisory workers in the private sector and net productivity of the total economy. "Net productivity" is the growth of output of goods and services less depreciation per hour worked.

Source: EPI analysis of unpublished Total Economy Productivity data from Bureau of Labor Statistics (BLS) Labor Productivity and Costs program, wage data from the BLS Current Employment Statistics, BLS Employment Cost Trends, BLS Consumer Price Index, and Bureau of Economic Analysis National Income and Product Accounts.

Economic Policy Institute

to work in it. In a 2019 video about the automation of call center jobs, labor correspondent Aki Ito interviews Laura Morales, an employee at OutPLEX, a call center in Santo Domingo, the Dominican Republic. Morales, formerly one of the center's agents, worked her way up and was chosen to participate in the company's foray into automation. Her new title is chatbot designer. After a tour of the call center and a visit to Morales's home, the two have an awkward conversation over drinks:

API dispatches a human to drive from point A to point B. And the 99designs Tasks API dispatches a human to convert an image into a vector logo (black, white and color). Humans are on the verge of becoming literal cogs in a machine, completely anonymized behind an API." Reinhardt worries that "as the software layer gets thicker, the gap between Below the API jobs and Above the API jobs widens."

ITO: Is there a small part of you that feels guilty for automating
   away the job that gave you your start in your career and is the job
   of a lot of your colleagues right now?

MORALES: Not at all, like zero guilt.

ITO: Zero hesitation?

MORALES (SIPPING DRINK): It's happening already.

ITO (RESIGNED): You might as well be a part of it.

MORALES: Thank god I'm part of it.

THERE IS A surreal part of *Labor and Monopoly Capital* in which Braverman describes the 1960s enthusiasm for Taylorizing the office. Having
worked both in the metalsmithing industry and at a publishing press,
Braverman is well suited to consider the transition, and in reading books
like *Scientific Office Management,* he finds time-study men measuring the
thousandths of a second it took to settle paper into a neat stack, walk to
the drinking fountain, or turn in a swivel chair. But the best is when they
time how long it takes to punch a time card:

PUNCH TIME CLOCK

Identify card	.0156
Get from rack	.0246
Insert in clock	.0222
Remove from clock	.0138
Identify position	.0126
Put card in rack	.0270
	.1158

Here, even mental parts of the process, like "Identify card" and "Identify position," were somehow timed (.0156 seconds and .0126 seconds,
respectively). This hints at another application of the division of labor,
which is the de-skilling of knowledge work. Writing in the 1970s, Braverman describes what now sounds like content moderation or other cognitive busywork: "The work is still performed in the brain, but the brain

is used as an equivalent of the hand of the detail worker in production, grasping and releasing a single piece of 'data' over and over again."

While I was researching this book, I was thinking about how social media users' time is also money for platforms and advertisers.* I googled "How does Instagram time seconds looking at post?" One of the top results was something titled "What is the average time someone spends looking at an Instagram post?" on a site called Wonder. It had been written by freelance researchers identified as "Ashley N." and "Carrie S.," apparently in response to a query from a paying customer. The service seemed like a cross between the question answering of Quora and the micro-tasking of Fiverr, where freelancers take on small jobs starting at five dollars each. On the main page, a testimonial read, "I love Wonder. It's like having an on-demand, personal, graduate level Ivy League research assistant who never sleeps (or complains)." In the midst of research myself, I felt like I was looking at the micro-tasking version of my own work.

On Glassdoor, an employer-review website, a Wonder Research worker in 2018 described their job's standard four-hour timer, which had a supposed option to extend the time by thirty minutes, if needed. However, Wonder Research was plagued by the same opacity as other platforms: "They must have just revamped their system, because no such extension exists." At the end of the four hours, whatever the researcher had completed was automatically submitted for review. (Another employee's review, from the following year, stated that each job paid between sixteen and thirty-two dollars.) If the work didn't get a certain rating, it would be sent back to the employee for revisions, and if the worker wasn't awake or online to make those revisions, the work would be discarded and the worker would go unpaid. "Truly a waste of valuable time," the reviewer wrote.

The lived experience of a person working with an anonymous, algorithmic, and inscrutable interface is one demonstration of the way

---

*On this topic, the sociologist Richard Seymour has called social media a "chronophage," something that "eats time."

automation doesn't so much replace work as reconfigure its content, conditions, and geography. In his history of Luddism, Gavin Mueller gives a useful overview of such reconfigurations, including "Potemkin AI," Jathan Sadowski's term for what he calls "services that purport to be powered by sophisticated software, but actually rely on humans somewhere else acting like robots." Making up the "human cloud," those humans can be recruited from anywhere and paid very little for their time. Mueller mentions the case of Sama (formerly Samasource), which recruits low-wage workers from Kibera, Kenya (believed to be Africa's largest informal settlement), to do the dull and endless work of entering data into a machine-learning system. While the legacy of Taylorism gets smarter, work continues to get duller, cheaper, faster, and more far-flung.

As a form of Taylorized mental work, something like content moderation has additional hazards beyond meaninglessness. In a 2019 Verge article about Cognizant, the content moderation company used by Facebook, Casey Newton writes about a very specific timetable on the job: Moderators were required to watch at least fifteen to thirty seconds of each video, which might well contain something unspeakably horrible. Employees were given nine minutes of "wellness" time per day to deal with this trauma. The workplace contained its own modern-day practices of "nibbling and cribbling": Employees were required to use a browser extension every time they used the restroom, and because Cognizant is based in Florida, the company was not required to offer sick leave. In a dark version of the Billows Feeding Machine, one woman told Newton that when she was ill at work and had used up all her bathroom breaks, a manager brought her a waste bin in which to throw up.

The job of content moderation exists somewhere uncomfortably between that of a human and that of a machine. On the one hand, being a human (with bodily needs and, just as important, emotional limits) is seen as a barrier to work, given that the sale of labor time implies a worker who is full of work time but devoid of other kinds of time, like biological or social time. But content moderation also requires human traits like empathy, morality, and culturally situated judgment. A psychopath would make a terrible content moderator. Mark Zuckerberg and others once imagined a day when AI would moderate content for us. But, as law professor James Grimmelmann has pointed out, "even humans have a hard time distinguishing between hate speech and a parody of hate speech, and AI is way short of human capabilities."

To some degree, content moderation could be considered a "cyborg job." Yet the fact that it requires its workers to be both robot-like and indelibly human raises questions about many other seemingly un-Taylorizable forms of work. If you have reason to—and many do—you can measure anything in a way that seeks to maximize a certain numerical outcome: words of content per day; test score increases and "learning outcomes" per semester; and clients, customers, or patients per hour. Social work, which requires incredibly high attention to context, nuance, and personality, is just as fragmented by bureaucracy as any other form of service work. Mueller quotes one social worker: "If I wanted to work in a factory, I would have worked in a factory." In those jobs that humans must still (or always) do, attempts to codify and intensify work continue to demoralize people just as they did those assigned the earliest Taylorized tasks. At Cognizant, where humans keep coming in for work despite the conditions, one worker tells Newton that they are merely "bodies in seats."

With regard to such bodies, there's still as much incentive as ever to make time become more money, providing a "closer filling-up of the pores of the working day." Many companies market gamification systems and leaderboards (like the ones produced at StaffCop and Teramind) to call centers, where they can be displayed on TV screens and mobile devices. When in 2021 I visited the website for one such company, Spinify, a window popped up that read, "Welcome to Spinify! Competitions that give you visibility and staff recognition," followed by three random images of people inside some circles, as though to assure me humans were involved. There was no X button to close the window, only the options "Chat with Us," "Request a Demo," and "I'm Just Browsing." In an attempt to get rid of the window, I clicked "I'm Just Browsing," but all this did was open up a chat box, where the chatbot replied,

"Enjoy browsing. Need anything I am here [*sic*]." This was immediately followed by an offer for "The Ultimate Sales Enablement Strategy Playbook," with my text field changed to accept only my email address.

Politely declining to enter my email, I scrolled through the offerings of Spinify and came to rest on a section called Gamify Your Team, in which things like contests and countdown timers were offered as ways to build employee excitement. One example of a contest was called Elimination, which "focuses on the bottom tier and randomly eliminates the person in last place." It's unclear what elimination would entail at any particular company. In the accompanying illustration, three cutesy figures were shown with progress bars, two of them green with the scores 55 and 63, respectively. But the third was red, and instead of a number, there was simply a trash can icon. At the bottom of the page, Graeme Johnston, co-founder and director of an unspecified company, offered a slightly sinister-sounding testimonial: "Spinify increased competition in the team and made people more accountable. There are no hiding places!"

Behind the upbeat phrasing, friendly cartoon characters, and pop-up windows were discordant implications: an intensifying work pace, not-so-friendly competition, and automated penalties. It made me think immediately of the workplace aesthetic in Keiichi Matsuda's short 2019 sci-fi film, *Merger*. Here, a nameless woman sits in a swivel chair at a wraparound desk, surrounded by various holographic screens, performing a job like customer service on steroids. As a version of Microsoft Word's paperclip user interface Clippy watches her from the desk, she nervously types and swipes as more and more screens, messages, and alerts appear, accompanied by dings and boings. Within this unsparing, all-seeing, and personalized assembly line, she visibly struggles to keep calm and keep pace.

At the beginning of the film, we hear a series of recommendations from someone we later realize is the employee herself, parroted to what sounds like an offscreen therapist: "Start your day with exercise. This increases concentration and contributes to your general well-being. Set up your work environment for maximum efficiency. No matter what you wake up to, try to stay in a positive mindset." Neither Bentham nor

Keiichi Matsuda, *Merger,* 2019

Taylor would likely have been displeased. This worker needs no feeding machine; instead, she takes a Soylent-like beverage and a set of pills, adding, "The aim is to achieve a laser focus and find ways to work around the limitations of the human body. There isn't always time to eat, but there are so many innovative ways to fuel your body and mind." Ironically, her time is so effectively absorbed that it's hard to imagine her having a free moment to read yet another listicle about how to be more productive in these trying times.

Merger is a 360-degree film, but rather than giving the viewer a sense of freedom, scrolling around simply imparts a feeling of claustrophobia inside the video sphere: There is no social or biological time, no physical environment, no personal identity, no humor, no co-workers, no human boss. There is only an algorithmically directed Cosmic Day, twenty-four hours of arbitrarily changing backgrounds and undifferentiated work time. At the end of the film, we realize that the employee's conversation is with someone (or something) that is about to transport her to "the other side." Counting down from ten, she closes her eyes with relief, becoming a disembodied algorithm. She has escaped to the realm of data, where time can finally be controlled. She has become the work.

The tragedy of fungible labor time lies first in its historical association with coercion, exploitation, and the imagining of people as machines. Time is the punitive dimension in which the wage worker is both measured and squeezed. But beyond that, an overemphasis on fungible time upholds an impoverished view of what time and labor are in the first place. The industrial view of time as money can see time only as work, the masculinized work of a machine with an On/Off button. Like a grid spreading outward from the Taylorized workplace, whether on the warehouse floor or on a gig platform's mobile interface, this framework contributes to a view of individuals who hold time like private property—I have my time, and you have yours, and we sell it on the marketplace. Now it's not just the employer who sees you as twenty-four hours of personified labor time; it's you, too, when you look in the mirror.

Chapter 2

Self Timer

INTERSTATE 880 AND STATE ROUTE 84

More important than any New Year's resolution is an annual report to
yourself. Your "fiscal period" can be any day—any day that you decide
the time has finally come to adopt this practice of self-accounting which
balances the status of your being in the world about you.

— P. K. THOMAJAN, "Annual Report to Yourself," in *Good Business*, 1966

Just because you're going forward doesn't mean I'm going backward.

— BILLY BRAGG, "To Have and to Have Not"

We exit the port southward onto Interstate 880, one of the Bay Area's less picturesque freeways and the one everyone loves to hate. Another double-stacked train runs parallel to us, the gaps between the containers revealing black cylindrical train cars full of oil. For a while, our car is like a lone beetle among the container-hauling trucks, but as the road ascends to pass over downtown Oakland, we're joined by commuters, trucks for FedEx, Walmart, and Amazon, and the occasional tech bus. (Pre-pandemic, there would have been more of them.) Between flat industrial buildings and a six-story county jail, cars from Interstate 980 join us to form a loose, anonymous morning mass. In the opposite direction, a white truck passes, its side reading, "Daylight Transport—Saving Time since 1977."

"Let's keep pushing through this Thursday!" says the deejay of Q102, the Bay Area's Throwback Station, before cutting to ads. Upstart.com offers to consolidate our debt with a loan; Sakara Life wants to boost our energy by delivering organic pre-made meals; Shopify suggests that moms everywhere can use its platform and go "from first sale to full scale." A podcast offers to enlighten us about meme stocks, cryptocurrency, and AI, "whether you're a seasoned investor like me or new to the game." Zoom pitches us something called a "united

bundle" for large enterprises, small businesses, and individuals. Up ahead, a
billboard announces that a tech company is hiring "self-driving people."

IN AN ARTICLE titled "Why Time Management Is Ruining Our Lives,"
Oliver Burkeman observes that when employment is insecure, "we must
constantly demonstrate our usefulness through frenetic doing." But
even when seeing time as money is not strictly necessary, this imperative
remains, and it often has a moral valence. For example, when I try to
imagine the extreme opposite of "frenetic doing," I picture one of my
favorite characters from *Futurama,* Hedonismbot, who is shaped like a
blobby Roman senator reclining on a chaise. In his occasional appearances, Hedonismbot requests that chocolate frosting be smeared on
him, asks whether the orgy pit has been "scraped and buttered," and
dangles grapes over his mouth, crying, "I apologize for nothing!" Far
from "improving every shining hour" and casting a prudent eye toward
tomorrow, he consumes hours (and many other things) on the spot,
seeming to epitomize sinful waste.

Particularly in the United States, it's not just busyness that's considered good—it's a specific image of industry, the result of a long romance
among morality, self-improvement, and capitalist business principles. It

owes much to Protestantism, an intensely rigorous and personal form of Christianity that sanctified hard work. Protestantism arose alongside a European bourgeoisie who had personal industry and commercial activity to thank for their social dominance. This is the same rhetoric that would be exported to the colonies, as described in the previous chapter. According to the Protestant work ethic, you weren't supposed to get rich in order to spend your money; work and the accumulation of wealth were inherently good, a way to serve God. And if you did manage to get rich, those weren't your riches to spend; they were God's, and they signaled your eternal salvation. Rich (but ascetic) was the way to go, the "business" of life a moral affair.

As a form of Protestantism, seventeeth-century Puritanism invited introspection and constant evaluation of the self against a high moral standard, a practice that included the use of daily journals where self-observation and measurement could take place. For example, in a close reading of the daily diaries of Samuel Ward, a Puritan minister who wrote between 1592 and 1601, Margo Todd finds Ward "fashion[ing] himself as both preacher and auditor, exhorter and penitent." This tension, Todd writes, explains his inconsistent use of pronouns even within a single sentence—referring, for example, to "thy gluttony at dinner, which distempered thy body; also my little care to pray." In these writings, Ward speaks both "for God in admonition, and for himself as sinner," inhabiting a space of both confession and rebuke. (He definitely would have been disgusted by Hedonismbot.)

In the industrializing United States, the Protestant work ethic would come under threat, especially as assembly-line jobs offered little room for advancement and were harder to find meaning in. Yet the feeling that a certain model of thrift and efficiency was intrinsically good survived, as did a sense of personal accounting. This made the rhetoric of "personal development" a potentially fertile ground for Taylorism as it spread throughout American culture. After all, as a system for ordering time and increasing profit, Taylorism never restricted itself merely to the workplace; that would have been impossible in a time when, as Taylor put it in *The Principles of Scientific Management*, "it is to the greater productivity of each individual that the *whole country* owes its greater prosperity." It was just one part of an obsession with rationalization, efficiency, and measurement that pervaded American Progressive Era culture at large.

What happens if you try to apply Taylorism to yourself? One possible answer shows up in Donald Laird's 1925 book, *Increasing Personal Efficiency*, "a practical and detailed manual to help the reader through step-by-step procedures to improved self mastery." Laird, a psychologist whose work prefigured modern ergonomics, personality testing, and self-tracking, spares no admiration for Taylor and laments that more of life isn't properly Taylorized: "Engineers have improved this world remarkably in the last century; but I cannot find any authority to state that man himself has improved in the last two dozen centuries. In fact, if we believe the arguments of the eugenicists[,] we might have to infer that mankind has actually deteriorated." Laird's reference to "the eugenicists" echoes his book's opening concern, which is the number of people currently institutionalized for mental illness. Seeing it from a systematic point of view, Laird interprets psychological breakdowns as a pitiful sign of lost productivity, an issue to be solved by better working practices.

An attempt to move Taylorist principles from the factory into the mind, *Increasing Personal Efficiency* promises that by becoming your very own "time-study man," you can greatly increase your output. After gesturing toward efficiency increases in offices, houses, and cars, Laird hazards "a personal question": *"Have you given as much attention to your personal mental efficiency? Are you laying your mental bricks with eighteen movements or with five?"* The book is shot through with the cultural moment's fixation on speed, mastery, and a single-minded mission to cut out the useless. After a section that tests you on speed-reading, Laird urges you to "avoid excessive eye movements" while reading and offers this rather puzzling advice: "Do not read on trains, cars, or busses. Nor should you look out of the windows. Instead, watch the other passengers and relax. Every minute of complete relaxation while riding can be subtracted from your sleep."

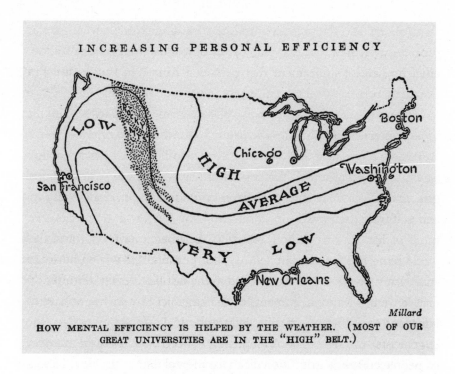

INCREASING PERSONAL EFFICIENCY

Millard

HOW MENTAL EFFICIENCY IS HELPED BY THE WEATHER. (MOST OF OUR GREAT UNIVERSITIES ARE IN THE "HIGH" BELT.)

One of the most morbidly fascinating aspects of *Increasing Personal Efficiency* is the way in which it introduces a division of labor into thought itself. Laird opens a chapter titled "Effective Thought" with an illustrative comparison. First, we see an executive sitting quietly alone, looking over typewritten sheets and a small map for a long while before making a call to a stenographer. This whole time, "he was apparently as motionless as a statue. But was he doing nothing? He was probably doing the hardest work of the week. The man we have just been observing was engaged in active thought." Next, we see a "maiden" in a comfortable chair, absorbed in a book as the curtains blow in the breeze. She, too, is motionless—until she looks up in reverie, imagining knights and fair ladies.

Laird admits that the maiden, like the businessman, wasn't exactly doing nothing. "She, too, was engaged in thinking," he writes, "but not in the active, constructive way that we observed in the earlier scene." The difference has to do with what the thinking accomplished, mostly in business terms: "She was accomplishing nothing more than the satisfaction of her wishes for romance. Our business man may have revolutionized his field of industry as the result of his hour of active thinking." With its emphasis on intentionality, Laird's active thinking might be mistaken for what we would now call mindfulness. But active thinking, which "separates man from the brute," sounds more aggressive to me. It evinces something that Samuel Haber identifies in his history of Progressive Era efficiency: a "turning toward hard work and away from feeling, toward discipline and away from sympathy, toward masculinity and away from femininity."

The end of the chapter on effective thought exhorts you to ask yourself, "Do I spend more time in active thought than I do in passive thought?" In other words, are you the boss of your own mind? The strange truth is that Laird's reader is implied to be both the time-study man and the timed, the businessman and the daydreaming girl. In place of the question "How Much TIME do They pay You?" we now get something like "How Much TIME do You pay Yourself?" Laird doesn't want to catch you slacking off on the job—*Don't look out the window!*—even in the space of your own mind. He wants to do you a favor and help you whip yourself into shape before you die in the competition, in the manner of a factory manager who has failed to read *Factory Magazine*.

While the forms and styles of personal development changed throughout the twentieth century, the legacy of personal Taylorism in many time management books is clear. Generally, their advice can be summed up as follows:

1. Keep an ever-more-detailed log of how you spend your time, in order to identify deficiencies and measure the increase in your productivity. (This part often involves filling in a time spreadsheet, with increments as short as fifteen minutes.)
2. Identify your most productive time in the day and arrange your work accordingly.
3. Militantly eliminate distractions, anything not having to do with your work. (What used to be stealing time from your boss is now stealing time from yourself-as-boss.)

Within certain bounds and with certain types of work, this is not bad advice. But when it is set in historical context, what's interesting is the kind of time we're meant to log in the grid. It's the fungible time we saw in the previous chapter, and the concept that each individual has an equal "supply" of these fungible hours to exploit is still the bedrock of mainstream time management. Though it is a patent mischaracterization of how we actually experience time, many people still subscribe to the adage that "everyone has the same number of hours in the day"—in essence, that everyone was born with the same number of hours in God's bank. Thus, Roy Alexander and Michael S. Dobson write in their 2008 book, *Real-World Time Management:*

> Your own time is not nearly as scarce as widespread wailing indicates. Say you work 40 hours a week for nearly 49 weeks per year (52 weeks less 2 weeks of vacation and six holidays). In a year your work time comes to 1,952 hours. Deduct that from your total inventory of time—8,760 (365 x 24) hours a year. Then deduct 488 hours for traveling to and from your job, 1,095 hours for meals (3 hours a day every day of the year), another 365 hours for dressing and undressing (1 hour a day), and 8 hours' sleep a night—count 2,920 hours for that. Your total deduction: 6,820 hours. Subtract 6,820 from 8,760 and you get 1,940 hours to do as you please. That's nearly 81 days of 24 hours apiece, 22 percent of the entire year!

The book is obviously not written for anyone tasked with caretaking or household duties, but we'll get to that shortly. Let's say for now that you really do have 1,940 hours with which you can do as you please. Just as in scientific management, you can always drill down from labor hours into labor minutes. In his book *15 Secrets Successful People Know about Time Management: The Productivity Habits of 7 Billionaires, 13 Olympic Athletes, 29 Straight-A Students, and 239 Entrepreneurs*, Kevin Kruse describes putting a giant poster up in his office with the number 1,440 on it: "I encourage you to try it yourself. Just draw a big '1,440' on a piece of paper and tape it on your office door, under your TV, next to your computer monitor— wherever it will best serve as a constant reminder of the very limited and oh so precious time you have each day." Once again, you're said to have the same number of minutes as everyone else. The only task that remains is running those minutes through your factory with greater and greater efficiency, as though you were using up especially clean-burning fuel. This is important because, as Kruse writes, "you can't make more time, but you *can* increase your productivity. Increasing your energy and focus is the most important secret to achieving 10x productivity in the same amount of time." Moments—your moments—are the elements of profit.

GRIDLOCK DOES NOT typically inspire a love of other people. As our cars sludge through the space between the freeway's faded red noise barrier walls, every movement made by one driver is either preempted or begrudgingly accommodated by another. Inside the cars, people pass the time by listening to things we can't hear, talking on the phone, eating, putting on makeup, or watching shows on phones tucked into the dashboard. While some seem resigned to the traffic, others weave restlessly through any little space they can get. Approaching the suburb of Hayward, we creep past the unforgiving gaze of the 880 Minion, a painted metal sculpture of a bacteria-shaped, overall-wearing Minion from Despicable Me *that someone has affixed to their roof so it can loom over the sound barrier.*

*   *   *

WITH ITS SUPPOSEDLY egalitarian quality, this approach to time seems to thrive in a bootstrapper culture. Appropriately enough, the modern meaning of *bootstrap*—to "better oneself by rigorous, unaided effort"— arose around the same time that books like Laird's were coming out.* Today's bootstrapper culture—informed by neoliberal values and intensified by the withdrawal of government services, fragmenting work, and the erosion of the social safety net—demands that each individual be responsible for her own destiny, ensuring her own security against that of others. To do so, she must invest her own time and effort, provide her own training, and calculate her own risk.

In the United States, the idea of the individual-as-entrepreneur exists as much in the cultural ether as it is does in labor statistics. A 2012 Pew study found that 62 percent of U.S. respondents disagreed with the statement "Success in life is determined by forces outside our control." The number of people who felt this way was lower in Spain, Britain, France, and Germany (where only 27 percent disagreed). Asked to choose between "freedom to pursue life's goals without state interference" and "state guarantees nobody is in need," the former won out in the United States 58 percent to 35 percent, with those numbers essentially reversed in the other four countries.† In a 2017 study, compared to Democrats, U.S. Republicans predictably attributed a person's wealth to their having "worked harder" versus their having "had advantages in life," and they attributed their poverty to "lack of effort" over "circumstances beyond [their] control."

This is an old debate, of course—effort versus circumstances. As I mentioned in the introduction, the question of how much latitude one

---

* Originally, "pulling oneself up by one's bootstraps" was a metaphorical description of attempting the actually impossible. In an 1888 physics book, for example, the question "Why can not a man lift himself by pulling up on his boot-straps?" came directly after the question "Can a man standing on a platform scale make himself lighter by lifting up on himself?"

† These results are consistent with a later 2019 Pew study, which found that a median of 53 percent of respondents in Western Europe and 58 percent of respondents in Central and Eastern Europe agreed with the statement, "Success in life is pretty much determined by forces outside our control," as opposed to 31 percent of American respondents.

has within "forces outside our control" is one of the abiding questions not only of sociology but of philosophy, in that you eventually get to questions of free will.* But, for the purposes of this chapter, one way to explore the topic is with a card game. I learned it as "Asshole," but it's also sometimes referred to as "President," "Scum," or "Capitalism," and it may have arrived in the West from China, where similar games (such as Zhēng Shàngyóu, or "Struggling Upstream") have long been popular.

For the most part, Asshole is a standard card-shedding game where you can play only certain cards at certain times. This game, however, has a form of generational memory. The winner of the first round becomes the "president," and the second-place winner the "vice president," while whoever lost becomes the "asshole," and the second after them is the "vice asshole." Before the subsequent round, everyone has to get up and rearrange themselves accordingly around the president. The asshole is tasked with shuffling and dealing.

Once the cards are dealt, the asshole then has to take two of their best cards and trade them with whatever two cards the president wants to get rid of; the vice asshole trades one with the vice president. Voilà! A miniature version of structural inequality.† The real torture of this game is that, when you're the asshole, no one sees the good cards you had to give up or the bad cards you were stuck with. Therefore, no one knows how much your poor performance had to do with the initial exchange

---

*For some explorations of "forces outside our control," see sociologist Pierre Bourdieu's concepts of field, habitus, and cultural capital in *Practical Reason: On the Theory of Action* and Harry Frankfurt's distinction between first-order desires (what you want) and second-order volition (what you want to want) in "Freedom of the Will and the Concept of a Person."

† Of course, in real life, the relationship between rules and practices is more complex and iterative than in a card game like Asshole. But this extreme example can be a helpful illustration of what it's like to occupy different positions in a network of advantages and disadvantages. In fact, a similar card game was used in a study of perceptions about inequality. When researchers at Cornell University taught "the Swap Game" to study participants in 2019, they found that winners were twice as likely as losers to describe the game as fair. While cautioning against generalizing card game results to actual socioeconomic inequality, the researchers nonetheless noted resemblance to "real-life stratification processes in which the distribution of opportunity matters for the distribution of outcomes."

and how much with a lack of skill in playing the cards you were dealt. And because the rules of this game are not negotiable, your only option as the asshole is to try desperately to be strategic. You must be the master of your own cards.

If we take this game as a metaphor, we can appreciate how much business there is in teaching people to play their cards right in a culture that systematically blocks avenues toward changing the rules. The resulting rhetoric of self-mastery, refashioned for the age of YouTube and Instagram, finds its apex in a group of people I will henceforth refer to as "productivity bros" and, in particular, in a pair of products designed and sold by John Lee Dumas. Among other things, Dumas runs a daily podcast called *Entrepreneurs on Fire,* where interviews with successful entrepreneurs are supposed to inspire listeners on their own entrepreneurial journeys. "If you're tired of spending 90% of your day doing things you don't enjoy and only 10% doing things you love, then you're in the right place," the podcast website reads.

In 2016, Dumas launched a Kickstarter campaign for something called *The Freedom Journal,* which promises to help you "CRUSH your #1 goal in 100 days." The leather-bound journal contains mostly the same two pages over and over again, asking the user to specify their goal and then evaluate their progress, spliced with "10 day sprint recaps" and quarterly reviews. Dumas's subsequent product, *The Mastery Journal,* doubles down on the quantification of tasks by breaking the day up into four sessions, with self-reported "productivity" and "discipline" scores. These scores are meant to be averaged and plotted out on ten-day productivity and discipline graphs. *The Freedom Journal* and *The Mastery Journal* were sold together as part of the "Success Pack for 2017." The pairing of freedom and mastery might be incidental, but the idea that one could be both the freed and the mastered speaks to the dual face of "empowerment."

Among productivity bros and many others, the Taylorist obsession with routines has morphed into an unhealthy fixation on morning routines. Craig Ballantyne, self-described as "the world's most disciplined man," has at least ten videos on this topic. In one, "This Morning Routine Will Increase Your Productivity and Income," he shows how he "dominates" his morning in order to live the life of his dreams and travel to five new countries a year. We see him waking up at 3:57 A.M., "twelve minutes after Mark Wahlberg and three minutes before The Rock." Unlike other entrepreneurs who while away their morning time with yoga and journaling, Ballantyne gives himself fifteen minutes to get to the computer and start working on his book *The Perfect Day Formula: How to Own the Day and Control Your Life.* And, of course, no such video would be complete without the making of a power smoothie.

Ballantyne's other videos and appearances involve advice on crushing or dominating the following: your goals, competition, sales, social media, chaos, and life itself. The freedom that productivity bros offer, however, is not simply a form of work-life balance within the status quo. Both Ballantyne and Dumas are acolytes of Tim Ferriss, author of *The 4-Hour Workweek,* which promises freedom from others—and from the necessity of selling your time at all. The idea is that by constructing passive income streams, you free yourself from the constraints of capitalism by recapitulating it within your very person. Books like Ari Meisel's *The Art of Less Doing* promise that "modern methods like the 80/20 rule, the three D's, and multi-platform repurposing let you build a high-powered, traditional-style 'success factory' that only requires one employee to run." Further illustrative of the genre is the site Screw the Nine to Five, whose founders share how they're "making money from our fleet of over 30 online businesses while living overseas."

In comparison to those other forms of "screwing the nine to five"— worker organizing, legislation, and mutual aid—the allure of the productivity gospel is supposed to be that you don't need anyone but yourself to achieve freedom. The problem is that, according to this plan, more freedom requires ever more (self-)mastery, ever-better playing of your cards. Increasingly unable to control any of her surrounding circumstances, the consumer of this kind of self-help risks turning on herself with displaced intensity, surveilling herself with spreadsheets and averages, docking points, and meting out punishment in a secularized space of "confession and rebuke." This approach perfectly fits the neoliberal worldview of total competition. Not only will you not find help among others, but everyone else becomes your opponent while you jealously guard and "supercharge" the time you possess. Whether you wring enough value out of it is on you.

THE SAN FRANCISCO BAY is not far, but you'll have to take my word for it, because at no point on our route is it visible. Presently, it's blocked by a distribution center for XPO Logistics, a company accused by unions of wage theft and attempts to "Uberize" freight. Occasionally, the monotony of the traffic jam is broken by the sight of a nearly all-white egret bound for the bay; a hefty red-tailed hawk waiting for rodents from atop a speed limit sign; or a turkey vulture making wide, wobbly circles in the sky. I recently learned that those vultures have the largest olfactory system of any bird and can smell things from a mile away. Here in the car, all we can smell is aged plastic and upholstery with a whiff of brake dust.

An electronic sign up ahead advertises the current driving time to Milpitas, the San Jose International Airport, and Menlo Park, the time shifting with the traffic. These numbers represent a fluctuating cost in life minutes, a different kind of toll than the one we're about to pay on the bridge. But, for the people living those minutes, the numbers mean different things. We look again at the other drivers, who are navigating their own temporal topographies, rushing to

meet some unseen demand. In the search for affordable housing and a place suit-
able to raise children, people have begun commuting to jobs in the Bay Area from
more than a hundred miles south and east, spending two or three hours in the
car each way. In some way or another, most people out here are just trying to
make it work.

"YOU WAKE UP in the morning, and lo! your purse is magically filled with
twenty-four hours of the unmanufactured tissue of the universe of your
life! . . . No one can take it from you. It is unstealable." That's a line
from Arnold Bennett's 1908 book *How to Live on 24 Hours a Day.* Henry
Ford gave five hundred copies of this book to his managers. The title
remains popular and was reissued by Macmillan in 2020 under the category of self-help.

Among people for whom the idea of twenty-four unstealable hours
makes little sense, working parents would have to be at the top of the
list. May Anderson, an admin for a Facebook Group for working moms,
told me that she had given up on mainstream time management books,
comparing them to the common financial advice "Just don't buy the
damn latte." Contrary to Kevin Kruse's "1,440" poster, not all minutes
are created equal. After listing the many tasks involved in her typical day
as an engineer and mother of two in rural Utah—where "work" meant
both paid work and unpaid childcare and housework—May described
how much the pressure of thinking about what she needed to do built
up in the back of her head and intensified the feeling of time scarcity.
"When you do sit down for ten minutes just to relax, it's not super relax-
ing," she said.

The pressure of obligations and the psychological variability of time
are just a few of the ways in which the idea of equal clock hours falls
apart as soon as you poke at it. Robert E. Goodin, a professor of philosophy and social and political theory, calls this supposed equality a "cruel
joke." First of all and most basically, some people control other people's
time. While slavery has been (officially) abolished, it's still the case that
the majority of people "rent their time to employers simply in order to
survive." Until the necessity to do that is addressed—for example with
universal basic income—"gross inequality" in temporal autonomy will

persist. Further, unless you are some kind of celebrity or high-powered consultant, the price for which you sell your time is liable to reflect aspects over which you have no control, like gender, race, and the current economic situation.

As we saw in the previous chapter, elements of time and pacing are also often outside workers' control. Given that time management books are marketed to the individual, this controlling presence is not usually acknowledged, but sometimes you can catch the ghost of it. Like the manager claiming that it was the "machines [that] kept the 'girls' at their desks," one 1990s time management book laments, "The computer chip didn't free us. It coerced us to produce at *its* speed." In *Real-World Time Management,* Alexander and Dobson provide an imagined Q&A session with an imagined reader, who protests, "You tell me to work on priorities. But *they* won't let me!" The authors' response is ominous: "You must control not only priorities but *them* (whomever [*sic*] they are)."

The question of *them* leads into a politics of time that is about more than numbers (even unequal numbers) of hours in the bank. Although time management often responds to the imagined feeling that one "doesn't have enough hours in the day," time pressure is not always or solely the outcome of a quantitative lack of time. A feeling of time pressure can result from constantly having to switch tasks or coordinate with external factors. Here, the German word *zeitgeber,* which translates roughly to "time giver," is useful. Frederick Winslow Taylor's detailed time charts would have been zeitgebers for industrial workers (or purchasers of Dumas's *The Freedom Journal*). For stay-at-home mothers, their children's moods, health needs, and school schedules might be zeitgebers. For a long time, the ten-week university quarter system and the ever-expanding Bay Area rush hour were my zeitgebers. For someone with a chronic condition, cycles of illness would be a zeitgeber. For an Instacart worker, both the whims of customers and the app interface are zeitgebers.

There's a pattern here: With a zeitgeber, someone or something is always giving time to someone else—not in the sense of gifting them minutes and hours, but in the sense of determining their experience of time. To follow a zeitgeber is to become entrained; your activities become entrained to patterns outside you; or others must become entrained to yours. But, as anyone with a chronic illness who works a nine-to-five job knows, different zeitgebers can conflict, and not all are created equal. Just as different peoples' hours have different "rental prices," some people are compelled by external structures to entrain to the lives of others.

In "Speed Traps and the Temporal," Sarah Sharma illustrates this negotiation with an experience she and her friend had when their train was delayed. Her friend—on her way to visit her twelve-week-premature baby, who was waiting at a hospital for her and her transportable cooler bag of breastmilk—had no time to spare. Scanning the crowd, the friend picked out a dark-suited businessman who appeared to be ordering an Uber. Based on his appearance and demeanor, she predicted that he was headed downtown, and she sped up to walk in step with him; although he agreed to share an Uber with her, he "never lost his pacing," walking fast and typing up until the moment their Uber arrived. Sharma writes that her friend used survival-mode intuition to exploit a power dynamic, having spotted "all the signifiers of the iconic and privileged protagonist of fast living in a culture that is dominated by the discourse that the world is speeding up":

> He had on a suit, [walked with] a quick step, and was tapping madly away into his smartphone. He was plugged in and on the go, using network time to navigate the space of flows so he could bypass the public transportation system that had ground to a temporary halt. He could maintain control over his time by ordering up a driver with his Uber app and get to work without losing a minute. He was in charge of his mobility and his time but also the time and mobility of others.

This kind of negotiation exists in stark contrast to the myth of equal hours. For an individual, time is not the measurement of something real but, rather, a "structuring relation of power." Just as your experience of a game of Asshole will depend on what happened in the previous round and where you are sitting, "individual experiences of time depend upon where people are positioned within a larger economy of temporal worth." It's an important clarification, and it reminds me of a Goodreads review of Kate Northrup's *Do Less: A Revolutionary Approach to Time and Energy Management for Ambitious Women*. In that book, one piece of advice is to sync your work schedule with your menstrual cycle (another zeit-geber) in order to exploit differing energy levels throughout the month. Reader Sarah K. observes that this makes sense only for someone with money or control of her own time. "Say I simplify by hiring a house-keeper," she writes. "What do I do when she is in her new moon phase, which should be dedicated to rest and reflection, and cannot clean your house? Better hope it syncs up with your waxing crescent/'putting in the work' phase!"

In the case of an on-the-go working woman who buys time for herself through the services of others, it's clear whose temporality is privileged. But within the workplace, more shades of power remain. Entirely apart from the phenomenon of the "second shift" and women's frequent role as the "default parent," multiple studies show that women in the workplace are expected to say no to work less frequently than men are. For example, one study showed that *both men and women* expect women to offer help and respond to requests for help; in the study, men would wait to volun-teer for favors when there were women in the group, but would raise their hand earlier if the group contained only men.

In the words of Sallie Krawcheck, a woman interviewed in an *Elle* story about women saying no, "We're all socialized to believe moms are helpful and dad is watching football." For BIPOC women, the disparity is worse. In a *Harvard Business Review* article, a manager at a technology company described the double bind: "If I don't accept the office housework, I'm considered an 'Angry Black' woman." Other professional women of color say that, when they attempt to control their time in the workplace, they're cast as "aggressive, out-of-character, or too emotional."

One of the things I talked about with May, the administrator of the working moms Facebook Group, was how everything from offices to cars was designed for men. (Car crash dummies are based on so-called average men.) She then told me about how, in a group for women engi-neers, someone had called out the women who were getting promoted for acting like men. "So now these women who have had to do this in order to get ahead are being criticized," she said. "And I'm like, I get it? I don't know. I've been on both sides." I nodded, realizing something out loud. "It's almost like the car seat, but as a metaphor. It's like you're trying to make yourself more man-shaped in order to not die in the car."

Becoming more man-shaped in order not to die in the car was my unwitting description of a *Lean In* type of feminism, and of time management aimed specifically at women. One straightforward example of this is Laura Vanderkam's 2010 book, *168 Hours: You Have More Time Than You Think,* which presents a Christian-tinted and slightly softened version of "crushing it" as both a career woman and a mom. Vanderkam's recommendations include finding your dream job, outsourcing tasks you dislike, and identifying your "core competencies" so as not to waste time on things you're not already good at. Vanderkam also provides her own handy time sheets with half-hour increments, allowing readers to view the 168 hours they have in a week as a "blank slate." Whether at home or at work, to be nimble means to be nimble like a corporation. A *Publishers Weekly* review of the book provided an apt summation: It contains great career advice at the same time that it risks "pummeling the life out of life."

The promise of *168 Hours* is that the reader—implied to be a working mother in the same socioeconomic class as its author—really can "have it all." Although this avoids the question of why women still do an outsize share of paid and unpaid work, suggesting instead that a woman's response to this reality should be a better allocation of resources, it's not an outright cruel book. Neither is it an empty, suspiciously pyramid-shaped offering like those of the productivity bros, who appear to make videos for people making videos for people making videos. What *168 Hours* offers is amelioration—making an uncomfortable situation more comfortable for those who have the same privileges as the author. In that sense, *168 Hours* has the same goal as a lot of self-help: It's targeted at an individual just trying to play her cards better.

Within a certain scope, that's just fine. Self-help has generally promised to revolutionize your life, not the social or economic hierarchy—and you can't really blame anyone for not fulfilling a promise they never made. At the same time, even seemingly practical self-help can read as an invitation to find a niche in a brutal world and wait for the storm to pass you over. The anthropologist Kevin K. Birth has described clocks and calendars, seemingly inert pieces of technology, as "cognitive tools that think for their users," reproducing "cultural ideas of time" and "structural arrangements of power." Just as a gridded schedule reproduces the idea of time as fungible units, advice for "becoming more man-shaped to not die in the car" reproduces the life of the wrong-shaped car. It is great advice to seek your dream job, but in many of these books, the implied answer to the question "Who will do the low-wage work?" is that it doesn't matter, as long as it's not you. That answer doesn't feel so good.

Time management illuminates the assumptions of the will-versus-circumstance debate because it takes the individual as the absolute unit and the near future as the time frame, at the expense of the collective good. Even Sharma understands time management's appeal, which also happens to be its danger. "It is an intoxicating concern: how to have a better relationship to time control and technology," she writes. "But this cultural fixation on time control and one's ability to modulate time, to manage it better, slow it down and speed it up, is antithetical to the collective sense of time necessary for a political understanding of time." It's precisely this political understanding of time that would allow one to look outward, imagining different "structural arrangements of power." This can't be done alone, and it usually can't be done in the short term. In the lengthy meantime, I'm reminded of a saying that a Spanish journalist shared with me, regarding the phenomenon of burnout: "Do you need a therapist, or do you need a union?"

At some point, you hit the limit of what an individual can do. Commercial time management recognizes this, advising you to "outsource" parts of your life, a market-based version of an old intuition about support networks. I wasn't surprised when May told me that she'd thought about getting together a group of seven other moms in which one would make dinner for everyone else one night a week. "I think a support system has got to be, like, the number one way to help with our time management," she said, citing informal networks of in-laws and friends. Taking this further, we might imagine, as Angela Y. Davis did in 1981, that "child care should be socialized, meal preparation should be socialized, house-work should be industrialized—and all these services should be readily accessible to working-class people." If anything, the pandemic showed us the opposite of this: the cost of every family (usually women) having to handle their own childcare, meal prep, and other household duties.

This is the perspective that allows us to consider the rules of the card game. If time management is not simply an issue of numerical hours but of some people having more control over their time than others, then the most realistic and expansive version of time management has to be collective: It has to entail a different distribution of power and secu-rity. In the realm of policy, that would mean things that seem obviously related to time—for example, subsidized childcare, paid leave, better overtime laws, and "fair workweek laws," which seek to make part-time employees' schedules more predictable and to compensate them when they are not. Less obviously related to time—but absolutely relevant to it—are campaigns for a higher minimum wage, a federal jobs guarantee, or universal basic income.*

And then there are all those time-consuming things that someone who has never experienced poverty or disability might not think about. In a piece on the "time tax" experienced by people dealing with gov-ernment services, Annie Lowrey observes that poorly run bureaucracy deepens gulfs between the rich and poor, white and Black, sick and healthy. She calls it "a regressive filter undercutting every progressive policy we have." Lowrey suggests the elimination of hurdles like asset tests and interviews, as well as the use of better tools—for example,

---

* One pilot UBI program in Stockton, which gave random residents five hundred dollars per month for two years with no strings attached, was found to reduce anxiety, depres-sion, and financial strain among recipients, specifically allowing "women who spent years prioritizing the needs of others over their own wellbeing . . . to focus on their health and subsidize gaps in family health care." As an example of guaranteed income (like UBI but targeted to specific communities), the Magnolia Mother's Trust gave a thousand dollars a month for a year to one hundred families headed by Black women in subsidized housing in Jackson, Mississippi. In a *Ms.* magazine series publishing the stories of each cohort, a participant named Tia gave a personal account of easing time pressure: "There's just knowing that if your kids get sick, it's going to be okay. That if needed to, I could take time off to care for my child without having to worry that my paycheck would be short."

well-designed forms easily readable in one's own language. But she also notes that the history of the time tax has deep, persistent roots in racism, skepticism of bureaucracy, and an old distinction between the "deserving" and "undeserving" poor.

Likewise, a true political understanding of time cannot be afraid to address the most general, widespread, and entrenched structures of power. For example, in a talk called "The Racial Politics of Time," author, activist, and cultural critic Brittney Cooper opens with the provocation that "white people own time." This has as much to do with how the colonized of the world are seen as existing outside history as it does with the fact that white people overwhelmingly set the pace of the workday and dictate the worth of everyone else's time. Plus, in many cases, you don't have to buy someone's time in order to waste it. As if in direct rebuke to Bennett's unstealable twenty-four hours, Cooper quotes Ta-Nehisi Coates: "The defining feature of being drafted into the Black race [is] the inescapable robbery of time." In place of the equal-hours myth, Cooper has this suggestion:

> No, we don't all get equal time, but we can decide that the time we do get is just and free. We can stop making your zip code the primary determinant of your lifespan. We can stop stealing learning time from black children through excessive use of suspensions and expulsions. We can stop stealing time from black people through long periods of incarceration for nonviolent crimes. The police can stop stealing time and black lives through use of excessive force.

If time is simply life, then, as Cooper makes clear, the question of "time management" boils down to a question about who controls whose lives. This is an example of the contrast Sharma highlights— between the political understanding of time and the dream of mastering our individual time units. The further question of what time *is*—an issue of language—is one I will come back to in chapter 6. My point for now is simpler: Only by acknowledging the real contexts in which experiences of time play out can we arrive at a different notion of "time management"—one that doesn't simply reproduce a cruel game.

*   *   *

AFTER AN EXIT onto 84 West, the flat terrain gives way to low-lying hills and scraggly, wind-tossed trees. A series of signs reads DO NOT STOP before a camera registers the FasTrak device in our car, which causes it to beep in satisfaction. The bay water brings a sharp sulfur smell through the car's vents. It's the anaerobic bacteria that live in the shallows on either side of us, where a few more egrets are stepping carefully. Through the haze, the Santa Cruz Mountains stretch as far as we can see in either direction, like a torn strip of blue paper.

The bridge launches us up past giant transmission towers and sets us down on the peninsula, a hostile-looking plane of salt marsh and concrete. In the distance, half hidden by Monterey pines, you can make out an odd conglomeration of buildings, mostly white with panels of red, teal, light blue, yellow, and gray. Only at the long light at the intersection where we're turning left do we finally see the sign: a giant blue thumbs-up and the text FACEBOOK: 1 HACKER WAY. The Instagram headquarters is in there, too. Normally, you might see people on blue Facebook-branded bikes crossing this road on their way to any of the campus's many buildings, but the massive parking lot looks emptier than usual, and many employees are working from home. Before the light turns green, I look down at my phone, ill-fitting in the old cup holder. A companion, yes, but also a device for measuring life.

*   *   *

FOR THE INDIVIDUAL, the opposite of time management would seem to be burnout. Things pile up; they can't be crammed into the time grid. Life becomes inelegant. For productivity bros as much as for Donald Laird in *Increasing Personal Efficiency,* burnout is a major concern, a failure of the machine.

I once found an embarrassingly spot-on characterization of my life in a paper on desynchronized work. The sociologist Hartmut Rosa is describing a hypothetical character named Linda, an overwhelmed professor who rushes through her day, never having enough time to fulfill all her obligations to students, co-workers, family, and friends; expected to be always available, answerable to everyone; with the feeling that she's always falling short and running behind. "Not enough time for cooking. Not enough time for her lover. Not enough time for household work. No time to go for a workout. She does not do enough for her health, her doctor tells her. At the end of the day, she is guilty because she is too stressed, not relaxed enough; she does not get her work-life balance right."

Within this state of conflict, Rosa duly highlights the role of digital technology in the expansion of "legitimate claims" inside and outside work—the idea that someone could be reachable by anyone anyplace and at any time. Linda does not have access to *Feierabend,* the feeling of leisure that peasants and farmers might have had when the cattle and children were in for the night. She might find *Feierabend* in rare situations, like staying in a mountain hut with no reception. Otherwise, constant demands mean that the mismatch between what Linda can do and what is asked of her is not an "abstract fact of life" but, rather, an "acute dilemma" she experiences in every moment.

Rosa then asks whether this story makes sense for anyone outside "a tiny jet-set elite of society." (Elizabeth Kolbert has likewise quoted economists' descriptions of complaints about busyness as "yuppie kvetching.") He compares Linda's situation to that of the temporality experienced by a truck driver, factory worker, nurse in a hospital, or shop assistant. People in these roles experience time pressure specifically *within* the job: The truck driver struggles to keep pace with his deadlines while obeying speed limits, the factory worker is pushed past capacity by her boss, the shop assistant deals with impatient customers, and the nurse is expected to give more care and attention while the hospital squeezes in more and more patients. "In work," Rosa writes, "underprivileged employees have very little time-sovereignty, pressure is put on them by the boss or by external authorities who regulate their time-budget. It is [these external factors] they can directly locate as the source of pressure. For Linda, the source of pressure is outside the [job] situation, it is herself she needs to blame."

With his concept of "discretionary time," Goodin makes a similar distinction between the Lindas and the non-Lindas, so to speak. Like discretionary spending, discretionary time is that which, strictly speaking, you don't *have* to use for something. You just choose to do so, for whatever reason. This idea allows us to distinguish between someone who truly has no free time and (for example) an ambitious person who voluntarily works long hours according to personal notions of necessity, only to wish she had more time. Goodin finds that some people, especially childless dual-earner couples, harbor a "time pressure illusion." These are people who, strictly speaking, have lots of free time—it's just that, according to their discretion, they don't see it as free.

If, for a burned-out Linda, the call really is coming from inside the house, then the question is: Why? To some extent, you can trace part of it to the increasing "flexibility" of work. If you don't know what's coming down the line, preparing for the future becomes an infinite task. There are certain forms of work (creative, freelance, or adjunct) that make it especially unclear whether someone is a Linda or a non-Linda. I have known many adjunct professors who have had to act like workaholics in order to "remain relevant" and secure work—and even when they do, a class (and therefore payment) could be canceled at the last minute. Adjuncts are typically not given benefits, and in 2019, one quarter of adjuncts in the United States were on some form of public assistance, while one third lived below the poverty line.

More broadly, the "discretion" of discretionary time can be difficult to assess in a culture whose injunction to "adapt or die" can be fearsomely convincing. Addressing the global rise of burnout, Rosa observes that drugs that slow people down are declining in favor of speed, amphetamines, and other substances that "promise 'synchronization' (like Ritalin, Taurin[e], Modafinil [sic], etc.)." Most forms of human "enhancement," he says, involve getting faster at something. Jamais Cascio, an author and futurist, shares a related anecdote in a documentary on human biotech-nologies. Given a legitimate prescription for modafinil, a wakefulness drug intended for his international travel, Cascio finds himself occasion-ally popping a pill at home when deadlines come around. Over some rather hilarious B-roll of suited businessmen sprinting on a racetrack, Cascio admits, "The real question is what happens if or when people that I'm competing with, people that I work with, decide to start using these cognitive drugs more often—and they start producing more and better work? It's not so much that my work is slipping, but their work is getting so much better. Will I be able to hold off? Will I be able to resist starting to use this kind of cognitive enhancement more often?"

No one is more aware of this situation than Laura Vanderkam. In *168 Hours,* she writes that the real reason to find your dream job is that being passionate about your work will make you more productive and creative: "This obsession is the only way to stay on top, because you can trust that your competitors are thinking about *their* jobs in the shower." While most people are aware of factory jobs being outsourced, increas-ing amounts of knowledge work can be, too. "To thrive in a world where someone else is always cheaper, you have to be distinctive at what you do," she advises. "In some cases, just to survive, you have to be world class." That means that you can't stand still and that, in theory, you could (must) always be improving.

And yet Linda's burnout has to be about something more than work and straightforward economic security, because even those who should be more than comfortable seem curiously apt to wear themselves down. In *The Burnout Society,* Byung-Chul Han suggests something even more general: that "the drive to maximize production inhabits the social unconscious," producing what he calls "the achievement-subject." Rather than be disciplined by something or someone exter-nal to them, achievement-subjects are "entrepreneurs of themselves," DIY bosses propelled from within. Although it answers to no one (else), an achievement-subject nonetheless "wears down in a rat race it runs against itself": "The disappearance of domination does not entail free-dom. Instead, it makes freedom and constraint coincide. Thus, the achievement-subject gives itself over to compulsive freedom—that is, to the free constraint of maximizing achievement. Excess work and performance escalate into auto-exploitation."

As any tech campus with free food, branded backpacks, and a climb-ing wall can attest, a society of achievement-subjects is great news for the bottom line. Vanderkam is right about obsession, it turns out. Han observes that "the positivity of *Can* is much more efficient than the nega-tivity of *Should*" and that "the achievement-subject is faster and more productive than the obedience-subject." This same limitlessness is what leads the achievement-subject toward burnout. Trained to set her sights on infinity, she never experiences the feeling of having actually reached a goal and, instead, exhibits the "auto-aggression" of the master and mas-tered rolled into one. She is forever "jumping over [her] own shadow," frustrated at the impossible gap between what is and what could be.

Unfortunately, that gap keeps expanding. Rosa writes that the capital-ist "logic of increase" infiltrates cultural notions of the good life, mean-ing that to stand still in the realm not just of work but also of money, health, knowledge, relationships, or fashions, registers as sliding back-ward or falling down in the social order. I would add that the language of comparison and competition is amplified by social media—a constant scroll through even just friends' photos is a never-ending tour through "what could be." Studies have documented the cruel cycle in which peo-ple with low self-esteem use social media for expression and connection, only to expose themselves to "upward social comparison information" that in turn restarts the cycle. Out on the feeds, the finish line never stops moving. You have twenty-four hours a day and must spend them in a better—and better, and better, and *better*—way!

We are taught early in life to compete with one another, most obvi-ously by being evaluated, timed, and graded in school. One response to a system of timing and grading is cynical: You game the system, as if with a cheat code (echoing the life-hacking approach of the productivity bros). My response was to internalize it. When I was in first grade, I cre-ated and submitted a "report card" for myself, *as a person*, to my parents.

Borrowing the "O" (outstanding) and "S" (satisfactory) grades from my elementary school report cards, I asked my parents to grade me on various criteria such as "being a good girl" and "cleaning up room." My parents, who probably found this gesture cute yet disturbing, obliged by ranking me outstanding in all categories. In a section for "what you need to say," my dad wrote "oh solo me oh," a reference to "O sole mio," a Neapolitan song he would often sing dramatically to make me laugh.

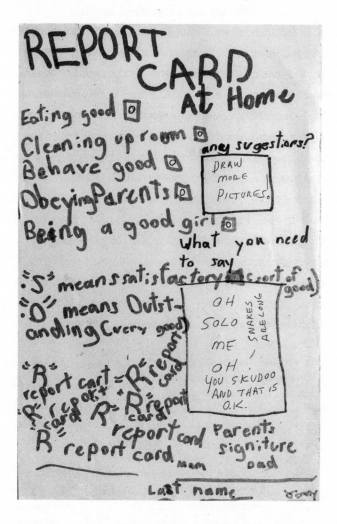

Decades later, I would find myself grading college students on art projects. An art project is about as easy to grade as a person would be, and I hated it every time. I hated the fact that giving everyone A's, even in a class where everyone had genuinely done well, had become stigmatized—and how grading was so addressed to the individual when the best classes had an ineffable group dynamic to which everyone contributed.

As conventional as it seems, the A–F grading system was not fully standardized in the United States before the 1940s. I was not exactly surprised to find that the grading system I had both used and been subjected to had taken shape within the early-twentieth-century "social efficiency" movement in education, influenced, in turn, by Taylorism. A socially efficient curriculum meant a more vocational, less strictly academic one that would be legible to employers or the military and would help shunt people into the jobs they were going to do. As a form of evaluation, grading requires you to invoke some kind of standardized scale on which qualities can be reduced to quantities—something I was reminded of every time I had to make a rubric for grading art projects. Social comparison is probably as old as time, but to compare a wide range of people using the same grades, you have to be able to turn those people into data and decide what you're optimizing for.* To fully appreciate what this has meant historically, and especially what it has to do with speed, we must turn to some older productivity bros.

Just two years before Taylor's *The Principles of Scientific Management* was published, Francis Galton, an English explorer and anthropologist who was also the forefather of eugenics, put out a memoir. Galton was

---

* American schools were already using ranking or grading systems before the turn of the century. The development I'm referring to has to do with the standardization of the A–F grading scheme, specifically within the context of the social efficiency movement and ideas about social utility. Franklin Bobbitt—author of *The Curriculum,* an influential book on structured curriculum—wrote in 1913 that the classroom teacher "needs a measuring scale that will serve him in measuring his product as well as the scale of feet and inches serves in measuring the product of the steel plant." Scholars have noted how the development of standard grading coincided with the desired objectivity of IQ testing, the need to control a dramatically increasing immigrant labor force, and the creation of grades for mass-produced commodities (such as wheat) as national markets expanded. While enthusiasm for social efficiency eventually waned in the mid-twentieth century, elements of scientific management have resurfaced in modern standards and tests, continuing the risk of de-skilling in educational work.

a lover of classes, quarters, and centiles. He was obsessed with measuring and ranking all kinds of things. In the memoir, Galton casually describes his efforts to make a "Beauty-Map" of the British Isles, for which he used a needle to secretly prick holes in a piece of paper listing "good," "medium," and "bad": "I used this plan for my beauty data, classifying the girls I passed in the streets or elsewhere as attractive, indifferent, or repellent," he writes. Other grading schemes of Galton's were more serious. In his book *Hereditary Genius*, he describes creating an A–G scale to measure intelligence and then uses it to compare the white and Black races. With the incredible caveat that "social disabilities" (by which one assumes he means racism and other legacies of slavery) made the data "rougher," he nonetheless found "a difference of not less than two grades between the black and white races, and it may be more."

Before Galton gets to the part on scientific racism, *Hereditary Genius* is mostly an attempt to construct genealogies of illustrious men: judges, statesmen, commanders, men of literature and science, poets, artists, and "divines." If, in Taylorism, the measurement of work was an attempt to intensify it, then, in eugenics, the measurement of people was an attempt to "mold" them in a specific direction, a mechanistic combination of Mendelian genetics and social Darwinism. "It would seem as though the physical structure of future generations was almost as plastic as clay, under the control of the breeder's will," Galton writes. "It is my desire to show . . . that mental qualities are equally under control." One way of getting there, he thought, was to "breed out" undesirable characteristics by evaluating marriages for genetic advantage. In this light, it's perhaps unsurprising that in a chapter on family life, Galton spends more time describing a highly efficient hiking snack (bread with cheese and a particular type of raisin) than he does his wife, about whom he mentions only the "hereditary gifts" of her lineage.

Just what were those qualities that Galton saw as capable of raising the grade? For him, intelligence was inherently linked to speed; at a testing center he established, he measured intelligence via reaction times to physical stimuli. But on the level of the human race, he measured worth by a different kind of "reaction time," namely, the ability to adapt to new social conditions. Civilization, by which Galton mostly meant colonization, was for him not a matter of human agency but "a new condition imposed upon men by the course of events," similar to geological events. The idea that humanity was destined for speed allowed Galton to view the "disappearance" of colonized people as "startling" and to view their fate as a warning even to him:

> In the North American Continent, in the West Indian Islands, in the Cape of Good Hope, in Australia, New Zealand, and Van Diemen's Land, the human denizens of vast regions have been entirely swept away in the short space of three centuries, less by the pressure of a stronger race than through the influence of a civilization they were incapable of supporting. And we too, the foremost laborers in creating this civilization, are beginning to show ourselves incapable of keeping pace with our own work.

In other words, it was high time to become more man-shaped so as not to die in the car—even for the men who designed it—by selecting against nomadism and "Bohemianism" (traits that Galton associated with barbarians). Long before Vanderkam warned about the competition thinking about their own jobs in the shower, Galton warned that a certain work ethic was marketable and thus adaptive: "No man who only works by fits and starts is able to earn his living nowadays, for he has not a chance of thriving in competition with steady workmen." On the ideal qualities of the modern British workman, Galton quotes a list from Sir Edwin Chadwick, a disciple of Jeremy Bentham (designer of the panopticon, with its human hamster wheel). This man would have "great bodily strength, applied under the command of a steady, persevering will, mental self-contentedness, impassibility to external irrelevant impressions, which carries them through the continued repetition of toilsome labor, 'steady as time.'"

Not even Charles Darwin (his cousin) was exempt from Galton's vision of steady performance. When Galton sent him a copy of *Heredi-*

tary Genius, Darwin politely responded that the book is "hard work" and that he'd read only fifty pages—"wholly the fault of my brain and not of your beautifully clear style." Quoting the letter in his memoir, Galton shares a private comeback: "The rejoinder that might be made to his remark about hard work, is that character, *including the aptitude for work,* is heritable like every other faculty."

Eugenics enjoyed broad popularity in the United States well into the twentieth century, especially in California, where beyond being discouraged to have children, tens of thousands of people deemed "unfit" were sterilized.* The rhetoric of optimization in eugenics also took root in the self-improvement literature of the era. (Recall that eugenicists were mentioned in *Increasing Personal Efficiency.*) One exemplary crossover happened in *Physical Culture,* a health and fitness magazine that ran from 1899 to 1955. Labeling itself the "personal problem magazine," *Physical Culture* mixed bootstrapper-type advice and bodybuilding content with Galton's passion for measurements and "racial improvement." The magazine once offered $1,000 to winners of a "Most Beautiful Woman" and "Most Handsome Man" contest, with entry forms showing a blank, vaguely Grecian ideal human body and places to list one's measurements.

Physical Culture was founded and, for a long time, edited by Bernarr Macfadden, who was in many ways the ultimate productivity bro. A proponent of bodybuilding and what might now be called "wellness culture," he carried out an early act of personal branding when he changed the spelling of his original name, "Bernard McFadden," to sound stronger ("Bernarr" was supposed to sound like the roar of a lion; "Macfadden" was supposed to stand out from the more common "McFadden"). Macfadden's articles had titles like "Vitalize with the Mono-Diet," "Make

---

*Though the term *eugenics* has negative connotations now and is not as obviously part of the political mainstream, its ideas continue to have influence. In 2021, California offered reparations for people who had been sterilized without consent in state institutions based on a law from 1909. These sterilizations, which often targeted individuals given the old eugenics labels "criminal," "feeble-minded," and "deviant," were still happening in the 2000s.

Your Vacation Pay Health Dividends," "Mountain Climbing in Your Own Home," and "Are You Wasting Your Life?"* Though the articles were mixed with some quackery, their objective was clear: Then as now, improvement meant getting a grip, getting faster, and getting ahead.

In a kind of editor's letter in the February 1921 issue of *Physical Culture*, Macfadden stressed that mental vitality was as important as—and inseparable from—superior health, which, in turn, was inseparable from financial success. He reflected all these ideas through the lens of social Darwinism, writing that anyone who did not "completely develop his or her physical organism" was "not a real man nor a complete woman." Macfadden gave this advice:

> This is a financial age. A struggle for wealth represents the main object of the average life. But a recognition of the importance of super-efficiency will soon awaken people everywhere to the importance of a splendid machine, in the struggle for life's great prizes, financially and otherwise. And a machine of this character must be complete in every sense. A body that teems with power, that is surcharged with energy, is capable of doing far more work and better work than one that is frail and undeveloped.

The degree to which this efficiency could be seen as part of the DNA of American culture is grotesquely illustrated in an issue of *Physical Culture* from 1937. Continuing Galton's line of thinking about genetically advantageous marriages, the magazine ran a piece called "What You Can Do to Improve the Human Race: In Gambling with the Unborn, How You Can Load the Genetic Dice to Bring Forth Superior Children and Raise Racial Level." Here, it wasn't enough that genetic combinations existed in a social (though eugenicists would have said "scientific") hierarchy of speed and progress. *The genes themselves* were productive or unproductive, "virile" and "sluggish," and they were laborers:

---

*The "mono-diet" refers to lessening the variety in one's meals. Macfadden reports eating only green beans and brown rice (separately) for a month and, improbably, "enjoy[ing] it as much [as], if not more than, the combinations used at the average meal."

We are far from knowing just how the genes do their work. But we know pretty well what they do. If we can forget their minuteness, we can conceive of them most clearly as workers. A single chromosome would be a string of these workers—literally a chain gang—for the genes are linked together, each forever in its designated place. Some of the genes are in effect architects, some chemists, some engineers, carpenters, plumbers, masons, colorists, dieticians [*sic*], etc.

Genes could also get faster, the author wrote. Instead of causing weakness, a mutation could act with the randomness of a lightning strike to improve the genes by "stimulating them or 'speeding up' their action." This resulted in genes for "physical superiority, brilliant mentality and genius." An accompanying illustration of different genes as stick figures showed a "champion" gene wearing boxing gloves, while a detrimental gene, labeled "black," holds two bombs.

This origin story hints at the depths of that moral equation I began with: Busy = good. In a study of "conspicuous busyness," the sociologist Michelle Shir-Wise finds that irrespective of work-life balance, busyness can become a lifelong performance of productivity, where "not presenting oneself as [busy] may be construed as evidence of an inadequate and unworthy self" (or, as Macfadden might have had it, "not a real man nor a complete woman"). We are taught from the beginning that faithfully squeezing the most value out of your twenty-four hours, "steady as time," is what a *good person* does; constantly expanding, pursuing opportunities, and getting ahead in every arena of experience is what a *good life* means. But if work inhabits the social unconscious, it is work according to a historically specific ideal: fast, muscular, unflagging, and white. Recognizing how deeply certain concepts of speed, efficiency, and progress are woven into our culture is yet another way of understanding Brittney Cooper's argument that "white people own time."

With this in mind, let's come back to Linda for a moment. If a Linda is burned out, it may not be in the same way that a non-Linda (a more socially and economically precarious person) is, and total burnout will likely not land her on the streets. But it would be a mistake to think that a Linda's burnout and a non-Linda's burnout are unrelated. Where a non-Linda is controlled and surveilled directly by external circumstances, Linda perceives herself to be controlled and surveilled by the cultural "logic of expansion." If Linda does not participate, she will be judged and have to pay a cost, whether it is social or financial.

The difference between the Linda and the precarious person is that the Linda can afford to pay that social cost. The *similarity* between the Linda and the non-Linda is that her "timer" (the culture of busyness) and the non-Linda's "timer" (wage labor and structural disadvantage) have common roots. They uphold the same system, one in which time can be only a means for profit and where someone else can appear only as your competition. Therefore, for her sake and everyone else's, the Linda should consider paying that cost—becoming *less* man-shaped in order *not* to fit into the car in which most people, in some way or another, are dying. I'm not suggesting that doing so is in itself any way revolutionary, only that it makes more sense. And it opens the door to an important recognition: not of shared consequences, but of a shared cause.

In a conversation transcribed in *The Undercommons: Fugitive Planning and Black Study,* Fred Moten models a useful way of thinking about such recognition. "The ones who happily claim and embrace their own sense of themselves as privileged ain't my primary concern," he says. "I don't worry about them first. But I would love it if they got to the point where they had the capacity to worry about themselves. Because then maybe we could talk." Then he paraphrases the thinking of Fred Hampton, one of the leaders of the Black Panthers:

Look: the problematic of coalition is that coalition isn't something that emerges so that you can come help me, a maneuver that always gets traced back to your own interests. The coalition emerges out of your recognition that it's fucked up for you, in the same way that we've already recognized that it's fucked up for us. I don't need your help. I just need you to recognize that this shit is killing you, too, however much more softly, you stupid mother-fucker, you know?

What would this mean on an immediate, individual level? Burkeman, who made that observation about our "frenetic doing," has suggested that as we work for policy changes, someone like the achievement-subject should accept their mortality and give up the impossible quest for total control and optimization. I would add that the "giving up" part is most true for those who can afford to do so, which means an honest and possibly painful reckoning with one's privilege. It all comes back to that *discretion* in discretionary time. Not all tasks are essential to survival, Burkeman writes, and it isn't universally "compulsory to earn more money, achieve more goals, realize our potential on every dimension, or fit more in." The hustle means different things to different people. But if you are truly an achievement-subject who is only wearing yourself down, then I suggest an adjustment of discretion: experimenting with what looks like mediocrity in some parts of your life. Then you might have a moment to wonder *why* and *to whom* it seems mediocre.

Of course, accepting a life with less of a certain type of ambition is not the same as settling for a life with less meaning. Deciding what can be (supposedly) mediocre entails asking what you want within the limits of your human life, not to mention the fact that it has a limit at all—topics I will return to in chapters 6 and 7. In the meantime, it is worth decoding the advice to "live your best life" as what it sometimes is: an imperative to "live *the* best life," in the sense of a high score. What about choosing to just "live *a* life"? Sometimes when I find myself grasping too much, I simply repeat to myself, almost in the tone of a parent toward a child, that *everything can't be everything*. Other times, I try to have a sense of humor about the American logic of endless expansion, in all its arbitrariness and absurdity, and also the quiet, comical dignity of refusing it. I think of a particular scene from *Beavis and Butt-Head*, where a customer makes an order at a fast-food restaurant where Butt-Head is working. "I'll have a double cheeseburger, large order of fries, small root beer, and an apple pie," he says. "Uh . . . what?" says Butt-Head. The man repeats his order, louder and more angrily. Butt-Head simply replies, "Uh . . . . could you, like, *get less stuff?*"

ACROSS THE ROAD *from the Facebook campus is a thirsty patch of grass and shrubs: coyote brush and toyon. I've never known what the deal was with that field, an "unproductive" tangle that just appears blank and gray on Google Maps. As we wait, a point of stillness above the field turns out to be a white-tailed kite (a bird that looks like a cross between a hawk and a gull): with the smallest flutters of its wings, it hovers in the exact same spot in near-suspended animation. Then the light turns green, and we turn left, toward the mountains, where the silhouettes of individual trees are now visible. We will be there soon enough.*

GIVING UP ON expansion in a world predicated on it can cause problems. In 2016, a young Chinese factory worker named Luo Huazhong quit his job and undertook a bike trip from Sichuan Province to Tibet, a distance of thirteen hundred miles, while living off of part-time work and his savings. Luo summed up his experience in a post on Baidu called "Lying Flat Is Justice." "I have been chilling," he said. "I can be like Diogenes, who sleeps in his own barrel taking in the sun." The post touched off the "lying flat" movement, which at the time of this writing is still going strong. In May 2021, an illustration circulated on Chinese social media of a reclining man, with the text "You want me to get up? That's not possible in this lifetime." Unsurprisingly, the Chinese Communist Party state media was less than charmed. "Struggle is always the brightest base color of youth," wrote *Nanfang Daily*. "In the face of pressure, choosing to 'lie flat' is not only unjust, but shameful."

When younger American Millennials picked up the trend in 2021, it was called shameful again—but translated out of the language of national duty and into that of bootstrapperism. In a *Bloomberg* op-ed titled "'Lie Flat' If You Want, but Be Ready to Pay the Price," Allison Schrager dismissed American "lie-flatters" as a privileged bunch whose choice to drop out was "a luxury they may come to regret." The clock was ticking, and the world was moving fast, with the same adapt-or-die stakes as before: "The economy is undergoing a big transition. Technology and globalization were changing the economy before the pandemic, and the aftermath will speed those trends up. This will create winners and losers among those who can embrace and benefit from the change. But it will be a messy and unpredictable process. One group that will certainly lose out is the people who opt out entirely."

Like Vanderkam, Schrager offers great career advice, pointing out that most pay increases happen before the age of forty-five. Given that all the important stuff—skill development and networking—happens in your twenties and thirties, this is "a terrible time to have a midlife crisis." But advice for winning the rat race assumes that you're running in it, rather than peeling away from a vanishing dream. On Twitter, "lie-flatters" replied to the article in acid tones. "It's just wild to me that everyday [*sic*] we read headlines about pandemics, climate change, famine, drought, fires, hurricanes, weapons programs and war, and folks at Bloomberg just want us to work through it for $36k a year," wrote one user. Another summed up Schrager's article this way: "Billionaire: 'Quick, newspaper that I own. Write a story about how young people are lazy for realising that they're just making me even more money while they can barely support themselves, will never own a house, and will need both parents working full time to support a family.'" Yet another asked, "Why work hard? I don't own my work."

I have tried to make the case here for both the difference *and* the link between those who can afford to "lie flat" and those who cannot; those who can say no to work and those who cannot; those who can take time and those who cannot. In other words: the self-timers and the timed (although, as I've mentioned, the boundary is not always clear). A recognition of that relationship—"this shit is killing you, too, however much more softly"—is important for several reasons. Most fundamentally, it opens up the possibility of solidarity, in the genuine sense of sharing a common cause ("this shit"). But it is also a safeguard against the reaction that privileged people sometimes have to their own burnout: fortifying walled gardens of slowness, minimalism, and authenticity. At best, such a reaction makes it easier for people to forsake the world and leave the status quo untouched. At worst, it actually deepens the status quo, creating a scenario in which slowness becomes a product that you buy off the backs of others. Nowhere is the danger of this happening greater than in the realm of leisure.

Chapter 3

Can There Be Leisure?

THE SHOPPING MALL AND THE PARK

Work dominates everything around it as a mountain dominates a plain.
—MICHAEL DUNLOP YOUNG AND TOM SCHULLER, *Life After Work*

O n our way to the mountains, we must stop first at a different kind of
campus. We get out of the car and skirt past the studied interiors of a
Pottery Barn and the salty-sugary smell of a California Pizza Kitchen. The
surroundings make us feel like render ghosts, those footless people you sometimes see floating in opulent architectural renderings. Green speakers are
embedded in planters of roses and snapdragons, joined by other speakers hidden above the magnolia trees. They're playing Rod Stewart's "Forever Young."
A Tiffany & Co. employee on break paces with her arms crossed and brows
furrowed, talking to someone on her phone. Behind her, in the window of a
store that has yet to open, a plain white panel reads, "Happiness can be found
in the smallest of things. It is our passion to turn everyday routines into more
meaningful moments."

Our walkway converges with another to form a faux town square, with an
antique-looking sign reading simply, PAVILION; it contains a smattering of café
tables outside a boulangerie called La Baguette. Around the corner, a wall has
been skillfully painted to look like another wall: a boulevard in Paris, with a
shaded "opening" onto the rue du Chat-qui-Pêche. Though the opening is not
real, the door handle on a painted door is, and so are the empty terra-cotta pots
hanging below painted windows. A (real) denizen of this town walks by, dangling an empty bottle of mineral water and letting out a small burp. Our eyes
follow him past a freestanding digital screen currently showing an enormous
image of a clock set at 10:10, the same time that is frequently shown in watch

ads.* It's an ad for Piaget, a company that sells a diamond-encrusted watch called Possession for $38,400.

AS PARTS OF the world were beginning to go into lockdown in late March 2020, the Instagram travel influencer Lauren Bullen posted a photo of herself emerging from a swimming pool into the rain, banana plants bowing in the background. Her eyes were closed, her mouth grinning. "All we have is now," the caption read. Five days later, she lounged in a nightgown on the edge of an infinity pool, against a lilac sky containing a well-placed bird. Her eyes were closed once again. The caption: "This too shall pass."

Bullen was not traveling. She appeared to be mostly at home, in the mansion she and Jack Morris had built in Bali that year with money they'd made as influencers. A photo captioned "Currently stuck in paradise" showed her in a white one-piece swimsuit, holding a Balinese woven hand fan. A video the next day shows Bullen's head and legs sticking out of a stone tub, with more banana plants in the background. She turns

---

*This time is used for several reasons, one being that the watch hands frame the brand logo.

around to reveal a smile and a charcoal face mask and lays her head on her forearm in a performance of relaxation. This time a caption is inside the video itself, in yellow text: "A moment of self-love."

On Instagram, where the unspoken goal of so many posts, influencer or not, is to influence, posts about slowness, self-care, and "taking time out" have a vague air of evangelism about them that is difficult to ignore. Like advertisements, they read either implicitly or explicitly as exhortations to the viewer: *You, too, could (should) be this slow!* This vision of retreat is often very pretty, rarefied, and existing somewhere else— somewhere the opposite of here, where obligations weigh heavily, the dishes pile up, and quiet moments are few and far between.

In "Slowing Down Modernity," the researcher Filip Vostal turns a critical eye toward the rhetoric of slowness, both in popular and academic culture. He argues that slowness is "not necessarily the equivalent of poise, deliberation, long termism, duration, maturing and consequently human betterment." The immediate irony is that when sold as a product, slowness is just another part of the logic of increase we tried to escape in the previous chapter. As an extreme example, Vostal describes a €260 "slow watch" with a minutely subdivided twenty-four-hour face where zero sits at the bottom. The enclosed manual reads, "I'll be your loyal companion as you journey through a whole new life—one where you learn to be slow." The problem, Vostal writes, is not just the "mere commodification of the slow brand," but also the fact that the temporal sense the watch face imposes is at odds with dominant clock time: "Indeed it might be a fascinating end to create the community of 'slowers' and, in the words of the founders of the brand, to 'slowify the world,' yet it seems that it would be a community of privileged individuals who not only can afford such accessories but, more importantly, who can adopt a different temporal reading of clock-time where punctuality and exactness are negligible."

The watch is an object lesson in how products and services become "paradoxically integral parts of . . . fast capitalism." In this world, slowness is not so much enacted as consumed: "Slow living is now 'for sale' and approaches a consumerist lifestyle mostly for middle-class metropolitan dwellers—the majority of whom are probably far from holding transformative, progressivist or even socialist agendas. Arguably, many would admit that 'it all needs to slow down,' but such slowness would then be, more often than not, consumed, and consumed privately."

Of course, no one expects a socialist agenda from someone like Bullen, whose literal job it is to sell the image of a privileged lifestyle for tourist destinations. If there is a problem here, it's nothing more than the latest elaboration of consumerism in leisure, where "slow products and services" happen to fit right in alongside fast (or any other kind of) products and services. Already in 1899, Thorstein Veblen's *The Theory of the Leisure Class* described how people used conspicuous consumption both to signal their status to those below and to aspire to a higher one. On social media, an endless wheel of comparison, you can do both with greater ease than ever before.

Elements of slow living, disengagement, and self-care have become favored products in the "experience economy." B. Joseph Pine II and James H. Gilmore coined the term in a 1998 *Harvard Business Review* article in which they theorized that "commodities are fungible, goods tangible, services intangible, and experiences memorable," the last being the most evolved form of economic value. In fact, they speculated that the more the experience itself was the commodity, the more it would make sense for non-theme-park businesses to charge admission.* This could turn time into money, but it would also create a sense of psychological envelopment that would help increase sales.

The article's celebrated examples included Niketown stores, Rainforest Cafe, and the Forum Shops in Las Vegas, where "every mall entrance and every storefront is an elaborate Roman re-creation. . . . 'Hail, Caesar!' is a frequent cry." In the fanatical pursuit of a sellable theme, no detail could be spared, no stone left unturned. "When a restaurant host says, 'Your table is ready,' no particular cue is given. But when a Rainforest Cafe host declares, 'Your adventure is about to begin,' it sets the stage for something special." And whereas trash cans at fast-food restaurants

---

* Insofar as it asks someone to pay for a particular period of time in a place, the experience economy overlaps somewhat with the idea of rent—another conversion of time into money that is admittedly beyond the scope of this book.

usually have a sign reading, THANK YOU, the savvy experience designer could "instead, turn the trash bin into a talking, garbage-eating character that announces its gratitude when the lid swings open."

Pine and Gilmore probably could not have foreseen how social media would supercharge the experience economy, the world itself becoming a twenty-four-hour, 3-D emporium of potential 2-D backdrops. While venues like the Museum of Ice Cream in San Francisco explicitly cater to Instagrammers, any ice-cream shop can be treated like a museum by someone with a camera and the right mindset: Susan Sontag's "acquisitive mood." In the context of the experience economy, Instagram, billed as "social," is better understood as a shopping app, a marketplace for both hawking and browsing those acquisitions, whether in actual ads or the pictured lives of friends.* While Pine and Gilmore seemed to think experience itself would be the souvenir, it turns out a photo (the communicable symbol of the experience) is still more than enough.

In 2017, the term *Insta-bae*—a word that combines "Instagram" and "haeru," which means to shine—originated in Japan as an adjective to describe something that would perform well on Instagram. A study that same year showed that two-fifths of American Millennials chose their travel spots based on their Instagrammability. In *The Independent*, Rachel Hosie opined that even though wanting to travel to picturesque places was not new, this was more specific, as "there are certain views, resorts, and infinity pools that are more likely to rack up the likes on everyone's favorite picture-sharing platform."† Given that each post acts like an ad, intentionally or not, the search for Instagrammability is easily contagious. (One of Bullen's more recent posts, where she poses enrobed amidst a lavender field, handily tells the viewer exactly what to

---

* When I originally wrote this, I was speaking figuratively. But in March 2022, Instagram announced that it would allow all users, not just creators, to tag products in their Instagram posts. The feature added to the app's other shopping features, including product pages and the ability to make purchases within the app.

† In his 2021 special, *Inside,* Bo Burnham deadpans that "the outside world, the non-digital world, is merely a theatrical space in which one stages and records content for the much more real, much more vital, digital space. One should only engage with the outside world as one engages with a coal mine. Suit up, gather what is needed, and return to the surface."

put into Google Maps to get to that spot.) The travel industry has taken note. Papers in the *Journal of Travel Research* have detailed the many uses of "social media envy" and "incidental vicarious travel consumption (IVTC)," noting that people with low self-esteem are especially susceptible and should be of special interest to marketers.

Slowness, it turns out, is extremely Insta-bae. One of Bullen's posts from a resort near the Dolomites shows her rising from bed, cradling a coffee, and looking appreciatively out a picture window at the mountains before lazily wandering offscreen. It is a well-crafted magnet for envy. But then, almost anything can be Insta-bae. In September 2021, actor and comedian Anna Seregina stumbled across a series of photos that travelers had taken inside a prison turned luxury hotel in Oxford. In the photos she collected and shared on Twitter, guests had written captions like "just spent the night in prison and I loved it" and "could get used to prison life." Although that sounds appalling—and it is—the Oxford hotel and the Dolomites resort actually have something in common. Just as the prison becomes an inert and sanitized tourist attraction (where *The Shawshank Redemption* is occasionally screened), "nature" at the Dolomites resort is a static backdrop. Offering "Peace as a new luxury," the resort "regards itself as a place of retreat surrounded by nature, where time can once again be sensed and filled with emotional value through physical and mental mindfulness towards oneself."

Landscapes, people, historical moments, and movements have all provided raw material for the experience economy. As the tourism industry has long understood, producing these experiences requires extraction and refinement—the removal of a husk of context, just as in any other commodity, like coffee beans or sugar, whose specificities and conditions of production are hidden away. People buying experience packages are not expecting them to be complicated—at least not in any way they didn't pay for. Pine and Gilmore understood this, too, suggesting that retail employees needed not just to work but to *act*, essentially becoming stage props. A pithy illustration of this dynamic happens near the beginning of the 2021 TV series *The White Lotus*, about a handful of mostly white tourists at a luxury resort in Hawaii. As he watches the guests arrive, the resort manager, Armond, gives advice to a new hire:

ARMOND: I know it's your first day on the job, and I don't know how
it works at your other properties, but here, self-disclosure is
discouraged. Especially with these VIPs who arrive on the boat.
You don't want to be too specific. As a presence, as an identity,
you want to be more generic.

LANI: Generic . . .

ARMOND: Yes. You know, it's a Japanese ethos where we're asked to
disappear behind our masks as pleasant, interchangeable helpers.
It's . . . tropical Kabuki. And the goal is to create for the guests an
overall impression of . . . vagueness . . . that can be very
satisfying. Where they get everything they want but they don't
even know what they want or what day it is or where they are or
who we are or what the fuck is going on.

If I harbor a special distaste for consumerist leisure, it has something to do with my upbringing in a Bay Area suburb full of theme-
parkified Rainforest Cafe clones. Working for an actual theme park
didn't help. For two summers, my job was to hang around faux public spaces with names like "Hometown Square," "Celebration Plaza,"
and "All American Corners" and convince passersby to sit down for
caricatures. Actually, what people were buying was not so much the
(frequently terrible) drawing—which I had ten minutes to complete,
preferably while making small talk and putting up with the occasional
creepy dad—as it was the experience of being drawn. We were incen-
tivized to pull in customers not only by the fact that we worked on
commission, but by the rule that prevented us from sitting down
unless we were drawing someone.

Depending on where I was stationed for the day, nearby speakers
blasted either songs from Paramount-owned movies and shows (including the theme from *Everybody Loves Raymond*) or nondescript patriotic
music, which in retrospect sounded like the result of feeding John Philip
Sousa's entire repertoire to an AI. One day, while standing under red
vinyl umbrellas, my co-worker and I discovered that the pavement was
so hot that you could nudge parts of it around with your shoe, exposing
boiling water underneath.

The break area, which contained a dispenser of Hostess products, sat directly beneath a roller coaster, and every five minutes or so, dangling legs and screams would pass by. At the end of my shift, I would exit through the center of the park's giant loop, a backstage area where you could see deflated game prize animals, the reverse side of plywood backdrops, and maybe even an exhausted-looking SpongeBob mascot taking a smoke break. Then, on the expressway home, I'd pass themed shopping centers that made me feel like I'd never left the park. Even though I liked some of my co-workers, my journals from that time express adolescent cynicism. "I don't think life has a whole lot of meaning when all you do is work, and your work is just basically ripping people off," I wrote.

This experience further soured the idea of "fun" being sold back to me in a carefully contrived package. I was skeptical, to a fault, of paid surprise, paid conviviality, and paid transcendence—the classic teenager's complaint that "the world is so *fake*." When an outdoor shopping development called Santana Row rapidly went up a few miles from my house in 2002, it promised to mimic the organic variation of an established downtown somewhere vaguely European. Without anything better to do, my high school friends and I would mill around on its cobblestone paths like bored movie extras, walking past the chain luxury retailers, giant outdoor chessboard, and new walls painted to look old—some idea of an idea of an idea of "the urban." Trying to encounter actual difference, surprise, or history in spaces like this made me feel like Jim Carrey's character at the end of *The Truman Show,* when the sailboat he's piloting runs into a wall painted to look like the horizon.

It is this old skepticism that informs my reading of the experience economy. I don't mean to suggest that there isn't an art to designing and acting in experiences, nor that there is some uncomplicatedly "authentic" experience hiding behind the screen of its commercial counterpart—if only I could grab it—nor that people can't have a genuinely good time at a place like a theme park. It's just that, as the experience economy expands to include commodified notions of things like slowness, community, authenticity, and "nature"—all while income inequality yawns wider and the signs of climate change intensify—I feel the panic of watching possible exits blocked. I keep wanting to *do* something instead of consume the experience of it. But seeking new ways of being, I find only new ways of spending.

In an article entitled "Why Millennials Don't Want to Buy Stuff," Josh Allan Dykstra, a member of the Young Entrepreneur Council, updates Pine and Gilmore's experience economy for a population hungry for meaningful connection: "The biggest insight we can glean from the death of ownership is about connection. This is the thing which is now scarce, because when we can easily acquire anything, the question becomes, 'What do we do with this?' The value now lies in the doing." Dykstra's recommendation is to "help connect people to other people through your business," and he adds that "sales isn't really about 'selling' anymore, it's about building a community." But, in other ways, it is still very much about selling: "We just have to think about the 'stuff' we sell in a slightly new way." I can't think of a better description of commercial social media, where the "stuff" is a sense of belonging. I have no problem with the idea of an online social network; I just don't want to buy a sense of community with my attention to ads, on a platform that implicitly encourages me to advertise myself, all while my data gets collected. It feels nefarious to me, like Nestlé selling us the public water supply in private bottles.*

Images and experiences are the leisure time counterpart of time management self-help. The same individual who is encouraged to buy time from others instead of having a mutual support network is also encouraged to consume periodic experiences of slowness instead of acting in ways that might reclaim her time—or help others reclaim theirs. In some senses, this could be considered not just conspicuous consumption, but compensatory consumption, where you buy something as a way of coping with a psychological deficit or threat. And right now, there's so much to cope with. Asked about the term *Insta-bae*, Hiroshi Ishida, who researches the "life course" of young Japanese, noted a high level of anxiety about the future. "Because of that, there is a feeling that they want to collect experiences of value now while they can," he said.

Whether conspicuous, compensatory, or both, consumption has long had a relationship to leisure, which can make leisure a strange kind of circumscribed freedom. Although leisure is typically defined in opposition to work, the cut that separates the two is also what has historically joined them. In a discussion of the internal paradoxes of the Protestant work ethic, Kathi Weeks describes how the ethic—originally warning against spending the wealth that one worked for—came to accommodate consumerism in the early twentieth century: "Consumption, rather than savings alone, emerged as an essential economic practice; as opposed to mere idleness, nonwork time was recognized as an economically relevant time, time to create new reasons to work more."

Because the Protestant work ethic is mostly about work, getting stuff was fine as long as you kept needing to work for it. In fact, leisure could even begin to stand in for work. Sociologists have observed that once assembly-line jobs made it difficult to see how well or hard someone had worked, what became visible instead was how much someone was able to consume. This consumption, in turn, became the new way to signal

---

*Instagram has allowed ads in Instagram Stories since 2017. I'll never forget the experience of seeing a friend's heartfelt story about finding out that someone close to her had passed away, immediately followed by an ad for VÖOST's "Effizzing Amazing Vitamin Böosts."

how hard one had worked. Weeks quotes Max Weber's classic study of the Protestant work ethic: "It is only because possession involves [the] danger of relaxation that it is objectionable at all."

These days, work can haunt the consumption of leisure even more directly, thanks in part to the "logic of expansion." Chris Rojek has observed that "leisure, without anybody planning it, [has] become a form of life coaching." An extreme example is Sensei Lānaʻi, an "evidence-based" resort launched by former Oracle CEO Larry Ellison, on Lānaʻi, a Hawaiian island he purchased nearly all of in 2012. Guests in the Optimal Wellbeing Program are asked to set physical and mental goals for their stay, with the spa tracking their sleep, nutrition, and blood flow. While one company note said that "guests have the luxury of limitless choice," another mentioned that *sensei* is the Japanese word for "master," with the master in this case being the data.

Most people will never be able to afford a Sensei retreat, but the rhetoric is familiar. Like the Progressive Era vision of socially useful public leisure (something we'll get to in a moment), individual consumption of leisure harbors similar notions of usefulness. We may not have a personalized wellness team, but there are hundreds of self-tracking apps from which to choose. One app, Habitshare, lets you set daily goals and make your progress visible to friends. But the much more common app for making your progress visible to friends is Instagram, a place to construct, improve, and groom one's self-image and receive constant feedback.

As anyone who needs social media for their livelihood knows, this involves work and can amount to acting as one's own ad agency. Rachel Reichenbach, an artist who began running an apparel shop as a college student and came to rely on Instagram engagement, wrote about a conversation she'd had with a media expert from the Instagram Partnerships team in 2020. "Think of it like the algorithm is grading you in a class," she wrote. "One test alone doesn't determine your whole grade—there's still participation points, homework, classwork, projects, and more. You've gotta participate throughout the class as a whole, not just show up for one test and get an A on it." The media expert she talked to had recommended creating three posts, eight to ten stories, four to seven reels, and one to three IGTV (now called Instagram Video) videos per week. Reichenbach's illustration for the blog post is an exhausted-looking frog with tired eyes laughing maniacally: "HA HA HA HA HA." After the head of Instagram announced in 2021 that it was "no longer a photo sharing app" and would emphasize video instead, some creators expressed fear, anger, and weariness, lamenting the amount of work and exposure that video requires.

Granted, Reichenbach and others are running a business, and most Instagram users don't have such concrete reasons to worry about getting their metrics up. But that is just the issue with social media: It's never clear where an individual ends and the individual-as-entrepreneur begins. This is especially true in an era that prizes "flexibility" and at a moment when "What makes you unique?" is a standard interview question. As a result, what once looked like leisure so easily becomes the arena both of the eternal self-upgrade and the search for some uniqueness to exploit. Marketing advice formerly given to companies—for example to "find your niche"—is now applicable to individuals during every moment of the day.

SURROUNDED BY PHOTOGENIC *families, we sit for a while in another plaza, this one between an Apple Store, a Tesla Shop, and a Macy's, and watch the occasional dog go nuts on the single patch of grass. Once, years ago, a swarm of bees took up residence in some vines growing on the Macy's wall, and someone put up a sign that read, in an austere font,* BEE ACTIVITY. *But there is no such activity now. Instead, Rob Thomas is singing to us from the planters that "maybe someday we'll live our lives out loud," and it's starting to grate on me. The time has come to head for the hills.*

To get there, we'll have to pass a giant golf course and a series of banks, hedge funds, and venture capital firms. Their unassuming office buildings are mostly hidden behind trees and small hills, but occasionally you'll catch a glimpse of their names—Accel-KKR, Lightspeed, Aetos, Altimeter, Schlumberger, Kleiner Perkins, Battery Ventures. We cross the freeway and snake through the trees on the other side, where similarly hidden houses currently go for three to five million dollars.

Soon, though, the houses fall away, and we turn in to the gravel parking lot of an open-space preserve. That torn strip of blue paper we saw from the bridge has resolved into something else: yellow, grassy hills and dark clusters of oak trees, all climbing westward into the more densely forested mountains. We're in the backdrop now. It blasts our faces with hot, impossibly dry air. The visitor's center contains a massive 3-D topographical map showing the three plant communities in the park (grasslands, oak woodlands, and riparian corridor), mortar bowls of the Ohlone Tribe, a brochure with a quote by the conservationist Aldo Leopold, and a button you can press to hear the sound of a meadowlark. While almost everything seems to be dying in the summer heat, the landscape is still beautiful, the edges of the oaks and the grass seemingly electrified by the sun. Most of all, it is quiet.

IT'S HERE THAT I want to return to Josef Pieper's *Leisure, the Basis of Culture*, which I mention in the introduction. In marked contrast to an experience to be consumed or a goal to be met, Pieper's leisure is something closer to a state of mind or an emotional posture—one that, like falling asleep, can be achieved only by letting go. It involves a mixture of awe and gratitude that "springs precisely from our inability to understand, from our recognition of the mysterious nature of the universe." It opens onto, and finds peace in, chaos and things larger than the self, the way you might feel when looking at an enormous cliff face—or a sunrise, for that matter. As "a form of silence . . . which is the prerequisite of the apprehension of reality," true leisure requires the kind of emptiness in which you remember the fact of your own aliveness.

You may recall the first of Pieper's distinctions: that leisure involves a fundamentally different attitude toward time than the one found in the world of work. Leisure is not refreshment-for-work but something completely different that exists for its own sake. The other distinction that Pieper makes is that, as "an attitude of mind" and a "condition of the soul," leisure cannot follow automatically from circumstances. He stresses, for example, that this mindset is "not simply the result of external factors, it is not the inevitable result of spare time, a holiday, a weekend or a vacation." There are many reasons someone might not be able to experience leisure while on vacation, including some of the internalized phenomena I have already mentioned (besides the awareness that you'll have to go back to work when the trip is over). At the same time, there are many ways that you could experience what Pieper calls leisure while not anywhere near a vacation.

In interviews about my first book, *How to Do Nothing,* I was sometimes asked about the kinds of activities I would choose in order to "do nothing." That Pieper's leisure was a state of mind and not a place, product, or service helped me understand the real reason it was hard to answer that question. I have experienced "leisure" while cooking, sorting socks, getting the mail, waiting for the bus, and especially riding the bus. If you have ever had a good trip on psychedelics, you know how something normally tedious and everyday, part of the horizontal realm of time, can switch into the vertical realm and become dizzying, fascinatingly alien.

One day during the pandemic, simply standing in a socially distanced line to get into a grocery store prompted me to see the street from an unfamiliar angle, revealing details I had never noticed: the new leaves appearing on the trees, the stucco on the wall next to me, the quality of light at that particular time of day. The people in front of me in line were not obstacles between me and the store but fellow travelers in a surreal historical moment. In short, I forgot about clock time and, for a moment before I went in, felt Pieper's "inability to understand" and his "recognition of the mysterious nature of the universe."

Yet, while it may be true that leisure is not the mere outcome of external factors, it is also impossible to say that it has *no* relationship to them. If not always literally or in deterministic ways, the mindset that Pieper describes involves time, space, and circumstance. You may not need a park to experience Pieper's leisure, but it sure is nice to live near a park and not to be harassed when you visit. You may find it outside vacation, but it does help if your entire life isn't subsumed by insecurity, anxiety, or trauma. If I exhibited the leisure mindset while in line for groceries, it was at least in part because I wasn't worried about paying for them.

Zooming out from Pieper's definition is hard, because even a state of mind is subject to the forces of a historical and political playing field. The difficulty of trying to account for this is not just the difficulty of reconciling individual agency with structural forces. It also means trying to see the vertical within the horizontal, the free within the unfree, and even peace of mind within a world marked by violence. Pulling on this thread leads me into what feels like an open field—where the entire concept of leisure, even "free" leisure, threatens to become a mirage. What does leisure mean in a world like this one?

By way of illustration, let me return to an argument from *How to Do Nothing,* this time from a different angle. There, I associate free time with public space, writing of a situation where "the parks and libraries of the self are always about to be turned into condos." My example of a noncommercial leisure space, a municipal rose garden in Oakland, was supposed to represent an escape from the productive and the commercial to something else—a place where you could be free from care and from work, including the work of self-optimization. A visitor to the park could, in theory, be just herself instead of a worker or a consumer. Comparing this with commercial, "scripted," and surveilled areas like Universal CityWalk, I wrote that "[in] a public space, ideally, you are a citizen with agency; in a faux public space, you are either a consumer or a threat to the design of the place."

In choosing the rose garden as a metaphor, I was also expressing nostalgia, however unfocused, for the ideals of public leisure that characterized the New Deal era. Oakland's Morcom Rose Garden was built with federal money during the Great Depression, a year before the Works Progress Administration (WPA) would begin building more than a thousand parks across the country. These projects reflected the idea that it was the state's responsibility to provide leisure resources to citizens, a notion informed by progressivism, the budding social sciences, and— something that sounds ridiculous now—concerns about more people having too much time. In 1930, the British economist John Maynard Keynes hypothesized that modernization would lead to a fifteen-hour workweek and wrung his hands over the prospect of freed-up time, "a fearful problem for the ordinary person with no special talents to occupy himself." An abundance of nonwork time became a reality for some during the Depression, given the high rate of unemployment and the National Recovery Administration's "blanket codes," which encouraged businesses to limit workweeks to between thirty-five and forty-five hours.

Rojek has suggested that "the birth of modern leisure is . . . indissoluble from the topic of the management of free citizens in civil society." With a sensibility similar to the social efficiency movement that codified school grades, early-twentieth-century reformers saw leisure time as both a risk and an opportunity to make citizens healthier and more useful. The National Commission on the Enrichment of Adult Life went so far as to suggest in 1932 that "what the American people do in their spare time henceforth will largely determine the character of our civilization." While public leisure was carefully marked off from the compulsions of consumerist leisure, its stated uses at the time could be starkly pragmatic: In light of the declining birth rate during the Depression, one study suggested that an important function of leisure was for people to meet, get married, and procreate; another mentioned needing to keep people healthy for potential military service.

A 1950 public education film called *A Chance to Play* gives a good sense of how useful something like recreation continued to look from an American institutional and economic view: It could keep youths out of trouble, keep men in better shape in case they needed to be drafted, keep mental patients out of the sanitarium (a taxpayer expense), and keep the nuclear family together. In general, leisure paid dividends in "health, happiness, and increased efficiency." The film points out that companies, too, had taken notice: "Many large industrial companies realize that no matter what his job, the American worker who has a chance to play in his off hours will invariably do a better job. Progressive companies today not only encourage their employees to take part in recreational activities but often cooperate in providing and maintaining floodlighted playing fields for their use." This final comment is the beginning of an entire section of the film about the need for floodlit recreation areas—less surprising once you learn that the film is presented not just by the National Recreation Association, but also by General Electric.*

---

* See also *Better Use of Leisure Time* (Coronet Instructional Films) from the same year. In this film, a narrator schools a young man on how easy he has it compared to his hard-

Alongside this pragmatism, of course, could also exist an emphasis on leisure as a salubrious arena of freedom and expression. It might provide the space and time for a visitor to act in a self-directed manner, as opposed to working and consuming. That anything designed purely for the enjoyment of life could be publicly funded is a beautiful idea. It is this concept, this version of freedom and agency, that I had pretty much accepted wholesale in my park metaphor.

WE ARE MOVING along what is normally the creek but is now an unhappy riparian corridor, its overstory suffering in the drought. It's so hot that, between shaded areas, it's difficult to think of anything except getting to the next one. From up in the hills on the other side of the creek, we can hear the sounds of maintenance trucks undergrounding power lines to reduce the fire risk in the area.

The trail, which used to be a ranch road, dives into a grove of large oaks and bay trees. I offer you a fallen bay leaf to smell: a mix of vanilla, clove, lemon, and black pepper. I have no idea how old these trees are. Looking at them, we drop out of the present and into the past of this place—how it used to be a ranch, how it was a different kind of home before that. The Ohlone mortar and pestle we saw in the visitor's center was from around 1750—not that long ago.

The road climbs up a hillside before topping out at an intersection with a private road. To our right, a large sign declares that the property beyond is private. Ahead of us: a gate, beyond which lies a spectacular view. This gate marks the boundary with a second park, one that until last year had been exclusively available to residents of Palo Alto, a mostly white city, since the 1960s. The ACLU filed a lawsuit on behalf of the local NAACP chapter, which argued that the restriction was, among other things, an echo of Jim Crow–era segregation. No evidence of the restriction remains, only a sign informing us that we are entering a different park.

---

working ancestors and how it's his responsibility to find a constructive leisure activity to fill his time. The young man chooses photography, which meets the requirement by being both a hobby and something that could become his profession. At the end of the film, the narrator addresses the viewer: "Will you let time slip away from you or will you use it well?" as the camera zooms in on the clock and the ticking gets louder.

IN AN AMERICAN sociology study from 1934 that asked people about enjoyable leisure experiences, several participants mentioned self-directed wandering. A forty-nine-year-old social worker, for example, describes a three-hour hike and lunch in what is presumably a public park in the mountains. His lengthy description of the day has many Pieper-ish moments of contemplation and appreciation. The man adds at the end that he "liked this day chiefly because":

1. I was on vacation and was carefree.
2. I had a congenial companion whose silences I enjoyed as well as her conversation.
3. There was the great natural beauty of clouds, trees, sunshine, dazzling air, etc.
4. And, most important of all: because our recreation had not been planned or directed by anybody. We went where and when we felt like it and had no predetermined destination.

I, too, have enjoyed undirected hikes on some beautiful trails, temporarily free of care. But in 1934, a huge segment of the American population would have found this description alienating. In fact, many still would.

In the 2016 essay "Walking While Black," Garnette Cadogan compares his childhood walks in Kingston, Jamaica, with later walks in New Orleans and New York. Varied and intoxicating, his walks in Kingston offered safety and respite from abuse at home. But New Orleans quickly proved to be different. From the moment he put together a "cop-proof wardrobe" in the morning until his return home, walking was no longer simple or liberating. Instead, it was a "complex and often oppressive negotiation":

> I would see a white woman walking toward me at night and cross the street to reassure her that she was safe. I would forget something at home but not immediately turn around if someone was behind me, because I discovered that a sudden backtrack could cause alarm. (I had a cardinal rule: Keep a wide perimeter from people who might consider me a danger. If not, danger might visit me.) New Orleans suddenly felt more dangerous than Jamaica. The sidewalk was a minefield, and every hesitation and self-censored compensation reduced my dignity. Despite my best efforts, the streets never felt comfortably safe. Even a simple salutation was suspect.

All of this lends an inescapable edge to Cadogan's walk and makes the joys of *flânerie* (strolling, idling) impossible. "Walking while black restricts the experience of walking, renders inaccessible the classic Romantic experience of walking alone," Cadogan writes, drawing parallels to the lives of women friends for whom this freedom is also elusive. Pieper's definition of leisure emphasizes wholeness; it is "when a man is at one with himself, when he acquiesces in his own being." But Cadogan is not allowed this relationship in either of the American cities he lives in. Rather than a wholeness, his experience reflects what W.E.B. Du Bois called "double-consciousness": "this sense of always looking at one's self through the eyes of others, of measuring one's soul by the tape of a world that looks on in amused contempt and pity."

In his essay, Cadogan recalls feeling whole only when he returns for a time to Jamaica: "I felt once again as if the only identity that mattered was my own, not the constricted one that others had constructed for me. . . . I strolled into my better self." The Lakota writer Barbara May Cameron describes a similar moment at the end of her essay "Gee, You Don't Seem Like an Indian from the Reservation," which is mostly about being misread and uncomfortable in a white-dominated world. It's not until she visits home in South Dakota that something like Pieper's state of mind appears:

> I rediscovered myself there in the hills, on the prairies, in the sky, on the road, in the quiet nights, among the stars, listening to the distant yelps of coyotes, walking on Lakota earth, seeing Bear Butte, looking at my grandparents' cragged faces, standing under [wakinyan], smelling the Paha Sapa (Black Hills), and being with my precious circle of relatives.
>
> My sense of time changed, my manner of speaking changed, and a certain freedom within myself returned.

In leisure, there is more to be free from than just the clock. Any consideration of leisure as a mindset—its definition, conditions, and purpose—is complicated by the history in the United States of the active destruction of anything and everything that many people have needed for wholeness, a sense of agency, and peace of mind. There are many people who, simply by walking down a street, whether public or private, are seen as "a threat to the design of the place" and for whom simply appearing in public at all is interpreted in some places as an invitation to violence. In 2021, a year that saw increased anti-Asian hate crimes, a Filipina American woman my mother's age was brutally attacked in New York City by a man who said she didn't belong here. I remember starting to see my mother's movements through public space as newly restricted by potential threats.

Just as social hierarchy pervades the experience of everyone who lives in it, it pervades the history of what has been purportedly public leisure. This is in direct contrast to its image as a neutral, apolitical, and noncommercial space to "get away from it all." At the same time that leisure as a public good was becoming a popular idea, the legal process of redlining was ensuring the spatial segregation of cities. Though it has a more diverse visitorship today, Oakland's Morcom Rose Garden may have been a de facto white space when it was built. (On a map from the 1930s showing the neighborhood grades used for redlining, the Rose Garden is in an area graded "B," versus West and East Oakland, which were graded "D" and carried a "high lending risk" because of the prevalence of nonwhite people.)

The 1930s concept of leisure didn't just exist within a field of social hierarchy; it also actively reproduced and entrenched that hierarchy. It had to, when the offer of safety and "freedom from care" to one group was composed of the tacit and violent exclusion of other groups. Safety and purity meant white and abled; improvement meant more white and more abled. Indeed, it was precisely *because* leisure spaces both public and private were associated with freedom that they invited anxiety over the specter of interracial mixing. The historian Victoria W. Wolcott writes that "even before the codification of Jim Crow in the 1890s, whites were more likely to enforce racial separation in recreational spaces than anywhere else." Barely any nonwhite people are featured in the film *A Chance to Play.*

At leisure facilities, time was just another tool of segregation. Some amusement park owners and staff in early-twentieth-century America allowed nonwhite people to visit only on a single day of the week, often Monday, or one day of the year (in at least one case, Juneteenth). In Ironton, Ohio, the sole municipal pool was open to Black visitors for only four hours on Mondays, despite its being built with WPA funds that included Black taxpayer dollars. The year itself became graded into premium and less-than-premium visiting days, as some parks allowed Black visitors only on so-called off days in less agreeable seasons. These strictures framed many peoples' experiences of "free" time. In his autobiography, Jackie Robinson recalls how his friend group of Black, Japanese, and Mexican kids chafed at these restrictions: "We were allowed to swim in the local municipal pool [in Pasadena] only on Tuesdays, and once we were escorted to jail at gunpoint by the sheriff because we had gone for a swim in the reservoir." Sammy Lee, a Korean American diver who went on to become a gold medalist in 1948 and 1952, was excluded from that

same Pasadena pool—except for Tuesdays—and had to construct a diving board and sand pit to train on the other six days of the week.

Ultimately, many facility owners who were challenged by protestors
or organizations like the NAACP simply let their facilities deteriorate,
shut them down, or sold them to private developers. In her book on segregated recreation in the United States, Wolcott suggests that nostalgia
for the lost "golden age" of public leisure, something I have unwittingly
exhibited, is whitewashed and forgetful of this history. Sometimes,
though, the history surfaces in unexpected ways. Wolcott writes that
in 2005, real estate developers in Stonewall, Mississippi, noticed some
concrete poking up out of the ground on some of their property, only
to discover a series of past events: "Further excavation revealed a swimming pool complete with tile work and underwater lights. The town
leaders had hastily buried the pool in the early 1970s rather than allow
local blacks to swim alongside their white children."

The idea that a safe leisure space is a white space continues to resur-
face in ever-new ways, including online. In 2020, after a white woman
called the cops on the science writer Christian Cooper as he was birding
in Central Park, Corina Newsome, Anna Gifty Opoku-Agyeman, and
others organized Black Birders Week. On social media and in events
and articles, participants shared stories of discomfort and harassment
in what has overwhelmingly been a white, male, middle-class leisure
activity. The artist Walter Kitundu, who'd had multiple run-ins with
the police while birding, told *The Washington Post,* "I can't really think
of anything more wholesome than standing under a tree and watching
a hummingbird build her nest, but I think if our activities fall outside
of the framework of possibility that's established for us by the white
imagination, then we're at risk."* But when #blackbirdersweek content (including the *Washington Post* article) was posted to online bird-

---

*After a man in a park called the police on him within earshot, adding, "The police will
take care of you," Kitundu posted flyers with a photo of himself with his photography
gear and the header "ADVISORY! Have you seen this man?" The text explains that "he
is a black man and also a bird photographer. While this combination may be rare rest
assured it is generally not considered DANGEROUS." The flyer also features "actual
photos taken by this man."

ing groups, it was sometimes reported or removed, or the person who posted it was banned altogether—a modern-day version of the buried swimming pool.

I thought of this phenomenon later that year, when the nonprofit organization Save the Redwoods League published a statement about one of its founders' roles in the American eugenics movement. Ridiculous as it sounds, Madison Grant associated redwoods with the Nordic race and equated the threat to their survival with threats to racial purity. He was the author of *The Passing of the Great Race,* a book that went on to directly influence the policies of the Nazi Party. On the Save the Redwoods League site, most of the comments on this acknowledgment were positive, expressing relief that the issue had been addressed at all. But one commenter was not so pleased. Calling the statement "out of place" and insisting that "skin color has about the same significance as hair color or eye color," he wrote, "For me, the Redwoods have been a special place of peace without regard to identity politics, and I feel that sanctity has been violated with recent posts on your site. . . . My hope is that the League will continue preserving the Redwoods as the sanctuary it was meant to be—free of the divisive identity rhetoric that has already divided the rest of society." The words *peace, sanctity,* and *sanctuary* prompt the question: Sanctuary for *whom?* This is to say nothing of the ahistorical vision of a place "as it was meant to be," as though it had always looked that way and did not contain histories of violence, plundering, and murder. Writers like Mark David Spence, author of *Dispossessing the Wilderness: Indian Removal and the Making of the National Parks,* have recounted how the establishment of National Parks and wilderness areas in the United States not only violated treaties with indigenous tribes, but also constructed wholesale the American idea of a " 'real' wilderness" or " 'virgin' landscape."

I found myself ruminating on that man's comment to the Save the Redwoods League site for a long time after I read it. There was something about it, the way it sounded like a head buried in sand, that made me have to reckon with what exactly leisure meant. What was leisure "for," beyond recovery from work? I, too, had used the language of sanctuary and peace of mind in *How to Do Nothing,* noting how the Oakland

rose garden was set into a hill and marked off from the surrounding
bustle. But this man's insistence on sanctuary made the idea sound mis-
guided and absurd—like a fridge in the middle of the desert.

*HAVING TURNED BACK from the gate, we stop at a pond I thought I knew. It's
totally dried up, something I've never seen, and in place of water, there's a miniature forest of some strange plant—maybe goosefoot. I tell you how I'm used to
seeing tons of birds here, and how I'm haunted by an image I saw months ago,
of dead fish in a similar pond about twenty miles south, a pond that was shrinking from the drought. We sit on a bench for as long as we can stand the tiny but
persistent black bugs. A white-breasted nuthatch (eater of bugs) visits for a brief
moment, lets out a few nasal eearrn calls, and then disappears. This bench,
dedicated to someone who recently died, was placed here for humans' enjoyment
of the pond. Whatever we're doing now feels like something else, given the view,
but I'd still rather be here than at the shopping center.*

WHAT LEISURE IS *FOR* is a question I would ask myself again on a bird-
watching trip near Pescadero, a small coastal town south of San Francisco. Right at the beginning of my leisurely walk along the rocks, I saw

a strange shape in the sand. It was a dead grebe, and it was not the only washed-up seabird I would see that day. Though I know many people see far worse things on a regular basis, it was a stinging sight. With the one bar of reception on my phone, I looked up "2021 Pescadero dead seabirds" and scrolled through articles about seabird die-offs across the country. Thoughts about climate change and loss consumed me for the remainder of the day. I noticed plants flowering earlier than usual and brooded about the lack of rainfall so far that winter. When the sun set, I sat on a log on the beach, overcome by grief, and looked at the ocean as if it could provide me with an answer. It responded only with its usual roar—another day, another set of waves.

Was this leisure? By traditional standards, probably not. The idea of leisure as unblemished sanctuary would have required screening off dead birds, not unlike screening off the "irrelevant" and unpleasant aspects of commercial leisure resorts. The beach would have had to be presented "as it was meant to be," atemporally, without the signs about invasive snails and declining trout populations. I would have had to be unaware of the Amah Mutsun, the local tribe whose descendants, forcibly taken to Mission San Juan Bautista and Mission Santa Cruz, were now working to restore balance to this very land. I would have had to be "colorblind" regarding the visitors I saw at the park. In other words, the situation would have to have been a postcard—and I the buyer of the postcard— rather than a living, breathing time and place subject to the same pain and injustice found anywhere else.

If my trip failed to provide peace of mind, it did provide affirmation and a sense of responsibility. Heartbreak did not make me love the birds less; it did not make the ocean less beautiful; it only suffused my see- ing them with a deep desire for things to be different. In that sense, my visit could not be characterized as that of a consumer buying a product, nor even of an untroubled patron to a public park, but of a troubled being meeting a troubled world. And, crucially, this encounter occurred *in time.* It was the opposite of a postcard, something that could not be pictured—because a picture would be immediately out of date, and because there was so much that wouldn't show up on the camera lens. It was complex and bittersweet, occurring in the interstices among ecological time, my personal memories, histories of injustice, and concern for the future—all washed over by a momentary pattern of light.

Perhaps this is precisely what Pieper meant with his "vertical" time—maybe it is vertical not just in that it's the opposite of horizontal, but also in that it reaches deep into the recesses of history even as it stretches up toward an infinite and utopian ideal. If the concept of leisure has any utility, for me it has to be this: an interruption, an apprehension, a glimpse both of the truth and of something completely different from what we normally see. This leisure is alien not just to the world of work, but also to the habitual, everyday world. Given the opportunity to slow down, what I find is not slowness per se, but simply what has been happening all along, just outside my perception.

During the Covid-19 pandemic, many people who were able to stay home expressed discomfort with their sudden status as homebodies. In some contexts, this was interpreted as their feeling they needed to be productive, but I think that, in at least some cases, it was something else. I think that people felt bad about inhabiting stillness and comfort while knowing that others were experiencing the opposite. Maybe what they wanted wasn't so much to be "productive" for the sake of productivity—as if they, like Charlie Chaplin's Tramp in *Modern Times,* didn't know how to stop making the motions of work—but, rather, that they simply wanted to *do* something and wished that their leisure time could be meaningful or helpful.

Thinking about a kind of leisure that pushes against rather than bolsters the current order makes it possible to consider it not as a rarefied escape, but as something vitally related to political imagination. If leisure has been an apolitical sanctuary for the people who are favored by such norms, it has always been political for those who are disfavored and for whom access to an enjoyable, dignified life is inescapably an issue of justice. I think of something that Mark Hehir, a Bay Area disability advocate, said when I asked him what he liked best about hiking. For Mark, too, the "classic Romantic experience of walking alone" is inaccessible: In 1996, he was diagnosed with muscular dystrophy and now uses a wheelchair and a ventilator, which require him to be accompanied on the trail. Mark told me, "It isn't unusual for me when I start a hike to say I am home." Yet Mark's feeling of being at home in nature has entailed *making home,* requiring years of trail reviews and feedback to park officials, at first unprompted and now as an official disability liaison for Santa Clara County Parks.

Past and current efforts by institutions to make public leisure spaces more inclusive are extremely important, and many of these organizations have taken significant steps in this direction. Even with their complicated history, I still feel strongly about public parks, which saved my life during the pandemic (and probably the lives of many others who didn't have access to outdoor space at home). But it is also worth remembering the history of those less visible, more inherently political leisure spaces: churches, kitchens, backyards, union halls, gay bars, community gardens, and activist hubs of all kinds. Sometimes fragile, short-lived, underfunded, and underground, these have been places not just for peace of mind, conviviality, and healing, but also for the building of power—at the very least because their existence is inherently at odds with their surroundings.

If these are sanctuaries, they are less places to bury your head in the sand than places where different languages of time and being are kept alive. They are the enactments of a kind of home, a "something else" for the sake of itself, like Pieper's state of mind realized on a collective level. In a 2021 interview, Saidiya Hartman, a writer and scholar focusing on African American studies, described one way of making a home outside both economic and social hierarchies: "Often the way people think of care is as an incredibly privatized thing. I mean, caring for ourselves, partly, is the way we destroy this world and we make another. We help each other inhabit what is an otherwise uninhabitable and brutal social context."

The work of the poet, performance artist, and activist Tricia Hersey is an example both of communal care and of a political slowness that, in stark contrast to "slowifying" your life, doesn't simply re-entrench the system. Her organization, The Nap Ministry, involves writing, workshops, performances, and collective napping experiences. "Rest is not some cute lil luxury item you grant to yourself as an extra treat after you've worked like a machine and are now burned out," Hersey tweeted in October 2020. "Rest is our path to liberation. A portal for healing. A right."

Hersey uses social media for work but is critical of the way it encourages a grind culture with historical roots in capitalism and white supremacy. Citing vicarious exhaustion for content creators producing "memes, infographics, reels, TiK Tok [*sic*] dance challenges, witty and funny skits and IG Lives," she tweeted, "Y'all be cranking out stuff every second and just seeing it drives me to the couch for a nap." Hersey is also aware of her words and ideas being co-opted by the white capitalist wellness movement (who must sense in her posts material for something Insta-bae). It's an especially cruel irony, given that The Nap Ministry specifically addresses the sleep deprivation of enslaved people and their status as commodified bodies. For Hersey, rest is simultaneously "a spiritual practice, a racial justice issue and a social justice issue."

In a four-minute spot on NPR's *All Things Considered*, Hersey was interviewed about her self-appointed role as nap minister. "What do you say to people about how to make that happen in their lives, especially if they feel like they can't rest right now?" the host asked.

Hersey responded, "Yes. You know, I love to reimagine rest outside of a capitalist and colonized system. So I love to think of resting as something that's subversive and inventive—closing your eyes for 10 minutes, taking a longer time in the shower, daydreaming, meditating, praying. So we can find rest wherever we are because wherever our bodies are, we can find liberation because our body is a site of liberation. So the time to rest is now. We can always—"

"I got to stop you right there," the host said at this point, cutting Hersey off. Their time was up.

LEAVING THE POND, *the trail comes up against a distinct line of planted oaks and redwoods. There is no fence, but just on the other side of the divide, the dry, scrubby grass changes abruptly into watered lawn that extends farther than we can see. It's the Palo Alto Hills Golf and Country Club. When we look it up online, we see that the club's website primly withholds prices up front, but does tell you what an initiation fee, monthly dues, and the right connections will buy you: golf, a pool, tennis, a fitness center, and a lot of stuff for your kids to do. I've never been to a country club, so I imagine most of them to be like the one on* Curb Your Enthusiasm. *Perhaps anticipating these associations, the club's site assures us that "change, innovation, fun and friends now rule the day, month in and month out at this multi-cultural club where diversity reigns and there is truly something for everyone." At the top of the page is a header image of a large outdoor clock on the club grounds, overlaid with the text "TIME WELL SPENT."*

THE CONCEPT OF leisure has always contained ambivalence. Historically, proponents and scholars of leisure have fallen into one of two categories: what Rojek calls pragmatists and visionaries. Pieper was a visionary, but the visionary par excellence was Aristotle. For him, the divide between the world of work and that of leisure was so fundamental that any activity undertaken toward practical ends, even play, could not be considered leisure. It could only be philosophy—contemplating, deliberating, and inquiring into the nature of things—which he saw as man's highest calling.

Yet Aristotle's definition of leisure also required work as its infrastructure: Ancient Greece was a slave society. Drawing a distinction between different types of reasoning, Aristotle believed that some people were born without the capacity for high-level deliberation, which made them "natural slaves." Specifically, he saw this trait as occurring in non-Greeks—which was a convenient observation, given that the majority of enslaved people in Greece were non-Greeks. Aristotle allowed that if a polis were to have autonomous laboring machines, it would not need slaves, but in the meantime, it was a good thing that natural slaves existed. It was good because the ideal polis would have leisure, and for some men to have leisure, someone else would need to do the work. What's more, enslaved people, unable to deliberate independently, benefited by working under the aegis of someone who *could* deliberate, and their lives could take on meaning through their contributions to the master's leisured pursuits. This model of natural inferiority and mutual beneficence would be invoked over and over again in the service of colonization, slavery, and the subjugation of women.*

If there was an infrastructure of leisure here, it was simply a social hierarchy, where the enslaved were seen as foreign to the leisure they made possible. Here is the kernel of the division of labor I mention in chapter 1, where some people's time is not only valued less, but also understood as existing *for the sake of* others' time. This sentiment persisted even as the landscape of work changed. During a surge in working-class self-education in the northern United States, a Jacksonian Democrat gave an 1830 toast in which he expressed his hope that "in the point of literary acquirements, the poorest peasant shall stand on a level with his more wealthy neighbors." In response, editors of the *National Gazette* in Philadelphia insisted that the division between classes underwrote high culture as well as stability: "The 'peasant' must labour during the hours of the day which his wealthy neighbor can give to abstract culture . . . ; the mechanic cannot abandon the operation of his trade for general studies; if he should . . . languor, decay, poverty [and] discontent would soon be visible among all classes."

In other words, leisure as an identifiable category of time can arise only through its contrast with work time, which is owed to someone else. Dismissing an argument that leisure was "invented" in Early Modern Europe by a bored leisure class seeking novelty, the scholar Joan-Lluís Marfany has proposed that "societies of hunter-gatherers or primitive agriculturists may have [had] no use for the contrasting notions of 'work' and 'leisure,' but it is hard to believe that the opposition failed to arise as soon as . . . some

---

* On this model, Aristotle wrote in his *Politics*, "Thus by nature most things are ruling and ruled. For the free person rules the slave, the male the female, and the man the child in different ways. The parts of the soul are present in all, but they are present in a different way. The slave is wholly lacking the deliberative element; the female has it but lacks authority; the child has it but it is incomplete."

Later, this hierarchy would also be interpreted through a Christian lens. In the United States, in an 1856 pro-slavery sermon, one southern Presbyterian reverend argued that the institution merely reflected the natural Christian order, quoting a line in the book of Genesis, addressed to Eve: "Thy desire shall be to thy husband, and he shall rule over thee." The reverend added, "There, in that law, is the beginning of government ordained of God. There is the beginning of the rule of the superior over the inferior, bound to obey."

Frank Blackwell Mayer, *Leisure and Labor,* 1858

form of socio-economic differentiation was introduced"—for example, indentured or wage labor. Whereas the source of boredom for the leisure class was free time, the source of boredom for everyone else was work, and working people had no problem deciding what to do with whatever leisure time they were allotted. Nor has this really changed. Marfany writes, "The really striking thing is that the most popular forms are still the same as five, six, seven centuries ago: playing certain games, drinking, dancing, simply chatting idly in the shade or by the fire. People play draughts in New York's Bryant Park just as they played them on the square in Bagà."

If true leisure really "runs at right angles to work," then the existence of leisure is at least the beginning of a gesture toward a life outside work, toward the consumption that justifies work, and toward the view of people as repositories of labor hours. Indeed, long before Keynes fretted about free time, questions about the primacy of work in workers' lives appeared with repeated urgency in American movements for a shorter workday. To demand leisure in the nineteenth century was also to ask a fundamental question about whether workers existed for capitalists or for themselves. How much of this one precious life was owed to capital?

Here, free time was seen as anything but inert. Combining demands for shorter hours with the call for an end to child labor, agitators imagined leisure time as inherently dynamic: It would be the domain not only of enjoyment but also of self-education and organization, which would lead to greater demands and greater political power. Unlike the social reformers of the 1930s, nineteenth-century labor leaders were not worried by the potential effects of this newfound time. Ira Steward, notable among labor leaders for embracing a racially inclusive vision of "the brotherhood of labor," was one of the foremost proponents of shorter hours in the late nineteenth century. He described leisure as "a blank—a negative—a piece of white paper." Getting an eight-hour day was not an end in itself but rather "an indispensable *first* step": It would give workers the time to figure out more ways to get free, and "make a coalition between ignorant labor and selfish capital on election day, impossible." Just as in the other instances of politicized leisure I have mentioned, Steward's "blank" was less like foam padding keeping a hierarchy in place than like a gas whose every increase carried the potential of more cracks in the system.

This expanding impulse and demand for freedom was evident again among American workers during the 1970s, before organized labor was weakened by neoliberal policies and globalization. Peter Frase has observed that the "Fordist compromise," in which labor accepted employers' demands in exchange for rising wages, turned out to be unsatisfactory on both ends; business owners had to deal with a powerful labor movement, and workers found they actually wanted more than just money and the trappings of leisure—they wanted not to have to sell their time in the first place. Referencing Jefferson Cowie's description of the "blue-collar blues," Frase suggests that this discontent pointed toward blue-collar workers' real desire: "for more free time, for control over the labor process, for liberation from wage labor." Or, as Harvey Keitel's character in the 1978 film *Blue Collar* says, "Got a house, fridge, dishwasher, washer-dryer, TV, stereo, motorcycle, car. Buy this shit, buy that shit. All you got's a bunch of shit."

In its least useful form, the concept of leisure time reflects an undignified process: working to buy the temporary experience of freedom and then faithfully breathing air in the little gaps that are allowed in the horizontal plane of work. Rest and recreation are applied like maintenance, the leisure machine to the feeding machine. Barbara Luck's 1982 poem "The Thing That Is Missed" articulates the absurdity of this "freedom":

> The thing that is missed is
> time without plans,
> time that invents itself
> like children with summer vacation,
> day after day of it,
> not one free square
> on your mark get set go
> Have FUN-dammit-FUN
> RUN-dammit-RUN
>
> Time's up.
> Back on the line.
> Well did you have fun?
> Not too much fun?
> Too hectic?
> More relaxing to work
> isn't it . . .
> heh heh heh heh

At its most useful, however, leisure time is an interim means of questioning the bounds of the work that surrounds it. Like a stent in a culture that can't stand what looks like emptiness, it might provide that vertical crack in the horizontal scale of work and not-work—that critical pause during which the worker wonders why she works so much, where collective grief is processed, and where the edges of something new start to become visible.

HAVING ADJUSTED TO *the pace of this place, we start noticing more signs of habitation and shelter, sometimes in collaboration with humans: deer trails,*

*bobcat tracks, bird boxes, snake holes in the ground, woodpecker holes in the
trees, woodrat nests (that we previously thought were just piles of sticks). Little
cylinders in the grass turn out to be protected oak seedlings, part of a local effort
to restore the oak woodlands. Other beings, other lives. As we pass, a brush rab-
bit watches us from inside a stand of dry, brittle-looking fennel and then scur-
ries in the direction of the creek bed. It doesn't know what "park" means, and
for a split second, we don't, either.*

WHILE WRITING THIS chapter, I had a conversation with Niki Franco, an
abolitionist community organizer and artist based in Miami. She told
me about the aggressive policing she and her friends had been subjected
to in National Parks and about sometimes being unable to enjoy her
own garden, as it was invaded by the sound of police sirens. Together,
we wondered how something like leisure could be possible in a world
saturated by patriarchy, capitalism, and colonialisms old and new. Then I
thought to ask her the same question from that 1934 leisure study: What
was a recent example she could remember of experiencing a leisured
state of mind?

In response, Niki recalled the weekly hikes she had gone on with
very close friends during a long visit to Puerto Rico. The experience
had been far from "politically void": She was always aware that Puerto
Rico was the world's oldest colony and was under U.S. occupation. She
couldn't dissociate Puerto Rico from the heartbreak of having watched
from Miami as the island was devastated by Hurricane Maria, a storm
from which, many claimed, the island would never recover. But on
these hikes—by sheer force of gratitude, of the company of friends she
deeply trusted, and of the sensory envelopment of the rain forest and its
birds—in some mysterious way, "it was like all of that didn't exist":

> Something has happened for me to be able to step back and feel
> the magnitude of our existence and also how small we are. Like
> coming back to our humanity . . . and that sounds kind of dra-
> matic, but that's how it feels. And, you know, just realizing that
> my existence is not just bound to my job, social media, and what-
> ever in between—these things that happen—it's like, *Oh, wow,*

I'm a living human in this moment in time. And wow, even with all the bullshit, I'm actually deeply grateful to be alive.

About a month after we had this conversation, I happened to visit the Mojave Desert during a Santa Ana wind event. The Santa Anas are incredibly strong, dry winds that blow down from the high desert toward the coast. They're capable of gusting at over forty miles an hour and are infamous in local lore for supposedly making people jumpy and violent. During the first two days of my stay, life consisted of wind: listening to it, being unsettled by it, and trying to stay out of it. But on the third day, the wind ceased and, almost immediately, the resident birds appeared: ladder-backed woodpeckers, white-crowned sparrows, roadrunners, LeConte's thrashers, and Say's phoebes. They filled the new quiet with their songs. I noticed a small mystery bird flying back and forth from a nest right outside the house where I was staying. The bird had chosen wisely: As opposed to the surrounding creosote bushes, it had built the nest in a young palo verde tree, whose branches were so dense and profuse that the interior had withstood the fifty-mile-per-hour winds that had knocked over all the patio furniture.

On the fourth day, the wind came back. It was as violent as ever. But I remembered that merciful pause and everything that I had heard and seen within it: I now knew what the desert was like without the wind. I thought of Pieper's vertical time "run[ning] at right angles to work," these fundamental interruptions that would always collapse back into horizontal time, and about the birds that Niki had heard in that brief allowance of gratitude and awe, before returning to the landscape of heartbreak. What songs are audible when the wind stops? What has been kept alive in the time snatched from work and sheltered from ongoing destruction—what moments of recognition, what ways of relating, what other imagined worlds, what other selves?

What other kinds of time?

Chapter 4

Putting Time Back in Its Place

A BEACH NEAR PESCADERO

But then, from the lower edge of the blank, black disc of the dead sun,
bursts a perfect point of brilliance. It leaps and burns. It's unthinkably
fierce, unbearably bright, something (I blush to say it, but here it comes)
like a word. And thus begins the world again.

—HELEN MACDONALD, "Eclipse"

Heading west from the park, we cross the ancient San Andreas Fault. Though we can't see the fault, something feels different on the other side of it. The rolling hills have disappeared behind us, and the road tunnels upward through hairpin turns into the shade of redwoods, Douglas fir, tanoak, bigleaf maple. The occasional hand-painted sign thanking firefighters reminds me of the giant burn scar, the one from last year we can't yet see. We pass a section of hill that has been armored in stone to keep it from collapsing, a road named Memory Lane, and a lone general store built in 1889.

Though the final stretch of highway is treeless and sunny, a blanket of gray lies ahead. It envelops us just as we reach the coast, and the slate-blue sea appears with its intimidating air of finality. We park and approach the edge of a set of cliffs where a hardy cover of ice plant barely moves in the wind. Despite the waves crashing against the crags below us and the sporadic cries of gulls, our eyes are drawn mostly westward, toward that changeless horizon unbroken by anything, even a ship. Out there, the ocean looks frozen.

IF SOCIAL MEDIA is any indicator, the beginning of the Covid-19 pandemic brought on an estrangement from common forms of marking time. As commuting declined, social events were canceled, and people worked from home, time became fraught: There was too much of it, but also, it

had become homogenous. A running joke developed about the seeming arbitrariness of time:

James Holzhauer @James_Holzhauer · Mar 17, 2020
So glad I set all my clocks an hour forward right before time became a meaningless concept.

jello @JelloMariello · Mar 28, 2020
quarantine messed up our concept of time so bad . . . really havin 2am-type breakdowns in 10am daylight now huh

Seinfeld Current Day @Seinfeld2000 · Apr 7, 2020
me running into the street at 3 am tonite after losing all concept of time [followed by an image of George Costanza in the street yelling, "It's June!"]

Mauroy @_mxuroy · Apr 9, 2020
incase anyone is unsure, today is Thursday, 47th of April

At the time, I was experiencing my own form of temporal weirdness. I taught two classes on Zoom in the same room where I slept; my

students were logging in from time zones as far away as Kenya, South
Korea, and the East Coast of the United States. Weekdays and week-
ends were mostly indistinguishable. The difference between work and
leisure often meant two different tabs in my browser. If I wasn't working
on class preparation or on this book or Zooming with a friend on the
same laptop, then my boyfriend, Joe, and I were on one of a handful of
maddeningly similar walks around our neighborhood. Every night, we
would eat dinner while watching TV, often an episode of a long series
like *The Sopranos*. I'm not suggesting that this was a particularly uncomfortable situation. But, just as in the pandemic memes, the time I experienced was repetitive, constant, and seemed to take place in a vacuum.
Crucially, at that moment, there was no end to the pandemic in sight. It
was just going to be this: boxes of time to be filled in my box of a room
forever and ever.

Amid all this, I started visiting a live webcam on Explore.org of a
nesting eagle in Decorah, Iowa. In March, the eagle had already laid
her eggs, occasionally standing up to inspect them or to chase away
some interloper offscreen. I soon began visiting a second webcam, of
peregrine falcons in UC Berkeley's Sather Tower; and then another, of
ospreys nesting on a shipyard crane in Richmond, about a half hour
north of where I live. The chicks in each nest hatched in late April and
early May, and they all grew precipitously from awkward, fluffy blobs
to something like their parents. The webcams became shortcuts on my
browser's Bookmarks bar; sometimes I would leave them open as windows in the corner of my screen, where they exerted a kind of comforting and talismanic quality. Late on some nights, I would look at the
darkened osprey cam and try to convince myself: *It is night. Time to sleep.*

In September 2020, when the nearby fires got so bad that checking the
Air Quality Index (AQI) and a wind map became a prerequisite to going
outside, I'd switch over to another tab: a pulsating map of wind currents
on Windy.com. Inevitably, I'd zoom out to see the local winds swirling
about in some larger pattern along the coast that fed into some even
larger pattern in the Pacific Ocean. Soon, I would find that I was looking at sixty-mile-per-hour winds on the coast of Antarctica. Of course,
I might naturally guess that it is windy on the coast of Antarctica, but

in this case, it mattered how I had gotten there: by following patterns from where I was sitting hunched at my computer. Zooming in and out, I would think about all that air pushing other air around. Those purple swirls had something to do with my local green swirls.

Then there was my actual window. Without yet being able to formulate a reason for doing so, I dug out my old tripod, set a camera on it, and pointed it out the window—just above the apartments across the street so that the view was mostly sky. Several times a day for a few months, I would walk over and press the shutter button, later flipping through the photos in what became a sort of DIY time lapse. As it happens, March is one of the months with the most varied skies around here. Time felt the same in my room, but in the photos, it rained, it stormed, and the fog rolled in from San Francisco. Sometimes the clouds were enormous and sharply defined; other times, distant and wispy. Midday, the sky could be a deep, dark blue; in the early evening, it would soften to an indescribable shade of something like purple or pink.

If work and online life were *Groundhog Day*, then whatever I glimpsed through these slivers in the day felt very different. It started to remind me of something I had noticed when I was seventeen. In my journals from that time, where I usually complained about being bored or having too much homework, I would also mention occasional sightings of something I referred to as "it." "It" was neither a thing in the environment nor an internal emotion (if such a thing is possible). Instead, it was some kind of gestalt, always unexpected and always fleeting—like catching a whiff of something for a split second and remembering something vast. Although my descriptions were sheepish and incomplete—it was always something that "wasn't here," or a time "outside time"—I documented these encounters nonetheless, writing that future me would know exactly what I was talking about.

NOVEMBER 3, 2003

Lately, this . . . unnamed, unidentifiable "otherness" for (severe) lack of a better word, has been showing its face more than usual. I might kill it with inadequate descriptions. It's like trying to describe a color you've never seen. I don't have the vocabulary.

NOVEMBER 8, 2003

This place is foreign not just in location but in time. It is either an infinite distance into the future or into the past. There is something so fundamentally different about it. Almost extraterrestrial but not on another planet.

NOVEMBER 21, 2003

I saw it at the intersection of Junipero Serra Freeway with Stevens Creek, in the left lane coming from school. Only for a split second when I wasn't paying attention. It is very much a reminding thing. A dejavou (sp?) induced by certain qualities.

DECEMBER 9, 2003

I found it in the newspaper—a crater near Bolivia and Chile—described by a person on the expedition as "the essence of earth" and "overwhelming, grandiose."

10:43 P.M., UNKNOWN DATE

I found it again on the drive to Saratoga library. It's incredibly sunny and the mountains look 5 times higher than usual. I am in two places at the same time. One is Saratoga and the other is very, very far away.

UNKNOWN DATE AND UNKNOWN TIME

It's something much bigger than I can explain, bigger than my perception. Something that shows through certain things, and it is so un-American, so un-anything.

IT WASN'T UNTIL I was an adult that I began to have a better idea of what "it" meant. The first clue that I found in writing was in the French philosopher Henri Bergson's 1907 book, *Creative Evolution*. For Bergson, time was duration—something creating, developing, and somewhat mysterious, as opposed to abstract and measurable. According to him, all our problems conceiving of the true nature of time stemmed from wanting to imagine discrete moments sitting side by side in space. He further noted that this "space" was not concrete environmental space, but something purely conceptual: Think of that green-on-black grid that sometimes shows up in the virtual nonspace of sci-fi movies, and think of moments in this kind of time as cubes existing in that space. (This conception also provided the grounds for the concept of fungible time I mention in chapters 1 and 2.) Bergson thought that our predisposition toward thinking of time in these kinds of spatial terms came from our experience manipulating inert matter; we wanted to see time in the same way, as something we could cut up, stack, and move around.*

Thinking of time in terms of metaphors from abstract space did not bear out for Bergson, who found the concept to be "extraordinary . . . a kind of reaction against that heterogeneity which is the very ground of our experience." Instead, his conception of time was one of interpenetrating and overlapping successions, stages, and intensities. In *Creative Evolution,* his model for this kind of movement is biological evolution, a process of branching and overlapping development where each step had to have been imminent in the last but where nothing about the process was deterministic. The other image I find useful for thinking about Bergson's idea of time is that of a lava flow across relatively flat ground, where the leading edge of the flow is alive and dynamic. Yes, at any point you could look back and perceive the continuous path the lava took to arrive where it is now, but this in no way means the lava was destined to end up there; nor does it let you predict exactly where it will go. Trying to isolate specific moments from one another in this process—like separating cubes in space—would be in vain.

Meanwhile, as you stand there thinking about it, the live edge of the lava is moving forward into the future, which is imminent in every present moment but also contains the history of everything that happened

---

*Abstract time and abstract space were ongoing concerns for Bergson. In *Creative Evolution,* he writes that "a medium of this kind is never perceived; it is only conceived." In *Matter and Memory,* he describes abstract time and space as something almost like a picnic blanket rolled out across an always-changing present, being a "diagrammatic design for our eventual action upon matter." Bergson grants that humans need these perceptual tools; the problem lies not in using them, but rather in taking them too seriously as the structure of reality.

before. Another example would be a seed that has fallen from one individual in a generation of plants and that contains the instructions for a future plant. Time as expressed in these processes—which Bergson explains using what he calls the *élan vital* (usually translated as "vital impetus" or "life force")—is not an abstract quantity to be counted up and measured. Instead, it's another irreversible turn of the kaleidoscope, something driving division, reproduction, growth, decay, and complexity. The old adage "You never step in the same river twice" also speaks quite well to what Bergson is describing, especially if you go on to consider the shifts in evolution of the riverbank, the canyon that the river may be slowly carving, and maybe even the cellular processes in your foot.

Yet, even as intuitive as this articulation of time seems to me, I often find it difficult to fully let go of those familiar, abstract, spatial metaphors for time—time as a linear path of uniform, separate, side-by-side time units. It is one thing (and already disorienting enough) to learn about the historical specificity of the default time sense that you harbor and quite another to be able to release your grasp on an old conceptual handle that is worn to the shape of your hand. This difficulty touches on something even more general than the time sense drilled into me by school schedules, timed tests, and report cards; those are simply parts or symptoms of a wider culture I grew up in and continue to inhabit. Beneath all these is a model of time that is linear and based in abstract spatial metaphors. It is a fact of my life on par with my assumption that if I turn on the TV where I am, most people will be speaking English. In fact, this is so deeply ingrained that anything different has felt to me to be "outside time."

Consider the graph on the following page, which measures sundial time against standardized clock time. This graph shows you which parts of the year the time indicated on a sundial would run ahead of a standard clock and which parts of the year it would run behind. The difference exists because, as John Durham Peters writes, "the sundial directly models natural facts, yielding stretchy days and hours that expand and contract as the earth makes its elliptical way around the sun, but the clock is a solar mood stabilizer, soothing the sun's annual

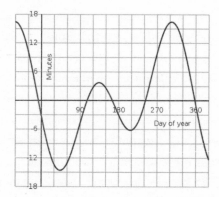

swings into twenty-four-hour average units and ticking away regardless of sun or cloud." It is the difference between place-based observation and the abstract, standardized system whose evolution we see in chapter 1.

The graph shows both readings of time, yet they are not equal. Sundial time is being described *in the terms of* clock time, which are the given grounds for comparison. It is as if, as anthropologist Carol J. Greenhouse describes it, "the clock were itself a materialization of some universal time sense." Clock time is not the only form of time reckoning we experience, but it is certainly primary in how many of us think about the "stuff" of time. And it was an allegiance to clock time that allowed colonists, anthropologists, and contemporary Western observers in general to view non-Western and indigenous cultures as being without, or outside, time.

In *Time Blind: Problems in Perceiving Other Temporalities,* Kevin K. Birth writes about the language barrier in how we think about time. In one 2011 study he cites, researchers from the University of Portsmouth and the Federal University of Rondônia in Brazil found that the Amondawa, an indigenous Amazonian group, used metaphors and language for time that made it inherently difficult for someone with a Western perspective to explain to a Western audience. In an attempt to forestall mistranslation when they shared it with the public, the authors of the study wrote, "We would strongly disavow any interpretation of the data that we present that would exoticize the Amondawa by suggesting that they are a 'People Without Time.'" Yet the media duly went on to twist the findings in order to present an image of timeless "primitives." The headlines they used are illustrative, worded as though something were missing rather than different: "About Time: The Tribe without Time" (*New Scientist*), "Amondawa Tribe Lacks Abstract Concept of Time" (BBC), "Amazon Tribe Has No Language for Time" (*Australian Geographic*).

The best description I have seen of this problem comes from a book by Tyson Yunkaporta, who straddles both worlds as an academic, arts critic, and member of the Apalech Clan in Queensland, Australia. In *Sand Talk: How Indigenous Thinking Can Save the World,* he puts it this way:

> Explaining Aboriginal notions of time is an exercise in futility as you can only describe it as "nonlinear" in English, which immediately slams a big line right across your synapses. You don't register the "non"—only the "linear": that is the way you process that word, the shape it takes in your mind. Worst of all, it's only describing the concept by saying what it is not, rather than what it is. We don't have a word for nonlinear in our languages because nobody would consider traveling, thinking, or talking in a straight line in the first place. The winding path is just how a path is, and therefore it needs no name.

The challenge of trying to overcome this barrier and conceive of time differently—not as some kind of exotic alternative or idle speculation, but in a fundamentally *felt* way—is difficult and fascinating. It is also urgent, a matter of political and ecological import. Conceptions of time are deeply related to how and where we see agency, including within ourselves. They matter especially now, when the world calls out not just for action, but also a less human-centric model of who and what is owed respect and justice.

For years, reading about this other conception of time, I could sort of grasp it on an abstract, intellectual level. But it took me longer to connect it to my personal experience of "it." If I have learned anything in the past few years of my life, it is the difference between thinking something and believing it. It is one thing to observe processes in nature, another to deal with the fact that, as Birth points out, the assumption of uniform time is present in our observations to begin with. Similar to the moment in which a 3-D form pops out of a Magic Eye image, it may just happen that, with effort, we can switch the prominence of the grid and the sundial. But how can we do that?

We can't do it from up here. We'll have to go down there, to the beach.

A creek is slowly washing out part of the cliff we're on, forming a small ravine. We carefully descend along caramel-colored walls studded with intrepid beach succulents. Wherever the walls block the wind, a hot seaweed aroma arises, and we have to bat away small beach flies. Occasionally, a huge explosion of white comes up from behind the black offshore crags, reminding us to beware of sneaker waves.

The hiss and crash of the ocean is many times louder down here. We meet the sand of the beach with a crunch—actually, it's something more like proto-sand. You lean down and see pebbles about one-quarter inch in diameter and in all different colors—dark red, black, gray, orange, sandy white, opalescent white, green—with a few tiny pieces of white shell mixed in. You pick up hand-fuls of them, sorting through the shapes: lumps, pills, shards, orbs, beans.

Among these are chert, quartz, siltstone, and sandstone. The identity of a rock is inseparable from both time and space. To become chert, for example, "you had to be there" many millions of years ago, far offshore, usually in an upwelling zone, in a shallow sea where microscopic marine organisms called radiolaria rained silica skeletons to the seafloor. During the Mesozoic era, this

material formed chert, which was then broken up, eroded, and recycled as pebbles several times over. By the most recent time that the seafloor was uplifted by tectonic activity—in the Pleistocene, an epoch when saber-toothed cats and dire wolves roamed the land above—the pebbles had become embedded in the seafloor's other material. More waves eroded that uplifted land, setting the pebbles free (to become the gravel we are standing on now), while the rest of the stuff was washed back out to sea. Of course, it's not as if this process were done. In front of us, the pebbles are slowly giving way to sand, being ground down by every subsequent wave.

Look again at the pebbles. Make no mistake: They are neither signs nor symbols of time. No—they really are *two things at once: seafloor from the last ice age, and future sand.*

Let's dig in the pebbles a little bit. Your hand feels something smooth. As you push the pebbles aside, you see a horizontally ridged surface:

These ridges line up with a series of larger striated rocks around us, where the stripes run between the cliff we've just descended and the sea. Each stripe is a layer of sediment deposited underwater between one hundred million and sixty-five million years ago, long before the pebbles' arrival. We're used to thinking of layers in a top-to-bottom formation, but unlike with the sediments of the cliffs

behind us, tectonic activity has folded this set of deposits sixty-three degrees
from its original position. That means that, in these rocks, time runs sideways
across the beach.

Resting here gives us a very different sense of being "on time." Rather than
avatars passing through an empty calendar square, we are actually on top of the
material outcome of processes that span millions of years into both the past and
the future. Suddenly, everything we look at is suffused with concrete time: not
just the pebbles, crags, and cliffs, but also the fog's slow movement to the south;
each wave's unrepeatable expression of tides and wind; the frenetic activity of
the beach flies; the dispersion of air and water through our bodies; and even the
chemicals flashing across our synapses as we think these very thoughts. They,
too, will never repeat, and they, too, make the world anew.

ROCKS WILL TEACH you the inseparability of time and space. (When I
say "space" in this instance, I am talking about environmental space,
not the Newtonian grid.) The geologist Marcia Bjornerud has called this
sense "timefulness," writing, "I see that the events of the past are still
present. . . . This impression is a glimpse not of timelessness but timefulness, an acute consciousness of how the world is made by—indeed,
made of—time." In a phrase that reminds me equally of rock strata,

tree rings, and layers of pearl inside a clam, Bergson writes that "wherever anything lives, there is, open somewhere, a register in which time is being inscribed." We can attribute some of our alienation from such registers either to ignorance or to lack of access to the natural world. But the difficulty here, crucially, also speaks to how we think about time and space. For Bergson, abstract time and abstract space are concepts that arise together; in the notion suggested by Bjornerud's timefulness, it is meaningless to try to separate time from space. I think Bergson, too, would look at a beach and see something full of time.

Like the concept of time as money, the abstraction and separation of time and space is a culturally specific and fairly recent event in human history. The idea was given its fullest expression in Isaac Newton's "clockwork universe," in which events and interactions between discrete and bounded entities played out: a sort of billiard ball universe of cause and effect that could be described and predicted in full if we had enough information. Yet, in the realm of physics, the concept didn't last; around two and a half centuries after Newton wrote his *Principia Mathematica,* Einstein articulated the existence of space-time, and thinkers like Bergson and Alfred North Whitehead eroded the concept of abstract time from various angles. Yet the Newtonian ideal has proven sticky. Standing Rock Sioux activist, historian, and theologian Vine Deloria, Jr., has observed that, despite the developments in quantum physics and philosophy, "most of Western society remained Newtonian in outlook while thinkers and philosophers abandoned the belief that nature existed 'out there.'"* It's worth noting here that such stickiness can't be attributed solely to cultural inertia: Abstract, Newtonian time is the kind of time that can be measured, bought, and

---

*In a 1992 article for *Winds of Change,* Deloria elaborates on how relativity in physics resonates with Native American ontology: "For American Indians . . . it was not necessary to postulate the existence of an ideal world or perfect forms untouched by space or time or to suggest that space, time and matter were inherent and absolute qualities of the physical world which, when properly described in mathematical terms, could accurately explain the universe." He adds that "for most Indian tribes it was enough that they understood the manner in which living things behaved."

sold. Wage work requires us to see time as "stuff" divorced from bodies and environmental context.

To understand the cultural specificity of abstract time and abstract space, it's also helpful to contrast the "Newtonian outlook" with the way Deloria and others have described reality in indigenous worldviews. Seasons give us one example of a context in which trying to separate time and space would be functionally meaningless. Whereas, as Giordano Nanni notes, the abstracting of time made it possible for Europeans to "carry the four seasons with them, superimposing them on local seasons wherever they went around the globe," most places did not (and do not) have four seasons. Instead, each has a series of stages corresponding to the ecological character of that specific place. For example, in what is now called Melbourne, the Kulin "recognized seven seasons, each of a different length, according to the appearance of specific flora and fauna": "Kangaroo-apple Season, corresponding roughly to the month of December, Dry Season (around January–February), Eel Season (around March), Wombat Season (approximately April–August), Orchid Season (September), Tadpole Season (October), and Grass-flowering Season (around November). Two longer, overlapping seasons were also recognized: fire (approximately every seven years) and flooding (approximately every 28 years)."

There is no inherent reason for a season to be any length of time, much less of four equal, mutually exclusive lengths. Until relatively recently, the naming and recognition of seasons or seasonal entities was an indicator of some action to be taken: collecting, hunting, harvesting.* Likewise, no element of a season can be considered in isolation from space, time, or other components—you will find no perfect billiard balls here, only dense meshes of interrelated or overlapping processes. Yunkaporta writes of Australia's silky oak tree, whose original name and medicinal use can be understood only in an extended spatiotemporal context: "That silky oak tree has the same name in

---

* Some sense of the utility of seasons is preserved in the English word *season*, which comes from Latin *satio*, meaning "to sow."

Aboriginal languages as the word for eel. Its wood has the same grain as eel meat, and it flowers in the peak fat season for eels, signaling to us that it is the right time to eat them. Eel fat is medicine in that season and can cure a fever."*

There is a common misperception that the Bay Area has, compared to other places, "no seasons." It probably feels the most this way from the perspective of someone moving from someplace like the Midwest or the East Coast, where winters are brutal, temperature swings are bigger, and climatic events are more likely to interrupt the pace of daily life. But even as someone who grew up here, I internalized this misconception about the Bay Area, which in turn rendered me insensitive to its seasons.

Recently, when I mentioned this to someone who had lived in the Santa Cruz Mountains for a long time, he theorized that, instead of abrupt changes, what we have is a constant and gradual "unfolding." Over the years, I've learned how to see it: Pacific hound's tongue always comes before Douglas iris, which comes before monkeyflower. Canvasback ducks come in the winter, elegant terns in the summer. As the drought wears on and fire season lengthens, I've become more attuned than ever to seasons of moisture—of rainfall in February and March and the layer of fog that settles over the coast in the summer. If, as Deloria puts it, each place exhibits a "personality," then it is made out of as much *when* as *who:* a string of overlapping developments like the tracks of a song. This song sounds slightly different in each place: Even within the Santa Cruz Mountains, I've observed how the chaparral-covered side of a mountain has a different procession from its redwood-covered side.

The corollary to fungible time is fungible space: the square footage of real estate or a nuisance to be crossed on the way to some destination. Whether due to lack of interest, lack of time, lack of access to safe outdoor spaces, or some combination of the three, many urban and suburban dwellers today might be hard-pressed to identify the liv-

---

*Deloria gives another example of such timekeeping: Tribes who lived along the Missouri River would plant corn crops and then temporarily abandon them for the high plains and mountains in midsummer. Having learned to use the mountain's milkweed as an "indicator plant," they knew that when the seedpods had reached a certain state, it was time to make the trip back to harvest the corn.

ing profile—what Deloria calls the "personality"—of the space they inhabit on a daily basis, or the ground below the shopping center. In "Indigenizing the Future: Why We Must Think Spatially in the Twenty-first Century," Daniel R. Wildcat, a Yuchi member of the Muscogee Nation of Oklahoma, wonders "what would happen if human beings once again took the places—the spatial dimension—where we lived as being constitutive of our histories as time or the temporal dimension." It's a vital question that pushes up against the grid of fungible time like tree roots under a sidewalk, especially as living in one place for a long time becomes difficult for more and more people. What would happen to our view of time if we could better see our *wheres*?

WE PROCEED FARTHER into this place, to a spot where tide pools have collected in pockets between the sedimentary ridges we first saw by digging in the pebbles. A sign in the parking lot has informed us that "the best way to see tide pool life is to sit quietly until animals emerge from hiding and resume their normal activity." We post up along one of the deeper pools and peer into what seems to be a mostly still scene: sand- and water-smoothed rocks covered in algae, small red and black seaweeds; and delicate, waving sea laces.

After a few moments, some dark, round things we thought were pebbles turn out to be snails. Some are still, while others bumble along the tiny, mountainous underwater terrain. A crab about an inch across scuttles into view. When it gets a little too close to a certain rock, a larger crab emerges, and a brief and quiet crab battle ensues, a miniature drama unfolding on a totally different scale from the waves still crashing in the background. The longer we watch, the more drama appears among the rocks.

IN A 1973 essay called "Approaches to What?" the French writer Georges Perec coined the term *infraordinary*. Media and the public perception of time, he wrote, focused on the *extra*ordinary—things outside the ordinary, like cataclysmic events and upheavals. The *infra*ordinary was, instead, that layer inside or just beneath the ordinary, and being able to see it involved the challenge of seeing through the habitual. This was no small task, given that invisibility is part of the very nature of habit. "This is no longer even conditioning, it's anaesthesia," Perec writes. "We sleep through our lives in a dreamless sleep. But where is our life? Where is our body? Where is our space?"

Clearly a person intent on defamiliarizing the familiar, Perec once wrote a three-hundred-page novel without using the letter *e*. For finding the infraordinary, too, he had his particular methods. In *An Attempt at Exhausting a Place in Paris,* he chose the Place Saint Sulpice, a large public plaza near the center of the city, as a place of study. Visiting it from a series of cafés and one outdoor bench multiple times a day for a few days, he sat and listed everything he noticed. The list sounds incantatory, with shades of police blotter:

A postal van.
A child with dog
A man with a newspaper
A man with a large "A" on his sweater
A "Que sais-je?" truck: "La Collection 'Que sais-je' a réponse à tout
    [The 'Que sais-je' collection has an answer for everything]"
A spaniel?
A 70
A 96
Funeral wreaths are being brought out of the church.
It is two thirty.
A 63, an 87, an 86, another 86, and a 96 go by.
An old woman shades her eyes with her hand to make out the
    number of the bus that's coming (I can infer from her
    disappointed look that she's waiting for the 70)
They're bringing out the casket. The funeral chimes start ringing
    again.
The hearse leaves, followed by a 204 and a green Mehari.
An 87
A 63
The funeral chimes stop
A 96
It is a quarter after three.

In the introduction to this piece, Perec briefly lists the normal points of interest in Place Saint Sulpice, like the district council building, a police station, and "a church on which Le Vau, Gittard, Oppenord, Servandoni and Chalgrin have all worked." By virtue of their identifiability, Perec was not interested in these. His intention, he wrote, "was to describe the rest instead: that which is generally not taken note of, that which is not noticed, that which has no importance: what happens when nothing happens other than the weather, people, cars, and clouds."

What happens when nothing happens. Perec was undoubtedly aware of the irony of this phrase, because it's never true that nothing happens. Weather, people, cars, and clouds are all things that move. Even if you were to stand on a vast, sterile concrete plaza in the middle of the desert, you would be surrounded by the swirling of air particles, the movement of the sun overhead, a drifting tectonic plate, and the aging of the mind and body you use to perceive these things. In the translator's afterword to a 2010 edition of *An Attempt at Exhausting a Place in Paris*, Marc Lowenthal emphasizes the "attempt" in Perec's title, writing that "time, unarrestable, works against [Perec's] project. . . . Every bus that passes, every person who walks by, every object, thing, and event—everything that happens and that does not happen ultimately serves no other function than that of so many chronometers, so many signals, methods, and clues for marking time, for eroding permanence."

For four years in a row, quarter after quarter, I used to give my design students the same in-class assignment, based loosely on Perec's text. I asked them to go outside the classroom for fifteen minutes and write down things they noticed. When they came back to the classroom, we would have a discussion not only about what they had noticed, but why they thought they'd noticed those things. Most of those times, my students were doing this exercise on campus, and they tended to notice human social interactions more than anything else. But when I gave this assignment in April 2020, many of us were not on campus. In most cases, my students were logging in to Zoom from their parents' or friends' homes and, for the assignment, either looked out the window or went out into a yard. After they returned to discuss their fifteen-minute study, we detected a striking theme: Many of them had noticed birds.

Moreover, they noticed that they had never really noticed those birds before—at least, not in those places.

Their observations were perhaps symptomatic of a larger national trend: For people staying at home during the pandemic, birds started to become more noticeable. In "The Birds Are Not on Lockdown, and More People Are Watching Them," *The New York Times* interviewed Corina Newsome, who pointed out that the beginning of lockdowns happened at the same time as spring migration. She suggested that it might "give us peace and calm to see that even though our rhythm is interrupted, there is a larger rhythm that continues to go on." The online database eBird was reporting a 37 percent increase in users posting observations in 2021, with record-breaking observations for a single day in May 2020. Dollar sales of binoculars in June 2020 had increased 22 percent from the previous year, and in August 2020, Lizzie Mae's Bird Seed reported a 50 percent increase in sales of birdseed and birdwatching accessories. Merlin, the Cornell Lab of Ornithology's bird identification app, had its largest-ever monthly increase in downloads in April 2020.

Some of this increase in participation is likely attributable to the ongoing effort to make birding feel welcoming to an audience more diverse in age, class, and race. But some of it must also have been because some people were stuck looking out their windows on a regular basis—or at the other side of a camera: Visits to Cornell's live bird cams had doubled by May. For some already in the habit of watching birds, the pandemic caused a shift from pursuing rare species in natural preserves to appreciating "what happens when nothing happens," or the minute activities of birds who had always been right nearby. In fact, eBird observation rates for suburban species increased significantly in some areas following stay-at-home orders. In Idaho, the statewide lockdown saw a 66 percent increase in eBird checklist submissions and a more-than-doubling of reports of "common residential species," including jays, chickadees, and the brown creeper, "a cryptic species that becomes easier to spot the more time you spend looking out your windows."

Indeed, the cryptic brown creeper is probably one of the better examples of something revealed by sustained and localized attention. About five inches long and weighing only 0.3 ounces, it has a chocolate-and-white, mottled texture that camouflages it extremely well against tree trunks. What's more, this bird doesn't often sit on branches like others do, instead clinging sideways to tree trunks, inching up or down in furtive, jumpy movements. I used to joke to friends that the only way to see a brown creeper was to accidentally point your eyes at a tree trunk at the exact right moment. The first time I saw one—by accident, of course—I momentarily thought I was hallucinating and that a piece of the trunk had broken off and was somehow moving upward. Now that I know how to pick out their tiny song, I can be a little more intentional, at least pointing my eyes in the correct general direction when I hear it. But, as the Cornell Lab's All About Birds online guide suggests in its description of the species, I still have to wait patiently and "keep [my] eyes peeled for movement."

Most living entities and systems on this planet obviously do not live by the Western human clock (though some, like the crows who memorize a city's daily garbage truck route, do of course adapt to the timing of human activities). To watch a brown creeper as it inches up and down, peering into crevices and extracting bugs with its little dentist beak, is thus a way of catching a ride out of the grid and toward a time sense so different that it is barely imaginable to us. In Jennifer Ackerman's book *The Bird Way*, I learned that the male black manakin, a South American songbird, can do somersaults so fast that a human can see them only in slowed-down video. Some birdsong contains notes that are sung too quickly or are too high-pitched for us to hear.* Veeries, a species related to the American robin, can predict hurricanes months in advance and adjust their migration route accordingly, and no one currently knows how. Birds' own bodies and their movements are an entanglement of time and space: If a loon is in the higher latitudes, it's summer, and the bird is mostly black with a striking pattern of white stripes. If the same loon is near my studio in Oakland, it's winter, and the bird is almost unrecognizably different, a dull grayish brown. (The only time I have ever seen a loon with the black-and-white pattern was when I was far enough north, in Washington State.) Thus, if you showed an expert

---

*For a particularly beautiful example of this, listen to BirdNote's slowed-down recording of a Pacific wren: birdnote.org/listen/shows/what-pacific-wren-hears.

birder an image of certain species mid-molt, they might be able to guess where that bird was on its migratory journey.*

In June 2020, eBird reported that new registrations for yard lists had increased 900 percent. On eBird, yard lists are a subset of "patch lists," examples of patches being "your local park, neighborhood walk, [or] favorite lake or sewage plant." The idea of a patch is instructive. Unlike roads, property lines, and city limits, patches often exist in the domain of the infraordinary, being unofficial spaces delineated only by attention. This attention is in turn responding to the fact that, as Margaret Atwood put it in an interview on birding, "nature is lumpy" in that birds have their own particular neighborhoods. I have my own patches around my neighborhood, like the side of an unkempt parklet where, in the right months, I know I'll see a Pacific-slope flycatcher. J. Drew Lanham, birder and professor of wildlife at Clemson University, has written achingly of a sparrow-filled patch along a public road in South Carolina that he'd spent "hundreds of hours cruising" before a racist encounter with a local farmer made him think twice. Before that, he'd "sit and just watch and listen—absorbing all the thickety sparrowness."

A patch is as small as you want to make it. The smallest one I have had is a single branch of a California buckeye tree in a nearby municipal park, a place I visited or passed through hundreds of times during the pandemic. Buckeyes are temporally notable around here: They go dormant in the late summer, their bare branches looking like an electrified brain, and they eventually grow hard, brown, poisonous seed pods the size of peaches. The scent of their white flowers in spring is my favorite smell, and I look forward to it every year.

---

* I first learned about loons from a class on diving ducks that I took in 2019 through the Golden Gate Audubon Society, taught by Megan Prelinger. In her essay, "Loons, Space, Time, and Aquatic Adaptability," Prelinger notes that Gaviiformes, the order to which loons belong, have performed another, much longer migration: After evolving in the Southern Hemisphere, they now reside only in the Northern Hemisphere. Along with other aquatic life, Gaviiformes have also made it through multiple worldwide waves of extinction. Noting that *Homo sapiens* lack the deep history that would allow us to apprehend or intuit these epochal time frames," Prelinger suggests that "we would do well to try, and to emulate the loon: that is, to imagine our species enduring millions of years on Earth."

Starting in late 2020, every time I visited the park, I made sure to check on what I began referring to as "my branch." In late December, the end of the branch had a small, reddish bud. In January, the bud got bigger and turned green. In early February, the bud opened to reveal small, tightly packed leaves. Over the next few weeks, the leaves and their stems grew briskly, and by the end of the month, they had opened fully, lost their ridges and waxy sheen, and become darker green and downright floppy. In March, I found that insects had eaten some holes in the leaves, and the branch had begun to grow a flower stalk. In April, the flower stalk doubled in size, and then some of, but not all, the flowers on the cluster opened—finally, that scent!—sending out long stamens into the sunlight. In May, all flowers were open, an invitation not just to me but to the explosion of bees whose buzz was audible on nearby streets. By early June, some of the flowers had begun to wither, and a bright yellow was starting to creep along the leaves from their tips. By mid-July, all the flowers had withered and the leaves had become thin, brown, and papery. The buckeye fruit became noticeable in August, mint green and fuzzy at first, then hardening and starting to turn brown in September, when the dead leaves were just barely hanging on. In October, the leaves were completely gone, but the buds of next year's leaves had already developed. In November, the buckeye fruit had fallen from the tree.

What's more, all this happened unevenly and within smaller and larger patterns of time. Within a single flower stalk, some flowers had opened and others had not, and at the very same time, some other trees across the walkway had only buds or fully withered flowers. The same was true of senescence: Some trees turned crispy before their neighbors did, and the encroachment of yellow proceeded unevenly even within one branch. Were you to cut off my branch, you would see the rings that were forming even now: a smaller number than in the trunk, as the branch was younger. One day, of course, the tree would die; buckeyes normally live from 250 to 300 years.

At some point, the tree I was considering must have been planted in the park, but the placement of wild buckeyes in the wider landscape is less easy to account for. Many plant species rely on birds and other animals to disperse their seeds, which makes the pattern of their place-

ment an echo of the animate movements of the past. But, as part of their adaptation to predation, every part of a buckeye tree is toxic. In a *Bay Nature* article, Joe Eaton notes that, unlike those other, animal-reliant species, buckeyes rely mostly on the large and heavy seed pods' falling and rolling down a hill. And yet, he observes, "the trees aren't limited to valley bottoms. Some grow on ridges, hilltops, even cliff edges." Listing the uses of buckeye for local indigenous tribes—who would roast and leach the seeds to remove the poison, or use them to stupefy fish in streams—he writes that the ridgetop buckeye trees might have grown from seeds discarded by those tribes at former processing sites.

Furthermore, the very existence of the buckeye—like any other species—belies an evolutionary moment. The unmistakable sight of its bare branches in late summer is a record of a change in the climate three million years ago, when the buckeye adapted to the newly dry summers that killed off its then contemporaries. In effect, it adapted by shifting its own calendar: Starting the growth cycle in late winter and shedding leaves in the summer, the trees lose less water by evaporation.

What is a clock? If it's something that "tells the time," then my branch was a clock—but unlike the clock at home, it would never return to its original position. Instead, it was a physical witness and record of overlapping events, some of which happened long ago and some of which are still occurring as I write this.

This exercise in observation is an example of what I have come to think of as "unfreezing something in time." To do this means releasing something or someone from their bounds as a supposed stable, individual entity existing in abstract time, seeing them not only as existing within time, but also as the ongoing materialization of time itself. Here, it's important for me to note the difference between seeing the tree as *evidence* of time and seeing it as *symbolic* of time. While it is certainly possible to derive some fruitful thoughts about time and fate from the branching structure of a tree, what I'm talking about is different: The literal tree in front of you is encoding time and change at this literal moment.

This exercise of unfreezing something in time is not hard to do. If you want to see time that isn't fungible, just pick a point in space—a branch, a yard, a sidewalk square, a webcam—and simply keep watch. A story is being written there. Like the larger and larger wind patterns on Windy.com, this story is inseparable from the story of all life, even yours. This story is, finally, the signature of "it": the restless, unstoppable, constantly overturning thing that *makes it all go.*

THE TIDE IS *coming in. It looks ready to subsume the pools that were, after all, only a moment in (tidal) time. The snails are hunkering down, the crabs are getting ready to wander, and the intertidal jumping spider will retreat to a barnacle shell that it seals with silk. These rocks, for a time, will disappear. And so will we, as we turn around to head back toward the cliffs.*

But before we do, we should appreciate this bottom-up view of them and the curiously flat land that lies up there, where we were earlier. The flatness itself is a record. It's a marine terrace, and it formed at a time in the Pleistocene when the sea level was stable for long enough that the waves cut into the side of the coast, leaving behind it a level area that was later uplifted by tectonic activity. Depending on how things go with the earth's next cold phase, where we are standing now may be atop a future terrace.

WHAT HAPPENS IN a world where Bergson's duration and Bjornerud's timefulness are palpable and where time is back in its place? Instead of things that the empty "stuff" of time simply washes over, you may begin to see "things" more often as patterns in time. The world, just like the architecture of a city, becomes a patchwork of outcomes from different weeks, decades, and centuries, all of it being built upon and eroded—pushing, trickling, and winging forward into the unknown.

Unfreezing something in time can convert it from a commodity into something else, a process that often involves having to acknowledge something—something related to "it"—that is uniquely unassimilable to the process of commodification. Robin Wall Kimmerer, a plant scientist and member of the Citizen Potawatomi Nation, includes in her history of mosses an illustrative chapter called "The Owner." Among other things, it's a story about someone who couldn't buy time in the way he wanted to, mostly because he didn't know how to observe it. In her capacity as a bryologist, Kimmerer is invited to consult on an estate whose owner wants to "create an exact replica of the flora of the Appalachians." In the name of authenticity, he wants to include native mosses as part of the overall design.

When Kimmerer arrives at the estate, a worker implies that she is late. "Looking at his watch, he commented that the owner monitors the use of consultants' time carefully. Time is money." A horticulturist gives her a tour of the property, during which Kimmerer looks askance at a gallery of African art. The objects are authentic, the horticulturist has proudly declared. But they are not only stolen; they are frozen in time. "In a display case, a thing becomes only a facsimile of itself, like a drum hung on the gallery wall," Kimmerer writes. "A drum becomes authentic when human hand meets wood and hide. Only then do they fulfill its intention."

The owner, it turns out, has similar ideas about moss. Shown a giant sculptured rock covered with beautiful mosses, Kimmerer recognizes the combination as unnatural—those species would never grow together in that way. When she asks how they achieved this, the horticulturist simply answers, "Superglue." But superglue will not do for the giant quarry wall for which the owner wants Kimmerer's help. She is told that the wall, a backdrop for the golf course, needs to look like it's been there for years: "The mosses will make it look old, so we need to get it growing." Kimmerer knows this is impossible: The only mosses that can grow on acidic rocks in full sun without moisture are not the luxuriant green ones the owner imagines. When she tries to articulate this, however, the horticulturist is unfazed, saying they could install a misting system or even "run a waterfall over the whole thing, if that would help." Money is not an issue, in other words. "But it was not money that the rocks required, it was time," Kimmerer writes. "And the 'time is money' equation doesn't work in reverse." When the two of them visit a nearby glen rich with rocks and mosses, the horticulturist says this is exactly what the owner wants up at his house. "One more time I launched my explanation of the relationship between time and mosses," Kimmerer says, observing that the moss beds are likely hundreds of years old. She is similarly skeptical of the owner's desire to transplant mosses to a rock terrace on the estate. Having specifically studied how mosses "decide" to grow on a rock, Kimmerer knows that the mosses that grow on rocks are "inordinately resistant to domestication."

A year later, Kimmerer is invited back to the estate, only to find that they have somehow gotten mosses onto the terrace. At first she's impressed, but she is later horrified to learn how this was achieved: The estate designers are selecting the "most beautiful" sections of that nearby glen and using explosives to extract the moss-covered rocks. The reason she's been brought in is that the stolen mosses are getting sick and turning yellow. Kimmerer, who at this point still doesn't know the identity of the owner, is furious. "Who was this man who would destroy a wild outcrop lush with mosses so that he might decorate his garden with the illusion of antiquity? Who was this man who bought time and who bought me?" Meditating on the "uniquely human behavior" of owner-ship, she wonders what the owner sees when he looks at his garden: "Perhaps not beings at all, only works of art as lifeless as the silenced drum in his gallery."

Kimmerer considers the cliff explosions a kind of crime, even if the owner legally "owns" the rocks. "Owning diminishes the innate sover-eignty of the thing," she writes. If the owner had really loved the mosses, "he would have left them alone and walked each day to see them." To see something in time is to allow that it has a life and to allow that this life entails more than the mechanistic cause-and-effect of a Newtonian world. In this way of thinking, mosses "decide" which rocks to live on, and even rocks have lives.*

In Barbara Ehrenreich's *Natural Causes*, a book I will return to in more detail in the final chapter, she articulates an understanding of agency that is atypical of the Western mindset and closer to Kimmerer's. Drawing on her doctoral studies in cellular biology, she describes cellular decision-making, positing that "second by second, both the individual cell and the conglomeration of cells we call a 'human' are doing the same thing: processing data and making decisions." Ehrenreich sees this on an even smaller scale as well, quoting the physicist Freeman Dyson: "There is a certain kind of freedom that atoms have to jump around, and they seem to choose entirely on their own without any input from the out-side, so in a certain sense atoms have free will." For Ehrenreich—who would, I suspect, agree with Kimmerer about the rocks and the moss—

---

* This story, along with the idea of "unfreezing in time," bears some similarity to the way that Bergson's notion of intuition—a way of seeing that can admit duration—was taken up in the Négritude movement's critique of European colonial way of seeing. Léopold Sédar Senghor, a Senegalese politician, theorist, poet, and co-founder of the movement, wrote that the European viewer "distinguishes himself from the object [and] keeps it at a distance, *immobilizes it outside of time* and in some sense outside space, fixes it and slays it" (my emphasis). In *African Art as Philosophy: Senghor, Bergson and the Idea of Negritude*, Souleymane Bachir Diagne calls this "the look that freezes."

agency simply means "the ability to initiate an action." Conceived of in this way, agency is "not concentrated in humans or their gods or favorite animals. It is dispersed throughout the universe." If we recall Bergson's "register" in which "time is being inscribed," it would seem that these actions and decisions are part of that inscription.

If you are feeling some resistance to the idea that rocks could be alive, I simply invite you to ask yourself why. Although the divide between the living and the nonliving might seem obvious—or "supracultural," as Sylvia Wynter would put it—it is inescapably cultural. In a study titled "Models of Living and Non-Living Beings among Indigenous Community Children," Mexican researchers interviewed Nahua children about whether different categories of things were alive. The answers often reflected the "biological" perspective taught in school, where being alive means eating, breathing, reproducing, and so forth. But other times, they reflected the "cultural" model where "being living implies that inanimate objects have the capacity to influence or affect the lives of humans and animals or being composed of a particular material"—a model similar to Ehrenreich's "initiat[ing] an action." This second model appears in the Mexican researchers' conversation with a six-year-old Nahua student:

RESEARCHERS: What do we put in the living area?
STUDENT: The ground.
RESEARCHERS: Why is the ground living?
STUDENT: Because we live there.
RESEARCHERS: Because we live there. Why is it living?
STUDENT: Because, for the animals.
RESEARCHERS: But, if we don't think about the animals, is the
    ground still living?
STUDENT: Yes (nodding his head).
RESEARCHERS: Yes, why?
STUDENT: For the plants.

For the Western mindset, rocks probably present the biggest challenge in terms of seeing agency. On Quora, answers to the question of whether rocks are alive are mostly in the negative, but some respondents grapple with the limits of the question. A rock might not be alive, but something like limestone was made out of the shells of living marine organisms, and it could support life in the form of lichen. One person wondered whether radioactive decay could be considered a form of rock death, and how differently the question might appear if we took a longer time span as our measure. Multiple people acknowledged that the definition of living and dead was itself a question of philosophy. And one simply noted that in some sense, we came from rocks and would turn back into them one day.

In "The Stones Shall Cry Out: Consciousness, Rocks, and Indians," the Osage scholar George "Tink" Tinker makes the case that rocks can talk. Pointing out that there is currently no agreement as to what consciousness is, Tinker finds it both paradoxical and arrogant that "the emergent world culture . . . of globalized capital and Western science is equally sure that rocks certainly do not have consciousness."

Learning how to hear rocks talk requires a Copernican shift away from anthropocentrism. Tinker describes attending a conference where a Kānaka Maoli artist was answering a question about how he found the boulders he sculpted. "I don't find them; they find me!" the artist had said. "I might be walking along the beach, and one would reach out and bite the heel of my foot." Tinker reports that a British professor of American studies immediately retorted, "That's what is wrong with you people. You are so anthropocentric! You think that everything in the world works the way you do." Tinker reflects on this critique, which he finds "emotional rather than rational":

[It] was rooted in nearly a week's worth of frustrated attempts to communicate across cultural barriers as well as in a lifetime of immersion in a culture that thinks of itself as somehow universal and normative—and thus inherently superior—a position of intellectual fascism, however naive. As he finished his short tirade, I rose to argue that exactly the opposite is actually the case. "I am sorry, Professor W., but that comment cannot go unchallenged.

You see, you are the ones who are actually anthropocentric. You believe that everything in the world works differently from yourselves."

In other words, a view like Professor W.'s sees Nature as fundamentally different from Man because it behaves deterministically. If time is involved here, it is not so much inscribed through a series of actions as a force driving materials as though they were inert. Tinker writes that "Euro-Westerners have come to divide the world into a clear hierarchy of the divine, the human, and nature—from greatest to least, in that order." In a development I return to in the next chapter, this divide was historically associated with the creation of race as a concept, where the people whom Europeans encountered in their expeditions and later enslaved were reimagined as the earlier stages in a Darwinian progression toward true civilization. Not only were indigenous people seen as "outside history," but they were frequently interpreted, as individuals and communities, to be lazy and lacking any interest in or understanding of the future. In short, they were seen as absent of true agency, where the model for agency was a European one.

One word that comes up frequently in Tinker's treatise on rocks is *respect*. For example, discussing the reductionist view of the mind as "the physical processes of the brain," he complains of the presumption "that a highly developed neocortical brain is somehow the ultimate achievement in terms of consciousness."* In contrast, Tinker observes that "the lack of a neocortical brain or even a limbic brain in reptiles, for instance, would not sway Indian people from a deep respect and appreciation for the intelligence and consciousness of lizards." Likewise, when J. Drew

---

* One example of this reductionism can be seen in the everyday phenomenon in which our memories seem to "live" in physical objects, locations, and landscapes. Similar to our use of writing and other memory aids, it is sometimes the case that you can't remember something (e.g., details about a certain era of your life) until you are in a certain place. For an example of how indigenous cultures have made use of this relationship, explicitly associating stories and memories with enduring features of the physical environment, see Keith H. Basso, *Wisdom Sits in Places: Landscape and Language Among the Western Apache.*

Lanham tells Krista Tippett on an episode of *On Being* that he "worships" every bird he sees, it is clear that this "worship" could not be more different from the estate owner's acquisitive "love" of mosses. The difference between respecting something and not respecting it is the acknowledgment that that something is not an automaton, that it is *registering* time by acting and not just existing in it.

Though I've associated it so far with colonialism, a version of this difference is visible in our everyday interactions with other people. Adam Waytz, Juliana Schroeder, and Nicholas Employ call it the "lesser minds problem," a cognitive bias that leads us to underestimate or overlook the emotional realities of others we perceive to be unlike ourselves, including a biased belief that those people are more biased than we are. We could interpret this to mean that we see people in these "outgroups" as being more automaton than human. The authors describe an incredible experiment in which participants were asked to consider "typically dehumanized outgroups" like drug addicts or people without housing. For someone outside them, thinking about people in these groups usually does not activate regions of the brain associated with theory of mind, the ability to imagine mental states in others. But "when [participants] are asked to engage directly with the minds of these outgroup members, such as by simply asking whether or not a homeless person would like a particular vegetable, then these neural regions become activated just as they are with higher status outgroup members."* The question about the vegetable presumes a person with desires. And desire, an attitude toward the future and a reflection of one's past, can exist only in time—the time inhabited by that person.

I think this is part of what Kimmerer means when she refers to "innate sovereignty" and what Tinker is asking us to respect. Conceiving of sovereignty in new places can require a pretty significant shift,

---

*Shonda Rhimes, the writer of *Grey's Anatomy* and *Scandal*, has made a similar point in a MasterClass lesson on writing realistic TV show characters. Having argued that compelling characters have fully formed hopes and desires—in other words, attitudes toward time—Rhimes adds that the risk of stereotypical, static, and boring characters is highest when people try to write characters whom they perceive to be most different from them.

and for a person accustomed to anthropocentrism (and, for that matter, Eurocentrism), unfreezing the whole world in time can be disorienting. I remember, for example, feeling turned upside down by a specific part of the 2001 documentary *Winged Migration*. Beginning and ending in the same location and season, the film shows us something simple yet profound: the lives and struggles of different migratory birds. Throughout the documentary, the filmmakers used lightweight cameras to move alongside Canada geese and other birds, attempting to show the landscape from something closer to their point of view.* The film's sparse soundtrack and narration make it easier to inhabit that perspective, or at least to feel one's own desire to do so.

The specifically disorienting part, for me, was when it showed Canada geese passing over New York City. Seen as part of a journey the geese had taken for thousands of years, the skyline looked suddenly alien to me; "New York" became an odd conglomeration of hard shapes and protrusions along a particular riverbank. The city existed for the geese, too, but they read it differently, perhaps as a signpost on a path of other signposts that may have included other rivers. Their flight path tied these places together into one big calendar. As the geese passed through the harbor, I can't say that I saw what they saw (most obviously, I have no ability to perceive the earth's magnetic field), but I didn't see what I normally see. Albeit very briefly, the sundial and the grid finally flipped, and I grasped something outside (my) time.

---

*In a 2009 issue of the *Journal for Critical Animal Studies*, Nicole R. Pallotta reviewed *Winged Migration*, including some of its more invasive filming techniques. She writes that "in an ideal world—at least in *my* ideal world—humans would not interfere with nonhuman animals and would leave them alone. However, our world is far from ideal and in *this* world, this film has critical potential and could serve an important purpose." Part of its importance has to do with the way that the "similarity principle" (which is similar to the lesser minds bias) can extend not only to human outgroups but to subsets of animals as well. For example, a 1993 study found that respondents rated birds below mammals and above reptiles and fish on a scale of perceived ability to feel pain; the ratings tracked with how similar the respondents felt the animal groups were to humans. With reservations, Pallotta finds *Winged Migration* successful at least as "an exercise in de-objectification" in which "the birds are transformed from 'dots' into 'characters.'"

★   ★   ★

THERE ARE STRANGE, spongelike rock formations behind us. They're called tafoni, and though it's generally accepted that they represent a form of salt weathering, they remain something of a mystery. Salt may be responsible for opening up holes in the rocks, but it's not the only factor: The actual intricate bowls, divots, and bridges may rely on differences in each rock's composition and still other factors. To really explain this signature of rock, salt, air, and water, we'd need to grasp multiple overlapping processes and feedback loops in this specific place. The tafoni are dramatic visualizations of something that is, in fact, happening everywhere, all the time, including in your body right now: things acting on other things. They are the traces of something like experience.

IN ENGLISH, THE word *experience* has a common origin with *experiment.* To experience something is to be present for it, to be the responsive co-creator of something that is happening—like the ducks and geese who make migration happen by sensing the weather and deciding when to leave. Mel Baggs, the late blogger on autism and disability, demon-strated their own form of experience in a generous and moving video called "In My Language." The video shows them using different parts of their body to interact with various objects in their home, producing effects, motions, and sounds as a recording of their singing plays in the background. The first few minutes of the video have no words (in the common sense of "words"). In the section titled "A Translation," a computerized voice reads the captions as Baggs's hand moves in circular motions beneath a faucet: "The previous part of this video was in my native language. Many people have assumed that when I talk about this being my language that means that each part of the video must have a particular symbolic message within it designed for the human mind to interpret. But my language is not about designing words or even visual symbols for people to interpret. It is about being in a constant conversation with every aspect of my environment."

Echoing the artist who observed that the rocks "found him," Baggs says that the water in the video "doesn't symbolize anything. I am just interacting with the water as the water interacts with me." In the video, the connection between experience and experiment becomes clear: To experience is to test, try, and respond to one's surroundings—a kind of call-and-response between different agents. But what also becomes clear is the political nature of who is afforded the capacity to experience the world. The very act of translation in Baggs's video (into English words) is necessitated by the position of nonbeing and nonexperience in which disabled people are typically imagined. "The way I naturally think and respond to things looks and feels so different from standard concepts or even visualization that some people do not consider it thought at all," Baggs's translation reads, "but it is a way of thinking in its own right." For Baggs to articulate their experience, even if in the dominant language, was also to declare themself an agent against the forces that would reduce them to an automaton.

Another illustration of the connection between experience and experiment—as well as ethics—appears in Ted Chiang's story "The Life-cycle of Software Objects." Ana, originally an animal trainer, is tasked with raising artificially intelligent "digients," a years-long process that ends up being much like raising a child. While technically software objects, the digients interact with and test their abilities in a virtual world and, when they're occasionally put into robotic bodies, the physical one. At one point, a company that sells household robots becomes interested in the digients, but negotiations stumble when the company learns that Ana and Jax, her digient, share the hope that Jax will achieve legal personhood. The executive says he understands why Ana would be so attached after all this time, and yet they're looking for "superintelligent products," not "superintelligent employees."

In a private moment of reckoning, Ana realizes the company "want[s] something that responds like a person, but isn't owed the same obligations as a person." She finds herself in a position similar to that of Kimmerer, who knew that hundred-year-old-looking mosses would take a hundred years to grow, and who reacted with protective anger when they were stolen. Money can't buy this kind of time:

> Experience isn't merely the best teacher; it's the only teacher. If she's learned anything raising Jax, it's that there are no shortcuts; if you want to create the common sense that comes from twenty years of being in the world, you need to devote twenty years to the task. You can't assemble an equivalent collection of heuristics in less time; experience is algorithmically incompressible.
>
> And even though it's possible to take a snapshot of all that experience and duplicate it ad infinitum, even though it's possible to sell copies cheaply or give them away for free, each of the resulting digients would still have lived a lifetime. Each one would have once seen the world with new eyes, have had hopes fulfilled and hopes dashed, have learned how it felt to tell a lie and how it felt to be told one.
>
> Which means each one would deserve some respect.

OUT ON THE *water-blackened rocks are the silvery lumps of resting harbor seals. An oystercatcher, an all-black shorebird with a cartoon orange beak, busily negotiates the smaller rocks, somehow never getting surprised by a wave. On a cliffside path through the faded remains of wildflowers, we encounter a series of stocky wooden signs describing the local geology and the plant communities' adaptation to the rough environment. But one sign, about the process of erosion, is harder to get to. The old path to it has eroded. In response, a new one has been worn into the cliff.*

SEEING MORE OF the world as constitutive of time, full of agency, and deserving of respect means abandoning that hierarchy that Tinker mentions, between the actor and the acted upon. Is this exhilarating or fearsome? Wildcat writes that "indigenous thinkers not only acknowledge contingency and human's lack of control in the world; they also see it as empowering and humbling, not something frightening." If "empowering and humbling" sounds like a paradox, it's because of how we normally conceive of power. In a worldview where power, agency, and experience are not bound by individual bodies but reside "in and through the relations and processes that constitute life," the paradox dissolves.

The real paradox is a mind that conceives of the world as inert but that may come to see itself as bound to the same laws of determinism as everything else—in a way, the ultimate self-own. In the autobiography I mention in chapter 2, the eugenicist Francis Galton recalls running experiments to test out his ideas of man as a "conscious machine" and a "slave of heredity and environment," one whose actions could largely be predicted. Ostensibly seeking a residuum of free will, Galton writes, "The more carefully I inquired, whether it was into hereditary similarities of conduct, into the life-histories of twins, or introspectively into the actions of my own mind, the smaller seemed the room left for this possible residuum." For his part, Bergson also admits that our actions exist on a scale between totally habitual and totally free—but the freedom he finds on the one end is of huge significance, opens onto infinity, and exists inside and outside humans. Comparing the life force to a rocket whose sparks are always falling back down as matter and form, he insists that it is not a *thing*, but a "continuity of shooting out." Nor is creation a "mystery," because "we experience it in ourselves when we act freely." Freedom is choice, and choice is scattered throughout the universe, pushing forward and acting upon what would constrain it.

For Bergson, the everyday experience of learning and recognition demonstrates both the freshness of each moment and the irreversibility of time. He describes walking through a familiar town where the buildings don't seem to change. But as he thinks back to the first time he ever observed those buildings, a comparison emerges that momentarily unfreezes the world: "It seems that these objects, continually perceived by me and constantly impressing themselves on my mind, have ended by borrowing from me something of my own conscious existence; like myself they have lived, and like myself they have grown old. This is not a mere illusion: for if today's impression were absolutely identical with that of yesterday, what difference would there be between perceiving and recognizing, between learning and remembering?"

Yunkaporta, too, speaks of learning and "creation events" in the context of *Turnaround*, an Aboriginal English word before settlers invented the more commonly known word *Dreamtime*. Describing the relationship between the abstract world of mind and spirit and the concrete world of land, relationships, and activity, Yunkaporta writes, "Creation is not an event in the distant past, but something that is continually unfolding and needs custodians to keep co-creating it by linking the two worlds together via metaphors in cultural practice." A "smaller but similar Turnaround event" happens during the release of dopamine in the brain whenever we truly grasp something new. A knowledge keeper, Yunkaporta writes, "is a custodian of miniature creation events that must continually take place in the minds of people coming into knowledge."

Like rocks pushing up out of the depths and the water that wears them down; like browned and ripened buckeye fruits falling off the tree and rolling down the hill; like poetry, which strains the boundaries of an ossified language; or like Bergson's cascading rocket that can never be arrested—the co-creation events of our lives do not play out in an external, homogenous time. They are the stuff of time itself. Grasping this fully can be like the moment when you actually have a conversation you've rehearsed in your head. Your rehearsal can never be complete because your imagination was missing not only the person you're talking with, but *yourself in each moment*—the person changing and responding as the conversation proceeds. When you remember this, the future can cease to look like an abstract horizon toward which your abstract ego plods in its lonely container of a body. Instead, "it," that irrepressible force that drives this moment into the next, is a thing that is speaking back to you always—even and especially from unexpected places. The task for many of us is to learn once more how to hear.

Chapter 5

A Change of Subject

THE PACIFICA SEAWALL

Alone, humanity has no future.

—ACHILLE MBEMBE, "The Universal Right to Breathe"

*W*e're thirty miles north, looking out from another cliff above the ocean. This time, though, the edge of the city is right behind us, and the fog is farther out over the ocean. A café selling "coffee, pastries, and cakes" has had the word cakes almost entirely scoured off the window by salt air, and the sidewalk gutters are full of sand. Ahead of us is a steep drop down to the beach, and just down the road, you can make out a flat, empty, fenced-in area right on the edge of the bluffs. An engineered vista point? No, it turns out. It's the footprint of an old house, removed before it would slide into the sea.

Nearby are two signs: one warning us of "dangerous cliffs," the other of rip currents and the absence of a lifeguard.

DEEP INTO THE 2020 fire season, on September 9, I woke up to a rust-colored glow coming from behind the blinds. I would soon find out that it was a mix of fog and smoke from the nearby fires, some of which had been sparked a week earlier by an ominous night of dry lightning. Solar panels, I read, were getting zero percent power. News and social media for the rest of the day would be apocalypse porn, an entirely orange time line. Here was Orange Bernal Hill. Orange Transamerica Pyramid, Orange Port of Oakland. Depending on where and when you are reading this, such an occurrence may not be an anomaly, but at the time, it felt unprecedented.

At nine A.M., it was still so dark that I had to turn on the lights in the kitchen. In a mindless bid for comfort, I fried garlic for a vegetarian version of *tapsilog*, Filipino breakfast, and watched it dry on a paper towel. The sky proceeded to get even darker, like a clock running backward, and it struck my animal body as deeply wrong. Rick Prelinger, a friend who co-runs the Prelinger Library, tweeted, "Morning is cancelled." But if morning was canceled, the workday was not. Across the street, my neighbor's lights were on; she was already at work, already Zooming. I myself had class prep to do and papers to grade. As I sat down at my laptop to work, I felt so humiliated by the contrast between my pedestrian tasks and the macabre surroundings that I couldn't decide whether to keep the blinds open or closed.

At the end of the changeless day, Joe and I went for our now-customary pandemic walk, away from our apartment building and into the land of single-family homes. For the first time, we could see into the interiors of the houses we were used to passing, because they all had their lights on. Outside, the air was winter-cold, empty, and odorless; the smoke was still too high up in the atmosphere to affect the AQI. It mirrored how my mind felt: eerily placid, dormant. But later that night, I had a dream about going to the dentist, where whatever they were doing hurt so much that I started crying and then screaming. The physical pain in the dream was real and overwhelming. When the dentist asked what was wrong, I said, "You were literally hurting me so bad I screamed!"

The smoke descended to our level the next day, like some kind of rent come due. The AQI went up to and then past 200; people complained of headaches, coughs, scratchy throats and eyes. Physical and emotional fatigue were difficult to distinguish. The sky turned white; the trees in nearby neighborhoods disappeared as though erased. No more walks. My nightmares continued, but now they were about fire: I was stuck in a traffic jam trying to get away from fire. I was on a trail with a group of people, running away from fire. I saw people fishing in a pond, but instead of catching fish, they were catching people who had drowned while trying to escape the fire. In these dreams, there was always a wall coming—a wall of fire or a wall of smoke—and it moved with terrifying, impartial finality, like the playhead on a video time line.

The fire dreams began to mix with dreams of my own death, which had increased during the pandemic. I wrote in my journal:

The future has disappeared—I want to say over a horizon, but there is no horizon, just this smoke-fog. I have never felt more distinctly that every year will be a worse year, that every minute is a minute closer to catastrophe and unrecoverable losses. Just like how you feel about your

*own aging body, but applied to everything in the world, and you don't
even have the comfort of knowing that it will flourish after you are
gone, like it's all actually ending.*

    *I keep thinking about my childhood and how I grew up not even
knowing about wildfires, and how I thought of myself as living in a
"normal time," and now everything in my past feels like it was traveling
along the surface of a folded piece of paper. And just now, we're going over
the fold, and everything after this is just about survival. Everything will be
different in ways that I can't imagine, and there is much reason to believe
it will be far worse, and the deep terror involved in that is the terror, I
think, that is driving my dreams. Not just of dying, but of suffering.*

READING THIS IN the midst of another nightmarish fire season—one that
started much earlier than usual—I recognize and sympathize with my own
sentiment. Yet I've also begun to see such nightmares as my internalization
of declinism, the belief that a once-stable society is headed for inevitable
and irreversible doom. As distinct from a clearheaded (and heartbroken)
assessment of our situation, declinism is probably one of the more dangerous forms of linear, deterministic time reckoning there is. After all, it is
one thing to acknowledge the past and future losses that follow from what
has occurred; it is another to truly see history and the future proceeding
with the same grim amorality as the video playhead, where nothing is
driving it except itself. In failing to recognize the agency of both human
and nonhuman actors, such a view makes struggle and contingency invisible and produces nihilism, nostalgia, and ultimately paralysis.

    Declinism is a close relative of nostalgia, and objects of nostalgia
are often atemporal, lacking aliveness. An example: Say you break up
with someone and many years later find yourself nostalgic for the relationship. Who is it that appears in this melancholic yearning? Assuming
they're still around, it is surely not your ex-partner *as they currently are,*
the one who has continued to age and evolve. Instead, it is a frozen, ide-
alized version of them, like a hologram that survives within and despite
the present. What's more, some relationships arguably end in the first
place because partners have stopped seeing each other *in time,* one partner having replaced the living, changing other with a static image that

can impart no surprises, only a comforting presence. As we learned with the moss, to think you love and appreciate something or someone is, unfortunately, not a guarantee that you can assign them their own reality or that you know them at all.

That's how it's been with me and "the environment" for much of my life. When I was a kid, my family took a few road trips up north, past the seemingly impenetrable Santa Rosa and Klamath mountain ranges. From the backseat of our car as we drove up Highway 101, I saw hundreds of miles of redwoods and Douglas fir. Admiring their unbroken density, I thought I was looking at forests immemorial. (Children can be nostalgic, too.) Even entering my thirties, I hadn't made much progress past "trees = good; fires = bad." I had yet to learn that California and, indeed, much of the world was actually in the midst of a fire deficit. I was not aware of how closely the local ecology had co-evolved with periodic fire, nor the extent to which indigenous people worldwide had used fire, nor how or when such practices were banned. In other words, I thought I was looking at natural history, not political or cultural history—as if the two could even be separated.

I have since learned more about the extent to which fire can be part of an ecology. Chaparral—the mix of grasses and scrubby evergreen shrubs you'll find variations of from southwest Australia to Chile to California, including where I live—is just one plant community dependent on periodic wildfires. Because this environment is so dry that few things decay or wash away, periodic small fires perform the function of removing dead underbrush, making space for new growth, and returning nutrients to the soil. The seeds and buds of certain plants will not sprout without fire and have evolved to be waxy and oily—basically extra-flammable. Uphill in the forests, species like the lodgepole pine need fire to release seeds from their otherwise-sealed cones. Lack of fire thus has cascading effects, like a decrease in wood-boring beetles that in turn imperils woodpeckers and other cavity-nesting species. The snag forest habitat that succeeds a large fire is surprisingly biodiverse—as I have seen myself on hikes through what I eventually realized were previously burned areas—and is preferred by some animal species.

People are often missing from the nostalgic view of nature, an omission detectable in the pandemic-era observation that "nature is heal-ing." Obviously, there is a difference between a healthy ecosystem and one stressed by people and pollution. But beyond that, a Westerner's attempt to arrive at the idea of how things are "supposed to be" is usu-ally fraught, because it doesn't take into account who is doing the sup-posing. Indigenous groups are sometimes said to be more attentive to an ecology's changes and temporal cues: flowerings, weather patterns, and migrations. Yet it's too easy to read this as passive adaptation, a total lack of footprint, rather than active construction and collaboration with the nonhuman world.

Indigenous practices, as much as any other, can speed and suspend—both on the minute level of individual plants and on the scale of entire landscapes and communities. Up until colonization, native tribes in many places used fire to maintain forests and prairies in a certain propor-tion and condition. In many parts of what is now California, the years after a burn would bring increased seed production, tall shoots attract-ing deer and elk, and plants in a bushy state ideal for basketry, rope, and traps. The flames from periodic burns under oak trees attracted and killed parasitic moths, who lived in the trees' canopy and jeopardized the food the trees provided. People, plants, animals, fire, land, and culture existed in a shifting co-evolutionary pattern that varied across California and many other places in the world.

At an event on fire management hosted by the Berkeley Center for New Media in 2021, one of the speakers was Margo Robbins, the execu-tive director of a council that facilitates burning on Yurok lands. Robbins used a pre-burn and post-burn photo to demonstrate the role of burn-ing in the very mountains I had gazed at as a child. With my untrained eye, I saw the first photo as a nondescript "natural area" like one you'd see on the side of a park trail. Robbins, though, described it in terms of process: Because the area hadn't been burned, the hazel (a serotinous plant, meaning that it is fire-adapted) was currently producing branches that would be useless for Yurok basket making. On top of that, other unburned brush was encroaching on the hazel, to the extent that ani-mals would not be able to eat the nuts off it and the plant would even-tually stop producing. Last, she pointed to a young Douglas fir tree, an ambassador of the forest. "This fir tree is starting to encroach on what is *supposed to be* an oak woodland savannah," she said (emphasis added).

According to Stephen Pyne in *Fire in America,* continually fired prairies and what nineteenth-century surveyors called "park-like settings" were not only common but expanding at the time of American colonization. By his measure, my childhood self had things backward: Because colonists banned indigenous fire practices, forests followed the footsteps of European invaders. Pyne writes that "the Great American Forest may be more a product of settlement than a victim of it." Robbins, too, emphasized this point: "Our landscapes were the way that they were when nonnative people arrived because of human intervention. . . . Native people, they purposefully made it that way to keep things in balance. It's just like if you have a big yard and you don't do anything to it. What's that yard going to look like in five years, in six years, in 10 years? Well, our yard is the forest and we took care of it just the way people take care of their fenced in yards." Yurok land, she said, had at one time been 50 percent prairie. Now, only slivers remained, and the elk had left. "One of the goals that we've set for ourselves is to expand those prairies in size so that the elk will come home," she said.

Far from immemorial, the forests I saw *were* memory materialized: created, marked, and later endangered by different fire regimes. Those regimes, in turn, reflected contests of power and different visions of what the land was. Initial bans on burning—by the Spanish in the eighteenth century and the incipient state of California in the nineteenth— were exercises of colonial power against indigenous tribes, tied up with other laws enabling subjugation, forced labor, and family separation.*

---

*Section 10 of the 1850 California Act for the Government and Protection of Indians prohibited the long-standing practice of firing the prairies. Tellingly, it sat between a section on punishment of indigenous chiefs if colonial laws were disobeyed and another section stating that an aggrieved white person could bring an accused indigenous person before the justice of the peace for punishment without due process. In a document prepared by the California Research Bureau, Kimberly Johnston-Dodds summarizes the overall act and its amendments thus: "[They] facilitated removing California Indians from their traditional lands, separating at least a generation of children and adults from their families, languages, and cultures (1850 to 1865), and indenturing Indian children and adults to Whites."

Although some frontierspeople learned from indigenous tribes and continued burning, the budding U.S. Forest Service was promoting a program of fire suppression by the early twentieth century. They saw forests as the nation's storage shed for wood during a time of exploding economic growth.

In this view, land became a mute receptacle of commodities. Fire, along with unregulated logging, appeared only as a threat to those commodities. In his 1871 *Report upon Forestry*, Franklin Hough, the first chief of what would later become the Forest Service, complained of a New Jersey fire that had burned "from 15 to 20 square miles, worth, before the fire, from $10 to $30, and after it from $2 to $4 per acre" and another in New York that "destroy[ed] standing timber beyond means of computation." For Nathaniel H. Egleston, who succeeded Hough and for whom "the history of our race may be said to be the history of warfare upon the tree world," trees were not just economically but also culturally valuable, providing a measure of aesthetic attraction that would keep young people from moving to the city.

Politically, this meant promoting the idea that all fire was dangerous, preventing the publication of studies that suggested otherwise, and dismissively referring to rural burning practices as "Paiute forestry." The debate over periodic burns versus total suppression was sealed during World War II, when the Forest Service put out propaganda linking fire prevention to the war effort. A 1939 poster showed something like a frontiersman version of Uncle Sam pointing at a forest fire and saying, "Your Forests—Your Fault—Your Loss!" Other posters put it more bluntly: "Forest Fires Aid the Enemy." This message persisted even after the war. One 1953 poster shows the relatively new Smokey Bear with a shovel and a forest ranger hat, a fire blazing in the background. "This shameful waste WEAKENS AMERICA!" it reads. "Remember—Only *you* can PREVENT THE MADNESS!"

In the decades that followed, California found itself at the forefront of a nationwide boom in suburban housing. I grew up in one of these houses, part of a shoddy, cookie-cutter subdivision built around the same time that Smokey poster went to press. Many of these suburbs ran along the wildland–urban interface, where fire risk was high, and attracted suburbanites who were less likely to be familiar with fire and more likely to accept Smokey Bear's all-American, zero-tolerance message. In the 1970s, when the Forest Service did change its ways, allowing fires to burn in wilderness areas (and, later, indigenous cultural burns),* the past decades of suppression had left a scar as cultural as it was ecological. At the fire event where Robbins spoke, Valentin Lopez, chair of the Amah Mutsun Tribal Band and president of the Amah Mutsun Land Trust, lamented that "non-indigenous people's relationship with fire is that fire is something to be afraid of, to see as very destructive." Robbins agreed, and hoped that younger people could help change the narrative and perception about fire.

That's a difficult task, given the spectacle of the megafires resulting both from decades of suppression and intensifying bouts of fire weather. In 2021, after an initially remote fire changed direction and destroyed a handful of homes in the Lake Tahoe area, the Forest Service caved to political pressure and suspended prescribed burns. Like so many things climate-related, this ignited a debate about temporal horizons. Wasn't a Band-Aid solution only going to worsen our accumulated "fire debt"? Ecologist Crystal Kolden said the ban meant "kicking the can down the road for those fuels to burn . . . in even hotter and drier conditions." Referring to a similar debate about Colorado wildfires, Jonathan Bruno, an operations director for an ecological nonprofit, said, "If we don't resolve how we invest our funding, and we continue to try to suppress our way out of the issue, we're not changing anything. . . . We're literally just throwing water on the hot stuff—and doing it again and again."

---

* Whereas, as Jan W. van Wagtendonk observes, the Forest Service was created in 1905 with "fire suppression [as] its reason for being," the shift away from fire suppression in California was gradual. In 1968, the National Park Service changed its policy to allow lightning-initiated fires in some parks to run their course if they happened within approved zones, and in 1974 the Forest Service did the same for lightning-initiated fires in wilderness areas. The Forest Service has also begun to allow indigenous communities to conduct cultural burns. In 2021, the Yurok Tribe provided guidance for a California law that eliminated the liability risk for private citizens and Indigenous people doing controlled burns.

*   *   *

SEEN FROM DOWN *on the beach, the cliffs are a chaotic tumble of rocks, pipes, tubes, orange plastic cones, bits of tarp, old fences, and the remnants of concrete pylons. In one spot, we can see the old foundation of a now-gone apartment building sagging over the edge of the cliff, rusted ropes of rebar twisting wildly in midair. The tubes are meant to redirect water down the cliffside without eroding it, but it's impossible to tell which tubes are still operational and which have been abandoned to the boulders. There's an unsettling ad hoc quality about it all, the air of an active burial, as people sit a few yards away enjoying a beach day.*

Various attempts to keep the cliff from moving are on display. All along the base of the cliffs are large imported boulders, sometimes referred to as "coastal armoring." In another place, some claylike stuff has been plastered to the sides like aspirational cake fondant. In still others, a fine-mesh net has been cast over the cliffside and bolted down. Below all this, someone has carved in the sandstone "OHLONE."

In one of the few apartment buildings left on the cliff—it looks to be the next to go—a series of dilapidated balconies juts out into the sun. On one, a shirtless man with a mustache leans against the railing. He looks out at the sea with an inscrutable expression, vaping.

*   *   *

"SUPPRESS[ING] OUR WAY out of the issue" is good shorthand for a range of things that structure the lived reality to which I and many other nonrural dwellers are accustomed. In California, the landscape of suppression is easy to find once you start looking: dams, seawalls, sand fences, netting, debris basins, concrete-lined creek channels, and the occasional plastered hillside—all dedicated to keeping water and rocks from moving in ways disadvantageous to people and property. With many such projects, especially those dating from the twentieth century, failure might be (in multiple senses) just a matter of time. In her otherwise politically agnostic book *Geology of the San Francisco Bay Region*, geologist Doris Sloan can't help passing judgment on Highway 1, a narrow road that, in the Bay Area, snakes between the Pacific Ocean and a fabulously unstable cliffside: "Continual repairs are required, and increasingly sophisticated (and expensive) engineering structures are built to maintain the road in an area where probably no road should ever have been attempted." Landslides onto and off the road are a frequent problem. In January 2021, a 150-foot chunk of Highway 1 near Big Sur fell off a cliff, closing the road until April of that same year.

This troubled road—closed at least fifty-three times between 1935 and 2001—would not be out of place in John McPhee's *The Control of Nature*, a collection of three accounts in which humans attempt to forestall the movement of water, lava, or rocks. Some of the most dramatic imagery in the book occurs in its final section, concerning the people in Los Angeles who live right up against the San Gabriel Mountains. The San Gabriels are a fast-rising and geologically young range that is "disintegrating at a rate that is also among the fastest in the world," with regular and enormous debris flows. Following a summer fire in mountainside chaparral, a large winter rain can send hundreds of tons of rocks, mud, and water into a canyon. Debris flows are part of the life of the mountain and actually built the flat plain where the rest of L.A. sits. But, in present times, by the time they reach neighborhoods, the flows may contain giant boulders, cars, and pieces of other houses. McPhee tells the story of a family whose house in Shields Canyon filled up with boulders and mud in the space of six minutes: "No sooner was the door closed than it was battered down and fell into the room. Mud, rock, water poured in. It pushed everybody against the far wall. 'Jump on the bed,' Bob said. The bed began to rise. Kneeling on it—a gold velvet spread—they could soon press their palms against the ceiling."

In the previous chapter, I suggest that one way to see time is to pick a spot and pay attention to it. This holds for larger spots and larger spans of time. Some folks I know who've lived in the Santa Cruz Mountains for fifty years told me about a time when you could walk to a giant rock at Pescadero that is now permanently out at sea. In the late 1980s, when McPhee wrote *The Control of Nature*, many who lived along the San Gabriels simply did not remember the last major debris flow, or did not see it as part of a pattern. Set by both transience and inattention, the cultural frame rate of "city time" could not register the geological: "A super-event in 1934? In 1938? In 1969? In 1978? Who is going to remember that? . . . Mountain time and city time appear to be bifocal. Even with geology functioning at such remarkably short intervals, people have ample time to forget it."

That's bad news for everyone except developers and real estate agents. One man who has been in L.A. since 1916 tells McPhee, "The people who buy the houses don't know that sooner or later stuff is going to come down through here like shit through a tin horn." That said, some of McPhee's subjects *do* know this. One response is to echo the city's aggressive building of debris basins by creating walls and fortifications around one's own house. Another family has installed overhead doors in the back of their built-in garage: "To guide the flows, they put deflection walls in their backyard. Now when the boulders come[,] they open both ends of their garage, and the debris goes through to the street." That's one innovative way to admit mountain time.

But isn't there something other than mountain time, captured in a series of recorded events, that needs to be admitted here? And if we're "suppressing our way out of the issue," what actually is "the issue"? On the material, everyday level, the issue in this case would seem to be a series of boulders that keep destroying property despite the city's increasing infrastructural reinforcements. But what I want to suggest is something more: that the "issue" is a failure to recognize the mountain itself. While the people McPhee interviews may appreciate what this rise in the landscape affords them—a vertical escape from the city, proximity to "nature," a masterful view of the valley, or even some neat rocks—the San Gabriels seem to appear to them only as a backdrop or a nuisance, a collection of lifeless stuff that just happens to be there. The mountain is inert and thus controllable—which explains the tragicomical hubris of a newspaper headline cited by McPhee: "PROJECT AIMS TO HALT EROSION OF MOUNTAINS; VALLEY AUTHORITIES VOTE LAND-SLIDES UNNECESSARY."

This hardheaded mentality arises from the same attitude that insists on total fire suppression. In their study of fire regimes in California and Greece, a group of Greek geographers describe a mindset that could just as easily be applied to boulders, floods, or mountain lions: "The general public perceptions are that forest fires should be controlled and not pose a threat to humans and property," they write. "It is interesting to note that what attracts people to live on the forest margins, that is a sense of living in a 'natural environment,' is done under the mystique of eliminating the 'wild' from wild lands."

The geographers suggest that what has been lost is a "vernacular" relationship with fire and a dynamic landscape. In Greece, before more people moved to cities, periodic burning was part of rural tenants' familiarity with and responsibility to their specific surroundings. Likewise, in California, Robbins noted that Yurok people out on their regular hunting rounds would take note in passing of areas that seemed to need a burn. And in Western Australia, Victor Steffensen, an indigenous land management expert educated by elders, explains how burning was related to the identity of an area: "From place to place, both old men would stop and tell the fire stories for each different landscape. They would talk about the right time to burn, how all the animals fitted in, what plants lived where, and the types of soils." Fire was part of a reciprocal responsibility between one subject (humans) and another (land).

At the age of nineteen, Steffensen was recruited by a national park ranger to help with fire management: "The rangers spread maps across the bonnet of the truck and started pointing their fingers at their plan of attack. 'We are going to burn this side of the road, but not that side,' they instructed. Their burn zones were broken up by roads and fences instead of reading the country and burning the right place like the old people do." In this particular instance, the fire jumped the roads, causing no small amount of drama. I wouldn't generalize this to all the interactions between state agencies and indigenous groups, some of which seem to be fruitful, good-faith exchanges. What this story does provide, however, is an extreme contrast between different ways of looking at land. In the first, land is the frozen stage on which those afforded identity might move; in the second, land *is* identity, expressed in time. Or, as Paula Gunn Allen writes, "The land is not really the place (separate from ourselves) where we act out the drama of our isolate destinies. It is not a means of survival, a setting for our affairs. . . . It is rather a part of our being, dynamic, significant, real. It is our self."

In the United States, the early Forest Service took cues from German scientific forestry, which dictated the planting of neat rows of the most economically valuable timber: a one-age, one-species commodity forest. James C. Scott observes that German scientific forestry sought to replace actual trees with "abstract trees" representing known quantities of lumber. While lucrative for one rotation of trees, this turned out to be a disaster—and not just for the German peasants who relied on the older forest ecology for grazing, food, and medicine. Monocropping made the forest more vulnerable to storms and disease, and the only reason the first generation of trees grew so well was that it could draw on the accumulated resources of the previous old-growth forest. After that, *Waldsterben* (forest death) became part of the German vocabulary, and attempts to artificially reintroduce everything that had been overlooked in the economic forest strategy (in the form of nest boxes, ant colonies, and spiders) had to contend with the unfortunate fact of the monocrop forest.

Scott includes this story at the very beginning of his book *Seeing Like a State: How Certain Schemes to Improve the Human Condition Have Failed*, as a parable:

> [This story] illustrates the dangers of dismembering an exceptionally complex and poorly understood set of relations and processes in order to isolate a single element of instrumental value. The

instrument, the knife, that carved out the new, rudimentary forest was the razor-sharp interest in the production of a single commodity. Everything that interfered with the efficient production of the key commodity was implacably eliminated. Everything that seemed unrelated to efficient production was ignored. Having come to see the forest as a commodity, scientific forestry set about refashioning it as a commodity machine. Utilitarian simplification in the forest was an effective way of maximizing wood production in the short and intermediate term. Ultimately, however, its emphasis on yield and paper profits, its relatively short time horizon, and above all, the vast array of consequences it has resolutely bracketed came back to haunt it.

I WAS IN college when the term *Anthropocene,* the geological age in which humans are the main influence on climate and the environment, started to gain currency in science and the humanities. The Métis anthropologist Zoe Todd and other indigenous studies scholars have already criticized the term for a number of reasons, including its wholesale adoption of a hierarchy of human and nonhuman, "flesh and things." In contrast to Allen's formulation, where the land "is our self," some versions of the Anthropocene see naturally exploitative humans exercising their will over something from which they are separate, something that exhibits no will in return. This is a very different story from that of the entanglement and co-evolution of "flesh and things" over thousands of years.

Seen through the lens of the Anthropocene, the nonhuman world is inert, but what's funny is that, upon closer examination, the innately exploitative humans don't have agency in this view, either. They just do what they do—mess up a "state of nature"—and they all do it. The *Anthro-* of Anthropocene lumps humanity together, as if one specific portion of humanity were not responsible for a culture of extraction, visiting environmental horrors upon the rest of the world. This blunt-edged framing has informed the unstoppable "video playhead" of my nightmares—a history with no actors, only mechanisms; no moments of struggle, only linear evolution. For example, many definitions of the Anthropocene trace it to the invention of the Watt steam engine in the late eighteenth century and see it unfolding from there, with the social and political simply bracketed out. In "Anthropocene, Capitalocene, and the Problem of Culture," Daniel Hartley writes, "Inherent to the Anthropocene discourse is a conception of historical causality which is purely mechanical: a one-on-one billiard ball model of technological invention and historical effect. But that is simply inadequate to actual social and relational modes of historical causation. The fact that technology itself is bound up with social relations, and has often been used as a weapon in class war, plays no role in Anthropocene discourse whatsoever." As Hartley observes, this sort of determinism reflects a view of history as a unidirectional and inevitable march of progress that can never be questioned or redirected, only sped up or slowed down. He cites two passages from a popular 2011 essay about the Anthropocene:

1) "Migration to cities usually brings with it rising expectations and eventually rising incomes, which in turn brings [sic] an increase in consumption"; and
2) "The onset of the Great Acceleration may well have been *delayed* by a half century or so, *interrupted* by two world wars and the Great Depression" (emphasis added by Hartley).

Hartley writes that "the first sentence seems almost willfully blind to the history of mass urban poverty, gentrification and accumulation by dispossession. The second seems to claim that the bloodiest century in human history—including Hiroshima, Nagasaki, the Dresden bombing, the Gulags, and the Holocaust—is a mere blip on the rising line of progress."

To think deterministically is to take things for granted, both forward and backward in time. Just as I misunderstood the forested mountains as a child, projecting them into a supposedly uniform past, the concept of the Anthropocene has the potential to make the outcomes of specific actions by specific people seem like a natural and inevitable condition.*

---

*This is not to deny that conditions can't over time take on a life of their own, becoming in a sense self-sustaining or leading naturally to other outcomes. The point is that actions did at some point play a role in the creation of a condition, as opposed to the condition being timeless, innate, or beyond question.

If the results weren't so horrifying, this phenomenon would almost be funny. Indeed, it has the flavor of one of my favorite sketches from the TV series *I Think You Should Leave,* in which a hot-dog-shaped car crashes into a clothing store without a driver in it.

"Somebody, call the cops! We need to find that driver!" says one bystander.

"Whose car is this?!" screams another.

The camera pans over to Tim Robinson, who is wearing a giant hot dog costume and an exaggerated expression of surprise.

"Yeah, come on, whoever did this—just confess! We promise we won't be mad!"

Despite being called out, the hot dog refuses to concede. "You know, I don't have to sit here and be insulted like this," he says. "I'm just gonna take as many suits as I can grab, get in that random hot dog car— *random!!*—and drive back to Wiener Hall."

In 2020, that sketch was frequently summoned to refer to Trump and his constant backtracking. But for my current purposes, it illustrates a broader disavowal having to do with what is a *given* in the most reductive versions of the Anthropocene. The Jamaican writer and theorist Sylvia Wynter has written about how the category of human came to be defined during the Enlightenment era: "Human" (the white economic man, the colonist, or the Man in "Man vs. Nature") was defined against "nonhuman" in a moment of colonial exploitation, the results of which were recast as biological, atemporal conditions that explained the supposed racial traits of "backward," "timeless," or "less progressed" peoples who were not quite human.* As it happens, this

---

*This definition is part of a shift that Wynter describes in her 2003 paper, "Unsettling the Coloniality of Being/Power/Truth/Freedom." Prior to this shift, a religious conception of Man placed the "True Christian Self" in opposition to the "Untrue Christian Other": heretics, infidels, and so on. Once Man was redefined as the rational and political subject of the state, a new group came to occupy the role of Other: "The peoples of the militarily expropriated New World territories (i.e., Indians), as well as the enslaved peoples of Black Africa (i.e., Negroes), . . . were made to reoccupy the matrix slot of Otherness—to be made into the physical referent of the idea of the irrational/subrational Human Other." Thus the very definition of what it meant to be human was based on a scientistic boundary between Man and those "dysselected by Evolution until proven otherwise."

conveniently obscured historical responsibility—a little bit like a bully pushing you and then suggesting that you're innately prone to crying. And it suggested new givens: The less-than-human were naturally inferior, while the properly human were naturally capitalist and individualistic. These were no longer outcomes of people's choices and beliefs but a priori qualities for which, in turn, no one was responsible. (Says the guy in the hot dog sketch, "We're all looking for the guy who did this!")

Drawing on Wynter's work, the writer serynada also points to the eighteenth century, when thinkers like Adam Smith contributed to the idea of a Western Man driven by the "imperative to survive":

> Human beings were rendered economic machines that seek to maximize their share of sparse natural resources. The inscription of a bio-evolutionary and thus inevitable impulse behind the ascent of Western Man—"we all want to grab more resources, Europeans just did it better than everyone else"—came to vindicate capitalism, white supremacy, and imperial expansion. The West invented Man and projected Him onto the past as natural and timeless, rather than historical and cultural.

From this point of view, the concept of the Anthropocene looks less like a descriptor and more like a symptom of belief in a "natural and timeless" capitalist Man and a helpless Nature. There is some irony here, given that long before the steam engine, the denial of subjectivity to so much of the world was part of what enabled processes of extraction and accumulation in the first place. In *This Changes Everything: Capitalism vs. the Climate*, Naomi Klein describes extractivism in terms that may sound familiar by now: "a nonreciprocal, dominance-based relationship with the earth," a "reduction of life into objects for the use of others, giving them no integrity or value of their own," and a "reduction of human beings either into labor to be brutally extracted, pushed beyond limits, or, alternatively, into social burden, problems to be locked out at borders and locked away in prisons or reservations." In other words, abstract people, abstract trees, abstract animals, abstract land; none of them subjects with agency; all of them ready to be mined, squeezed, bracketed, or simply destroyed.

WE CONTINUE SOUTH, *toward a promenade that is actually a seawall. When I was here in winter, the area near the pier was a mess of orange cones, barricades, and sandbags. Outside one of the dusty-hued houses along the promenade, ensconced in a carpet of what was at the time a purple-flowering succulent, I had noticed that a mermaid statue's head and outstretched arm had broken off. Today, however, people are out en masse and enjoying the unfoggy day. Although the motion of the sea remains below us, we can hear its constant assault against the concrete, and a yellow sign reads* CAUTION—WAVES MAY BREAK OVER SEA-WALL. *Farther down, a sign on a tall pylon shows a satellite image of this area and invites us to* JOIN THE CONVERSATION *about the local infrastructure resiliency project. I tell you that I recognize this sign because, when researching the area, I'd come across the minutes for an earlier meeting in this series. Residents couldn't seem to agree on whether the city should pursue a seawall or a restoration with plants and other natural elements. One person had said they were not interested in the "living shoreline" and "managed retreat" options; what they really wanted was a seawall that would last for at least fifty years. I wondered how they'd settled on that number.*

AS I WROTE this chapter, I'd occasionally look out the window at the ashen sky in which the nearby mountains had disappeared and think of something Stephen Pyne writes in *Fire in America*. He's just explained that, at the turn of the [twentieth] century, "the controversy was at base a conflict between two sets of fire practices, one set learned largely from Indians and sustained by a frontier economy of hunting, herding, and shifting agriculture; the other set, better suited for industrial forestry." Then he adds: "There was no a priori reason why American forestry should have rigorously excluded all forms of broadcast burning." No a priori reason. No inevitable march toward the technocratic "wisdom" of fire suppression. Instead, there were only differing worldviews and a messy knot of political machinations that began long before I was born. Now the fire debt was in my lungs, and I was tired.

Those whitened days felt like purgatory, and purgatory is enervating. The danger, to me, is that it can become so enervating that there's no energy left to see past the boundaries of the present. But this creeping sense of inevitability doesn't just feel bad; it obscures the actors continuing to tighten the ropes and everyone who has fought, and is fighting, to get free. The story of Enlightenment Man teaches me an all-too-common truth: that the people who stand to gain the most from determinism (in others) are typically the people doing the determining. This strategy is detectable not just in the long historical sense, but also in the current maneuvers of those energy companies now driving climate change.

In *Overheated: How Capitalism Broke the Planet—and How We Fight Back,* Kate Aronoff describes a moment in which the energy industry learned how to sell inevitability. In the 1960s, a couple of Shell executives were invited to a Hudson Institute seminar on scenario planning. Developed by futurist academics and defense planners during the Cold War, scenario planning involved imaginatively spinning up and filling out different future scenarios in order to get a jump on one's opponents. This was supposed to be an explicit break with linear thinking (which might involve, for example, computer modeling). The scenario-planning seminar sought to transfer the practice to the multinational corporate world, and the idea found fertile ground with the Shell executives—especially the eccentric "idea man" Pierre Wack, who sounds a lot like Hank Scorpio from *The Simpsons:*

> Wack, Newland, and their colleagues became evangelists for scenario planning within the company. In the early days, Shell's brightest minds went "into the green" at chateaus in the South of France to write early scenarios, enjoying wine with long meals and walks between heady, marathon sessions mapping out the changing face of geopolitics and evolutions in the oil and gas business. . . . Wack, famously, split his time between the West and the East, where he had sought out spiritual guidance in scattered ashrams and monasteries from the age of twenty. His office smelled of incense. One member of the planning team recalled that his

final interview for the job was with Wack, who conducted it "in a complicated yoga position."

Aronoff points out that scenario planning was much more—or maybe much less—than a vaunted philosophical exercise, given that "you didn't need to be any great genius . . . to see that linear forecasting models would only work for so long for the oil industry of the late 1960s." This was a time when Shell faced pressure from the Global South and the implications of reports like the 1972 *Limits to Growth*, which highlighted the unsustainability of fossil fuels. As economic historian Jenny Andersson tells Aronoff, Shell needed a way of "dealing with the future" that would avert a self-defeating determinism and to seek out "other versions of the future that were not catastrophic to them." That's just good business sense. It's also a reminder that, as Aronoff writes, Shell "has a constitutive block that keeps it from being an ally in the climate fight: an inability to envision a future without Shell. The company's overriding mission is to ensure an indefinite life for itself and its profits."

Since then, Shell has merged scenario planning into what is more straightforwardly PR. Having shifted from funding climate change–denial ads in the 1970s to "painting [themselves] green" in the 2000s, those same companies who faced down "self-defeating determinism" are selling the public their own strain of determinism. Energy companies have every incentive to make *their* future be *the* future. In a sweeping 2021 study of ExxonMobil's climate change communications since the mid-2000s, Naomi Oreskes and Geoffrey Supran find language that portrays extraction and consumer demand as inevitable:

> [A] 2008 ExxonMobil Corp advertorial stat[es]: "By 2030, global energy demand will be about 30 percent higher than it is today . . . oil and natural gas will be called upon to meet . . . the world's energy requirements." Another, in 2007, says that "increasing prosperity in the developing world [will be] the main driver of greater energy demand (and consequently rising CO_2 emissions)." A 1999 Mobil advertorial is even blunter: "[G]rowing demand will boost CO_2 emissions." In other words, they present growing

energy demand as inevitable, and imply that it can only be met with fossil fuels.

It was BP that popularized the notion of an individual carbon footprint, for example, by releasing a carbon footprint calculator in 2004. This was one of several ways energy companies would imply that the responsibility for solving climate change lay with the consumer. It is certainly true that consumption habits need to change; Klein suggests that the well-off 20 percent in a population are the most responsible for making those changes. But she also points out that if we want reductions to span beyond "earnest urbanites who like going to farmers' markets on Saturday and wearing up-cycled clothing," we need "comprehensive policies and programs that make low-carbon choices easy and convenient for everyone."* In the meantime, energy companies' emphasis on consumption is disingenuous. This rhetoric echoes Big Tobacco's effort to portray itself as a neutral purveyor of what consumers just can't seem to help but demand. In other words, *We just sell the cigarettes; you're the ones smoking them.*

A framing like this one portrays climate change as solely "our" fault, where the "our" is an aggregate of consumers who should attend to their carbon footprint calculators. All the while, as Aronoff writes, "every shred of evidence suggests the [energy] industry is moving full speed ahead in the opposite direction, pushing more exploration and more production as temperatures rise, seas swell, and fires burn." One smoky day while I was writing this chapter, a Wells Fargo ATM asked

---

*Likewise, in *Overheated,* Aronoff notes that "if there is to be such a thing as a low-carbon society, it will be the government's job to build it." Of course, personal choice within the structures we have is still important. Douglas Rushkoff, in *Survival of the Richest: Escape Fantasies of the Tech Billionaires,* has this suggestion: "Instead of debating whether to buy electric, gas or hybrid, just keep the car you have. Better yet, start carpooling, walking to work, working from home, or working less. Like Jimmy Carter tried to tell us during his much-ridiculed fireside chats, turn down the thermostat and wear a sweater. It's better for your sinuses, and better for everyone." Toward the end of her book, considering the possibility of doing less, Aronoff ties her arguments to the proposed benefits of a shortened workweek. In some ways, these recommendations evoke the end of chapter 2 of this book: the idea of giving certain things up, as well as Butt-Head's request: "Could you, like, get less stuff?"

me if I wanted to donate to help with the wildfires. I stared back at the screen. Wells Fargo is one of the largest funders of fossil fuels, having invested $198 billion into the coal, oil, and gas industry in the four years following the Paris Agreement.

Just as the industry of individual time management resells the idea of time as money to the isolated bootstrapper, energy companies sell the idea of the carbon footprint to conceal larger and more significant avenues of change. These include both technological and political tools to which we already have access. For Klein, Aronoff, and others, some of those tools would be public regulation and oversight—things like the Green New Deal—and standing up to the global trade agreements that favor the suicidal time horizon of energy companies. Indeed, Klein has an entire chapter titled "Planning and Banning."

Klein acknowledges that this is an uphill battle in the United States, where both planning and banning are currently decried as government overreach. Nevertheless, she writes, "we should be clear about the nature of the challenge: it is not that 'we' are broke or that we lack options. It is that our political class is utterly unwilling to go where the money is (unless it's for a campaign contribution), and the corporate class is dead set against paying its fair share." For her part, Aronoff takes great pains throughout her book to remind us that the hill in the uphill battle is historically specific: "In positing all of human existence as an endless striving toward market society, neoliberals had to erase not just the possibility of a future but all memory of a past when humans managed to organize themselves in other ways. The kinds of tools needed to navigate out of the climate crisis—things like public ownership, full employment, or even just tough regulations—have receded into memory." Aronoff is talking mostly about policies of the New Deal era, before a globalized economy took hold and the perception of government regulation soured in a neoliberal atmosphere. But one could extend this notion of political amnesia even further back, as an echo of what serynada describes: rewriting the history of Man as an economic machine.

Again, purgatory is enervating. Like a fog machine spewing a priori dystopia, energy companies are still selling their certain future, still designing targets and portraying us as drifting helplessly toward them. I think back on my nightmares, about how the future looks there. Who wrote that scenario?

A PIER JUTS *out from the promenade into that uncompromising ocean. The moment we move past the wall, the violence of the waves against it becomes clearer and louder. Navigating the crowded pier, we pass groups of crabbers who have set up on the sides with tables, buckets, umbrellas, and stereos.*

From out here, we look back at the seawall. Now you can really see it—the north end of the wall is failing, has probably been failing for a long time. It sags downward, seems to thin out, and then just sort of disappears. In fact, the whole sequence of events is easier to grasp from this distance, where the houses and streets look like a precarious dusting of civilization upon the restless, churning cliffs.

I DON'T REMEMBER what year it was that I started noticing apocalyptic language in the art classes I taught at Stanford. I just remember the student who made a detailed, animated triptych based on *The Garden of Earthly Delights*, by Hieronymus Bosch. The three collages got darker and bleaker from left to right. "It's kind of like . . . the sunset of humanity," the student said, laughing nervously. Or the student who, standing in front of a projector screen showing his 3-D project and tasked with explaining it, said in a small but tortured voice, "Well, I just feel like the world is ending and all that," at which everyone mutely nodded. I remember thinking it seemed vulgar to continue on talking about vectors and shaders after that. And I remember wanting to run across the classroom and give that student a hug.

Years later, I would see my previous book recommended in an online forum for people heartbroken by climate change and worried about civilizational collapse. In a post typical of the forum, one person wrote, "I know that I should be grateful that everything is still here but every single thing seems to be a reminder that one day it will disappear in a miserable way." "I could just unexist without hurting anyone else," wrote another.

On the forum, these expressions were typically met by good-faith suggestions for transcendence: pursuing Buddhist ideas of impermanence, finding the small joys in life, and in one case, reading *How to Do Nothing*.

It is important to grieve, especially to grieve in common. I'll take wailing sorrow over denial or unrealistic optimism any day. But dangling there, unattached to anything else, the sentiment felt similar to my nightmare and what it had come to represent—a non-future where people's beliefs and behaviors are as determined as the earth seems inert and helpless. Without suppressing grief, there has to be a different way of thinking about time than the one in which we're simply strapped in all the way to the end. One way, which I've tried to outline so far, is to recover the contingencies of the past and the present. Another is to shift your temporal center of gravity by looking to those whose worlds have already ended many times over.

In 2019, Thom Davies wrote a study of a place in Louisiana informally named Cancer Alley. He interviewed residents of Freetown, once part of the Landry-Pedescleaux Sugar Plantation, founded as a settlement by formerly enslaved people during Reconstruction and now overrun by the petrochemical industry. While Davies was writing, the Bayou Bridge crude oil pipeline had yet to be completed, a process that saw sixteen protestors and one journalist arrested and charged with felonies. But things were already so bad that a resident told Davies the air some days was "so full with gas you can hardly breathe."

For Davies, what happens in Cancer Alley illustrates the concept of *slow violence,* a term coined by Rob Nixon, of the High Meadows Environmental Institute, for harms that remain below the level of public perception because they're too gradual and lack a spectacle. But Davies makes one important clarification: "Instead of accepting Nixon's oft-cited definition of slow violence as 'out of sight,' we have to instead ask the question: 'out of sight to whom?'" A spectacle means something different for those who view it on the news for a week than it does for the people who live in it. "Having spent almost a decade investigating the lives of communities in various toxic geographies—including Chernobyl, Fukushima, and now 'Cancer Alley' . . . the last thing I would describe these spaces as, is lacking in spectacle," Davies writes. "Communities who are exposed to the slow violence of toxic pollution are replete with testimonies, experiences, and bereavements that bear witness to the brutality of gradual environmental destruction."

In other words, seeing the future can be more a matter of looking around than looking ahead. Because of my family history, I am inclined to look across the ocean. As with other Southern Pacific countries, the Philippines has seen increased tropical storm activity since the 1970s, and between 1960 and 2012, the sea level in the Manila Bay Area rose by nine times the global average rate. In Sitio Nabong, in an area just north of Manila, locals told Channel News Asia that it had been decades since they'd walked on the paved streets there. They attend church in boats. Yet differences in perspective do not have to be geographically distant, from me or from one another. For example, the title of an August 2021 *New York Times* op-ed struck a future-dreading, hypothetical tone: "Lovely Weather Defined California. What Happens When It's Gone?" In contrast, a month prior, California farmworker Martha Fuentes told an Al Jazeera reporter that, out in the fields for thirty-one years, she was already well aware of shifts in the temperature thus far.

Here we arrive again at that question of the Anthropocene and what you take as your inflection point. Kathryn Yusoff, in *A Billion Black Anthropocenes or None,* writes against the way the Anthropocene is "configured in a future tense rather than in recognition of the extinctions already undergone by Black and indigenous peoples." Māori climate activist Haylee Koroi, when asked how she felt about the contemporary idea of climate depression and fatigue, replied that "without invalidating those that feel that way, the reality is that through colonisation, we've been experiencing the symptoms of climate crisis for generations." Likewise, Elissa Washuta calls her people "post-apocalyptic"; for them, annihilation is not in the future but, rather, in the past, continuing into a white American present that seeks "to exterminat[e] the Siwash they see in me."

I am citing these perspectives not in order to shame those, like me, whose worlds only now seem to be ending. Rather, to the nihilist who cannot imagine the future, I am highlighting a perspective that has survived, and continues to survive, the long-ago end of the world. There are many people and places that could accept neither Enlightenment Man's march of progress nor the billiard ball declinism of the Anthropocene—because that narrative was inherently premised upon their destruction, commodification, and relegation to a state of nonbeing. For those people and places, the historical past can never be an object of nostalgia, and the future has always been in jeopardy. If you don't want to kick the can down the road, look to those who never recognized the road in the first place.

*   *   *

BACK ON THE *seawall, there is a circle of five wooden posts that looks like a miniature Stonehenge, that most iconic of calendrical tools. In the center of the circle, we can just make out the text of a plaque embedded in the ground and obscured by sand:*

ANCHOR OF THE BRIG ROLPH
A FOUR MAST SAILING VESSEL
SANK OFF POINT SAN PEDRO 1910
ANCHOR RAISED BY THE SEA LIONS CLUB IN
1962 AND DONATED TO THE CITY OF PACIFICA
WEIGHT 2000 POUNDS

The plaque reads more like a memorial to a memorial than to an artifact, because for some unknown reason, the anchor is gone. Nor does the plaque tell us where the ship was going. We squint at the horizon, and I tell you about the ship's destination, how it was carrying lime, hay, and lumber to a sugar plantation in Hana, Hawaii. The plantation was run by Theo H. Davies and Company, one of the "Big Five" commercial conglomerates that held most of the land and monopolized the economy on Hawaii. In an effort to build a suf-ficient and steady labor force—Native Hawaiians protested work conditions* and were also devastated by foreign diseases—the companies imported people from China, Japan, Norway, Germany, Puerto Rico, Russia, Korea, the Philip-pines, and Portugal, arresting laborers if they didn't work fast enough. To carry the sugar away, the Big Five also ran the Matson Navigation Company—the same Matson we saw at the Port of Oakland.

There's a Hawaiian saying that translates to "The land is chief, man is its servant." As it happens, commercial interests on Hawaii caused their own cli-mate change: By cutting down ancient trees and grazing cattle, they may have altered the local rainfall patterns to their own detriment. The territorial gov-ernment, under the thumb of sugar interests, frantically worked to reforest the

---

*Ronald T. Takaki, in *Pau Hana: Plantation Life and Labor in Hawaii, 1835–1920,* writes that the native Hawaiian workers on one of the island's first plantations refused the plantation owner the "control and loyalty" he expected. Illustrating racialized ideas of work and time, the plantation owner hoped there might be steady, obedient men among the natives (*kānaka*) who might "turn themselves into 'white kanakas.'"

slopes; unfortunately, they did it with fast-growing nonnative eucalyptus that formed less-complex forests with fewer species.

That night in 1910, on its way to the plantation, the freight-bearing ship fell victim to dense fog and strong currents. It crashed in the exact spot that another ship had wrecked just six years earlier, under similar circumstances. No one died, but efforts to remove the ship were unsuccessful—the rocks held fast. Now the ocean looks calm, and on the horizon, a wisp of fog slides by so slowly it doesn't appear to be moving.

THE WEATHER SPEAKS, though not in English. Many of the spectacular events of climate change are the unprecedented versions of this old language: fire, storm, flood—just louder and in new places. While we "suppress our way out of the issue," mountains shed, fault lines slip, lava flows where it wants, and chaparral has "an always developing, relentlessly intensifying, vital necessity to burst into flame." Creeks swell their banks, and rivers periodically like to change their course. In the first section of *The Control of Nature*, about the losing battle to keep the course of the Mississippi River from being "recaptured" by the Atchafalaya River (an effort requiring higher and higher levees in New Orleans), nonhuman agency is briefly acknowledged in an unexpected context. It happens in a conversation between a contact pilot for the river and a civil engineer: "Cano was speculating about the Atchafalaya's chances of capturing the Mississippi someday despite all efforts to prevent it from doing so. 'Mother Nature is patient,' he said. 'Mother Nature has more time than we do.' Rabelais said, 'She has nothing but time.'"

It is a sentiment out of place with the arrogance that would "DECLARE . . . LANDSLIDES UNNECESSARY" and live insulated from whatever it was that caused them. In *Technics and Civilization*, Mumford observes that this insulating, time-outsmarting quality is precisely what industrialists had liked about coal, "which could be mined long in advance of use, and which could be stored up, [and] placed industry almost out of reach of seasonal influences and the caprices of the weather." As it would turn out, this was one of the first times we kicked the can down the road. Already, in 1934, Mumford surmised that industrialism might actually cause "a long cyclical change in the climate itself."

Around the time of that sunless day in September, I added a book-mark to my browser bar for AirNow, an air quality forecasting website. That day, the circle was red, with the AQI hovering around 153. The burnt remains of trees had arrived as $PM_{2.5}$—the "PM" for "fine particulate matter"—and they'd be sticking around all week. "Have flexibility in your schedule?" the website said. "If the forecast is Red (unhealthy), there may still be times during the day when air quality is OK for outdoor activities. Check current air quality to see if now is a good time for you to be active outdoors." Centuries after coal was first commercially exploited in the United States, the caprices of the weather were now laid out on the screen in front of me. *Hear me,* they said. *Ignore me at your own peril.*

I take no pleasure in the lives destroyed by megafires and megastorms, especially when the world's poor have disproportionately borne the cost. I also cannot deny weather events an aspect of speech and an element of return, even haunting. I think of the sign I sometimes see on Northern California beaches, which have rip tides, sleeper waves, and no lifeguards, and where unsuspecting people are sometimes swept away. It says NEVER TURN YOUR BACK ON THE OCEAN. That sign always puts me in my place. It reminds me that the beach is not an amenity for humans—that I can be there, but I'd better learn the laws of the ocean if I want to stay alive.

These days, more and more of us are compelled to seek "flexibility in [our] schedule," becoming periodic people who must learn the language of fire and flood. There have been legal acknowledgments of the non-human world: New Zealand in 2017 gave Mount Taranaki the same legal rights as a person; Bangladesh did the same with all its rivers in 2019; and in 2022, a Florida lake was a plaintiff in a lawsuit against developers.*

---

*In a chapter of *The Metaphysics of Modern Existence* called "Expanding the Legal Universe," Vine Deloria, Jr., writes, "Nature has no rights of its own in our legal system. If our legal system reflects our view of reality, then we believe that we exist over and apart from the physical world." He also discusses Christopher D. Stone, a law professor from the University of Southern California who used a theory of legal standing in 1972 when arguing *Sierra Club v. Morton* and went on to write the book *Should Trees Have Standing?* Since then, similar legal undertakings have happened in Ecuador, Argentina, Peru, Pakistan, India, New Zealand, Canada, and the United States. In 2019, the Yurok Tribe (the same tribe that provided guidance for the California law on controlled burns) granted legal personhood to the Klamath River under tribal law, hoping it would aid legal actions on behalf of the river.

But, for the most part, what could not be admitted in the Enlightenment fantasy of elimination and control has yet to be widely (re)admitted: the subjecthood of the nonhuman. I know that not every reader is going to follow me here, but this is how far I need to go. Especially in the midst of climate change, not to acknowledge this feels like living with a roommate and pretending they don't exist—that you're not killing them, and killing yourself in the process. It is thus a question that blends the practical and moral. Māori writer Nadine Anne Hura (Ngāti Hine, Ngāpuhi) gives this diagnosis: "We are unwell because Papatūānuku is unwell. What's coming is worse. How can we talk about solving this sickness if we don't acknowledge its fundamental causes? Greed, waste, the accumulation of individual wealth, an arrogant belief in the superiority of 'man' over every other living organism, and the perception of land as a resource to be wrung out like a dirty cloth and then discarded."

Likewise, the energy policy and climate scholar Seth Heald warns against talking about climate adaptation and resilience "without mentioning what it is we are adapting to or working to be resilient from." He cites a study finding that most Americans conceive of climate change within an environmental, scientific, or economic lens, but not a moral or social justice one. For Heald, this is a form of "partial climate silence." While it is undeniably good that more people are polling as concerned about climate change, partial silence will bring partial solutions. I can imagine, for example, a future in which the increasingly burning, storming, sliding world is vilified at the same time it is suppressed and denied agency—the same way all objects of colonialism have been. This perspective would (and does) see waves of migrants in the same objective terms as a hurricane, declaring them as "unnecessary" as a landslide and substituting technocratic interventions for a long-overdue reckoning.

THE SEAWALL PATH *gives way to an armored berm protecting a golf course— and the threatened species of frog that now lives in the ponds there—from the ocean. To our left are cypress trees permanently sculpted by the wind, and ahead is a bald set of hills into which we can see walkers disappearing like dots. We decide to follow them.*

The trail thins out, passing under a small cypress grove. Improbably, things are still blooming here: coast Indian paintbrush and a flower with a temporal name, "Farewell to Spring." At the top of a bare hillside, we're met with yet another sign telling us not to approach the unstable bluffs. From here we can see it all: the pier, the failing wall, the cliffs to the north and south, and that infinite ocean, seemingly more infinite the higher one gets.

Out there, we see something: an explosion of mist. It's just far enough away, and the sun bright enough, that it seems like a trick played on the eye, but then it happens again. It's a whale.

For a moment, I am speechless. Then I say something silly about how I'd forgotten that whales were really real, not symbols on bumper stickers. What I'm actually thinking is that it's not just the whale but the ocean that suddenly looks more real. All this time, it's been a universe bordering ours, an unfathomable Umwelt that is not for us.* The shift in our center of gravity reveals the whale and the ocean as sovereign, the cliffs the edge of their world as much as ours.

---

* Umwelt is a German word for "environment" or "surroundings." In the early twentieth century, the Baltic-German biologist Jakob von Uexküll began using Umwelt to refer specifically to the world as it is experienced by a particular organism. For an exploration of this concept, see Ed Yong's An Immense World: How Animal Senses Reveal the Hidden Realms Around Us.

*   *   *

WHEN WE ALLOW the climate crisis a moral dimension, certain things lost in the haze become clearer, including its relationship to other fundamental injustices. For example, the seemingly utilitarian reasoning of energy companies and investors can be compared to that of the apologists for slavery in nineteenth-century America, who also saw it as an apolitical, economic issue with technocratic solutions. Only by viewing enslaved people as nonsubjects could someone like Henry Lascelles, Second Earl of Harewood, have spoken plausibly of a "progressive state" of "improvement in the slave population" at an 1823 meeting about his West Indian plantations. Amelioration was technical, a question of how to use objects better; abolition was moral, a question of who was a subject. Energy companies cannot imagine a future without the objects of extraction and, therefore, must promote and fund a worldview in which earth remains an object. Plantation owners could not imagine futures without the objects of slavery and, therefore, promoted and funded a worldview in which enslaved people remained objects. This connection is more than an analogy: Multiple scholars have emphasized, for example, the role of plantation cotton in the textile factories that drove the Industrial Revolution.

For a modern subject, there is so much about this historical moment that can seem helplessly convoluted—but some things are cut and dried. Whenever I see the future being frittered away in cold calculations; whenever someone says it's ecological and economic but not moral or political; whenever a technocratic framing hides and continues the arrogance of centuries past; whenever the colonized and objectified fail to appear as plaintiff; whenever those who profit fail to appear as defense; whenever I start to lose sight of the horizon and forget why the smoke is there—I play out the argument in my head. *It's a complicated subject,* says one side. *Not really,* says the other.*

The alternative to saying "this is it" is the idea that this was never it. The trees I saw as a child were not timeless. Like the forests sprung from

---

* These are lines of dialogue from a conversation about climate change between a priest and the owner of a polluting factory in Paul Schrader's 2017 film, *First Reformed.*

fire suppression and the question of whether land was a *who* or a *what*, I grew up on a false plateau I took for infinity. And until I learned otherwise, all I could perceive was the loss of what was familiar and comforting to me. Now I endeavor to release my grip. To look into the future is to look around; to look around is to look into history—at not the apocalypse coming but the apocalypse past, the apocalypse still unfolding. Observing that the Greek word *apokalypsis* meant "through the concealed," Washuta writes that "apocalypse has very little to do with the end of the world and everything to do with vision that sees the hidden, that dismantles the screen." Likewise, French feminist poet and philosopher Hélène Cixous wrote that "we need to lose the world, to lose a world, and to discover that there is more than one world and that the world isn't what we think it is." The current meaning of *apocalypse* is modern; in Middle English it simply meant "vision," "insight," or even "hallucination."

The world is ending—but which world? Consider that many worlds have ended, just as many worlds have been born and are about to be born. Consider that there is nothing a priori about any of them. Just as a thought experiment, imagine that you were not born at the end of time, but actually at the *exact right time,* that you might grow up to be, as the poet Chen Chen writes, "a season from the planet / of planet-sized storms." Hallucinate a scenario, hallucinate yourself in it. Then tell me what you see.

AND WHAT OF the meantime, where nightmares still happen? The future is not written, but there is the loss that has already occurred, the loss happening now, and that portion of loss that is already locked in. Writing this chapter, there were moments when I felt like I was drinking poison—or, perhaps more accurately, letting several tons of San Gabriel boulders through the small house of my self. I wasn't always sure if the walls would hold.

Grief on this scale can kill the lone mourner—if not physically, then in other ways. It's just another curse of an isolated *Homo economicus:* What consumers do is buy green, not hold each other and cry. If we've been robbed of "all memory of a past when humans managed to organize themselves in other ways," that must extend to our emotional lives as well: Your problems are personal and pathological, their solutions circumscribed to your own life choices and a couple of self-help titles.

I remember telling two close friends, at a dinner right before the Covid-19 pandemic, that I thought I might be depressed. The way I talked about it, you'd think it was a broken limb, a nutrient deficiency, or even a personal failing, and not the heartbreak of a person existing in a world. "Well, Jenny," one of them pointed out, "there is a lot to be depressed about." The other simply put his arm around me.

The present cannot and should not be borne alone. Grief, too, can teach you new forms of subjecthood. I think of a kind of double-ness, a mutuality with the power to witness and not turn away. That which pulls me through to another day has always been another body, whether that of a friend, a flock of birds in a shrub, or the east-facing side of my favorite mountain. I draw near them, draw from them some kind of *something* that doesn't quite reside in me. A review of *How to Do Nothing* once said that I "employ[ed] the annoying term 'bodies'" when, clearly, I must have meant people or humans. But I don't mean "people" or "humans." I mean *bodies*: double bodies, triple bodies, alliances and amalgamations that can shift and bear the weight, brace the walls. This moment requires that we be pressed together, pressed against the world. Now is not the time to turn your back on the ocean.

Back in September 2020, most of my nightmares ended with me looking at the advance of fire. But there was one notable exception: In one of them, I ran up to a stranger who had a dog with him and asked him for help. He grabbed my hand, and the three of us ran for our lives to a grocery store parking lot. As the fire surrounded us, we stood and watched it together. The world had ended, but the dream hadn't. *Now what?* I asked.

Chapter 6

Uncommon Times

THE COMMUNITY LIBRARY

We live according to the sun, not the clock.

—Seville woman quoted in the BBC article "Spain Considers Time Zone Change
to Boost Productivity," 2013

Heading northeast from the cliffs, we cross back over the San Andreas Fault and into another traffic snarl: this time on Highway 101, near a San Francisco hospital named after Mark Zuckerberg. Just as it starts getting really congested, we exit left and head down into the South of Market, alighting on a wide but busy street that vanishes in the direction of the Financial District. Here, boxy four-story condos mix with old industrial buildings that now host businesses like Leather Etc, a leather and bondage supplies shop.

We duck into the entryway of one of these old buildings. In the 1920s, this place housed a commercial laundry that was listed in the phone book under an ad for the Anti-Jap Laundry League, at a time when businesses proudly advertised "white labor" as if it were a fair-trade label. A hundred years later, we're dialing "P" for "Prelinger" on the building's intercom. An elevator takes us away from the noise of the street and toward the second floor, where a pole dancing studio emanates thumping bass. A warm light beckons at the end of the hall, with one of a set of double doors propped open. Inside: three corridors between floor-to-ceiling steel bookshelves, two smiling librarians, and a couple of people gathered at a large table poring over various books and maps.

HOW DO WE make a home for desire? This is a formidable question for anyone living in the kind of bootstrapper society that casts their dissatisfaction as little more than a private shame, and where *what you want* and *the way things are* can seem completely unrelated. Cynicism and nihilism

will make you dry up, like soil compacted by neglect and abuse. But soil holds the memory of life, and with some water and a garden fork, you might be able to bring it back. It helps to remember that you're not alone. Look around. Is it really true that everyone sees time as money? Or is it true that everyone spends all their time wishing it didn't seem like money?

I'd like to poke at this soil by trying another thought experiment. As I describe in chapter 2, time management often sees units of time in individual time banks: I have mine, and you have yours. In this world, when I give some of my time to you, I have less. Our interactions can be nothing other than transactional. If that is not true—if you and I exist in a field of mutual influence where time is neither fungible nor commodified—then what could "time management" mean?

I think it would have to mean, at least in part, some kind of mutually beneficial agreement between you and me about when and how we want to do things. It could be on a very minuscule scale. One friend and I have an explicit agreement never to apologize for delays in our epistolary email exchange; the understanding is that you'll get to it when you get to it. My boyfriend and I have an unspoken rule that whoever cooks dinner doesn't have to do dishes. But all of us live within a much larger, more serious version of this negotiation. Instead of "marching in lock-step toward the abyss," where the admonition to be realistic refers to an increasingly untenable reality, we at least have the right to imagine, and to imagine in common, whose time should be worth what, whose time is worth anything, and what our time is for.

In trying to imagine other temporal landscapes, we might take some lessons from what Allen C. Bluedorn refers to as the "temporal commons," the social agreements that construct and define participants' experiences of time. Bluedorn is especially concerned for endangered temporal phenomena, such as the Spanish siesta, which is declining. If laws fail to protect the siesta, or if people stop observing it for other reasons, it will perish as a temporal form. Like all other commons, then, a temporal one requires stewards. "The idea is not to save time in the time management sense; rather, it is to save *times*," he writes. "Or at least to preserve some of them."

But a temporal commons doesn't exist in a vacuum, and it can often be at odds with its surroundings. At one point, Bluedorn recounts Leslie Perlow's 1999 experiment introducing "quiet time" at a Fortune 500 software company. The engineers there were frustrated at not being able to get work done effectively in the face of constant interruption. Perlow's "quiet time" meant one period of the day (sometimes two) in which "spontaneous interactions and interruptions" from co-workers were not allowed. Bluedorn reminds us that "quiet time did not just happen; like so many times it was constructed, socially constructed, and in this case it was socially contracted too."

Perlow was able to glean important insights from the study about different kinds of work time. But what ultimately happened at the software company is telling. Although quiet time was popular, and some of the engineers wanted to continue it past the end of the experiment, the structure could not hold once Perlow was gone: "Apparently key elements of the organization's culture, such as the *criteria for success*, had not changed and these aspects of the culture motivated behaviors that led the practice of quiet time to 'disintegrate'" (emphasis mine).

What would it have taken for the engineers to "steward" the quiet time after Perlow left? Probably, it would have required more than an informal agreement—that is, some codification of new "criteria for success" that would protect everyone from the old ones. This kind of tension exists, to some degree, even in the small examples I've just given: The agreement with my friend refutes the wider expectation that one be constantly available for email. The agreement with my boyfriend refutes the wider expectation that women do all the housework.

It was in Bluedorn's book *The Human Organization of Time* that I learned the German word *zeitgeber,* something that organizes and patterns your time. If you recall from chapter 2, one zeitgeber can conflict with and overpower another. This "capture" suggests a frame through which we might revisit the subtitle of Naomi Klein's book *This Changes Everything: Capitalism vs. the Climate.* Klein describes how international trade agreements like NAFTA are used to hobble countries' individual efforts to regulate the sale and extraction of fossil fuels or to build renewable energy infrastructure. Multinationals can even use them to try to overturn grassroots victories, such as Quebec's moratorium on natural gas fracking. In other words, we have international trade bodies, and we have climate summits; each has a different set of temporal targets in mind, but those targets have never enjoyed equal enforcement. Klein quotes a WTO official saying in 2005 that the organization makes it possible to challenge "almost any measure to reduce greenhouse gas emissions," adding that although there was little public outcry at the time, there should have been.

Centuries after Sandford Fleming dreamed of a Cosmic Day utterly divorced from an earthly context, the zeitgeber that dominates our lives seems to be not the Doomsday Clock but the quarterly earnings report. This would explain the strange temporal bifurcation I experienced one day this summer, when the AQI was too high to go out and I was lurking in the earnings call transcripts that BP makes public on its website. In the shareholder Q&A section of the 2018 transcript, an analyst from Santander Bank politely asks about Tortue, the company's proposed offshore natural gas field on the border between Mauritania and Senegal. In 2020, with the gas field under way but suspended by Covid-19, an analyst from Panmure Gordon, a British investment bank, asks about it again:

Yep. Thanks for taking my question. It was back to gas. One of the projects, Bernard [Bernard Looney, CEO of BP], you didn't mention was where things stand on Tortue and perhaps more broadly, if you could comment about how further developments in Mauritania and Senegal play into the 25 million tons per annum and 30 million tons per annum 2025 and 2030 LNG [liquid natural gas] targets. And in particular, what triggers them into becoming proper, full blown, FID [foreign income dividend] projects? Thank you.

Looney assures the analyst that despite Covid-related delays, everything is *on schedule*. Outside my dull horror imagining thirty million tons of liquid natural gas, there is nothing at all remarkable about this exchange. As Marx writes in *Capital*, "*Après moi, le déluge* [After me, the flood] is the watchword of every capitalist and of every capitalist nation."* The most important criterion for success for most companies is growth. Bernard Looney was doing his job; the bank was doing its job. When they design ads selling natural gas as "clean," marketers at BP are doing their jobs. In another quarter, there would be another meeting, and it would be just as casual. All I was seeing was a window into another day at an extractive industry operating on a time horizon advantageous to itself. And yet what I saw, as a *giver of time* and determinant of my time horizon, concerned me directly. At the end of the day, I live on their clock.

WE ENTER THE stacks. This library does not use the Dewey decimal system, instead relying on the librarians' intuitive, psycho-geographical layout. Starting on the left-hand side with subjects local to San Francisco, things proceed outward and along the aisles into the American West, World Geography and Natural History, Extraction, Transportation, Infrastructure, Housing, Art,

---

* This expression occurs in the same chapter of *Capital* ("The Working Day") that I referenced in chapter 1, and it follows up on Marx's comparison of exploitation of laboring bodies to that of the earth: "What interests [capital] is purely and simply the maximum of labour-power that can be set in motion in a working day. It attains this objective by shortening the life of labour-power, in the same way as a greedy farmer snatches more produce from the soil by robbing it of its fertility."

Film, Networked Media, Material Culture, Language and Gender, Race and Ethnicity, U.S. Political History, Geopolitics and Un-American Activities, and finally, a section called Abstract and Off-Earth.

I walk you over to the remaining section, Oversize, and pull a book out of a series of bound periodicals. It's Factory Magazine. We flip through the ads for time clocks and efficiency systems. "Geared Up to Time," says one. "Successful men realize that in business there is one element that controls the others—Time." Another, which shows workmen at a table, reads, "Human Efficiency Determines Factory Efficiency."

We turn a corner and find some issues of Physical Culture. "Oh, You Unfit!" reads a World War I–era full-page ad by Lionel Strongfort, who poses in his underwear while seriously sucking in his abs. "Unfit, flabby, weak, useless—doing nothing for yourself, your family or your country, at the one time in the whole history of America when the Nation expects EVERY man either to FIGHT or WORK." And further down: "Why Don't You Better Yourself?" There is something funny about looking at these originals. Even though we recognize how much these ideas have infiltrated the culture we live in, this language looks, on the page, desperate, arbitrary, and fragile.

*   *   *

IN HIS MISSION to "save times," Bluedorn often sounds like an ethnologist concerned about extinctions of the world's languages. Indeed, any shared time sense is deeply related to language, being itself a system that orders and parses the world, its contours upheld by words, phrases, and ideas about what time is. The sociologist William Grossin wrote of a "correspondence" between "a society's economy, the way it organizes work, the means it uses for the production of goods and services, and the way time is represented in the collective consciousness, a representation that every individual receives, internalizes, and accepts almost always with no problem."

Almost always. What happens when there is a problem?

Language is dynamic, unruly, always splintering. It has to be, because in order to use it, we take words and constructions we never chose and make them do what we—as collectives, however big or small—want them to do. In March 2021, deep into the pandemic, Kathryn Hymes wrote a story for *The Atlantic* about "familects," dialects and shorthand that develop between those who spend a lot of time in a shared space, speculating that the pandemic lockdowns may have accelerated the process. One person gives Hymes the example of *hog,* meaning less than a full cup of coffee: "She explained that this comes from 'a smaller-than-the-others coffee mug with a little hedgehog on it that my roommates and I found one day.' *Hog* has become an established unit of measurement in her house: 'I've now also asked for and been offered *half a hog.'*"

We might understand agreements within a particular temporal commons as a time familect—like the ones I've described regarding email and dishes, or my Filipino relatives' observance of "Filipino time," a general acceptance of what in other contexts would be considered lateness (a topic I'll return to . . . later). You could even imagine creating an arbitrary time familect, like deciding with just one friend to observe some kind of ritual every eight days. Whenever dealing with anyone but each other, of course, you would incur the cost of maintaining a temporal language that ground against the normal seven-day week.

If eight-day cycles sound strange, they can't be any stranger than the historical phenomenon of different religions choosing different days on which to have the Sabbath, for the sole purpose of distinguishing them-selves. The Filipino side of my family is Seventh-day Adventist, a Chris-tian denomination that grew out of the Second Great Awakening in the early nineteenth century. One of this denomination's main identifiers is that it holds Sabbath on Saturdays. Thanks to missionaries, it gained a toehold in the Philippines in the early twentieth century, around which time someone managed to convert my great-grandfather—but not my great-grandmother, who remained Catholic. Until my great-grandmother converted, the Sabbath was a source of tension. Family lore has it that she would create a mess in the kitchen for her daugh-ters to clean up on Saturdays, purposefully making everyone late for the Seventh-day Adventist service.

Holdouts against new temporal orders have their reasons, which range from the petty and practical to the symbolic and separatist. Those most likely to feel (and sometimes resist) the imposition of new time zone standards are often those for whom such changes introduce the most dissonance. For example, being 7.5 degrees of longitude away from a time zone meridian means living with a standard noon that is up to half an hour off from your observed noon. Standard time zones could also be seen as a "blasphemous interference with the divine natu-ral order"; Eviatar Zerubavel notes in his study of standardized time that Muslim countries insist on solar time (based on the apparent position of the sun versus the reading of a clock) for scheduling prayers. In a move similar to Seventh-day Adventism's Saturday service, Twin Oaks, one of the 1960s communes I mention in *How to Do Nothing*, purposely set all its clocks one hour ahead of "outside time," observing what they called Twin Oaks Time (TOT). And until 1911, the French stubbornly refused to observe Greenwich Mean Time—based as it was in England—and even when they did, they called it "Paris Mean Time, retarded by nine minutes twenty-one seconds."

As that final example attests, standard time has often been the right-hand man of state identity. In 1949, citing national unity, Chairman Mao Zedong put all of China on Beijing time; the single time zone persists to this day, with one exception I will revisit shortly. During World War II, Germany adopted Daylight Saving Time (DST) and also imposed it on parts of Nazi-occupied Europe. And in a signal of solidarity with Hitler, Spanish dictator Francisco Franco moved Spain to Central European Time (CET) in the 1940s. For that reason alone, Spain currently shares a time zone with Germany and is a full hour ahead of Morocco, which is due south. In 2019, the European Parliament voted to get rid of DST, but its actual abolishment was, ironically, postponed by Covid-19 and disagreements over whether to stay on summer or winter time.

In the United States, the story of DST is rather ridiculous, influenced by that especially American blend of wartime morality and blatant commercial interest. In the surprisingly hilarious book *Spring Forward: The Annual Madness of Daylight Saving Time,* Michael Downing writes that soon after the United States adopted DST in March 1918, "the lofty humanitarian goals of Daylight Saving—to get working girls safely home before dark, to reunite dads and moms with the kids before shadows fell on the backyard garden, to safeguard the physical and mental health of industrial workers by increasing their daily opportunity for sports and recreation—also resembled an innovative strategy for boosting retail sales." Clock companies ran thousands of ads for alarm clocks, new "five-to-midnight" outfits were marketed to working women, and discounts abounded on gardening tools, sporting goods, and vacation homes.

Downing is able to devote an entire book to the subject of Daylight Saving Time in the United States because of how messy the switch was (and continues to be). In an account likely to please James C. Scott, Downing writes that, by the 1960s, DST had created a nation "absurdly out of sync with itself":

In 1965, eighteen states observed Daylight Saving, so that their clocks ran one hour ahead of Standard Time for six months of the year; eighteen other states half-heartedly participated, which meant that the clocks in some cities and towns in these states ran one hour ahead of Standard Time for periods ranging from three to six months every year and some didn't; twelve states did not practice Daylight Saving at all, keeping their clocks one hour behind the clocks in the observant states; and in areas of Texas and North Dakota, local residents adopted "daylight in reverse," so that their clocks ran one hour behind Standard Time, and two hours behind Daylight Saving Time. In that year, *The Nation* estimated that "100 million Americans were out of step with the other 80 million" and quoted a U.S. Naval Observatory official who had dubbed the United States "the world's worst timekeeper."

Practical concerns continue to stand in the way. Arizona does not observe Daylight Saving Time because, as two Arizonans put it in 2021, "when you live in the desert, daylight is way overrated. . . . So no, we don't want to save it." Moving to DST would put summer sunsets an hour later on the clock, merely "prolonging our heat-based agony." Within Arizona, however, the Navajo Nation *does* observe DST—an exigency of running a legal territory that crosses from Arizona into New Mexico and Utah. (Whereas the Hopi Reservation, which sits within the boundaries of Arizona and is surrounded by the Navajo Nation, does not observe it.) Given the patchwork nature of some parts of the Navajo territory, it's possible for someone to drive into and out of Daylight Saving Time several times on a single stretch of highway in Arizona.

The example of DST and time zones might at first seem trivial, a simple question of hours and daylight rather than what time itself means and is for. But the very idea of time zones and standardization implies dominance—the subsumption of one zeitgeber (e.g., local solar time or locally embedded agricultural cues) into another (international time and standardized commercial agriculture). The question of official versus unofficial time becomes a variation on the question I ask in chapter 1: Who is timing whom?

In China, the one holdout from Beijing time is Xinjiang, a mountainous and desert region in the west that partially observes Xinjiang Time (or Ürümqi Time, named after the capital of Xinjiang). Situated on China's border with Kazakhstan, Xinjiang is home to the Uyghur, whose pan-Islamic and pan-Turkic identity has never sat well with the Chinese Communist Party. Although Xinjiang was designated an autonomous region in the 1950s, China began trying to assimilate it politically, a project that included an effort to officially abolish Xinjiang Time in 1968.

On the one hand, Xinjiang Time appears merely practical: Xinjiang is more than a thousand miles west of Beijing, which puts its solar time two hours behind that of the capital. A sanitation worker in Ürümqi told *The New York Times* he thought they must be the only people who eat dinner at midnight (by which he meant Beijing midnight). But Xinjiang Time is fundamentally cultural, running along ethnic lines: Local TV networks put their schedules for Chinese channels in Beijing time, while Uyghur and Kazakh channels are in Xinjiang Time. In a period when the Chinese Communist Party's efforts have moved from assimilation to anti-Islamic annihilation, observation of Xinjiang Time could not be more political. Uyghurs have been subjected to sterilization, forced labor, detainment in reeducation camps, and bans on Uyghur cultural materials and practices.

At the beginning of his book on Daylight Saving Time, Downing jokes about the time he adjusted his clock well before the official two A.M. start time of Daylight Saving because he was tired and wanted to go to bed. "You were breaking the law," a neighbor tells him the next morning, offering to "lie for [him] if the Feds came around asking questions." But in Xinjiang, temporal noncompliance is no joke. A former Uyghur political prisoner told Human Rights Watch about a man who had been detained for setting his wristwatch two hours back to Xinjiang Time. It was evidence, Chinese authorities said, that he was a terrorist.

LIKE ANY OTHER language, a system of time speaks of a shared world. If you and I had some practical reason to observe our eight-day ritual cycle, it wouldn't be arbitrary; it would be a natural outcome of our relationship—to each other and to our shared situation—and every bit as relevant to us as the other temporal forms of the world. If all the roommates in a house know about the hedgehog cup, and regularly need to specify an amount of coffee, then a *hog* of coffee makes perfect sense.

In *Seeing Like a State,* James C. Scott quotes a Javanese proverb: *Negara mawa tata, desa mawa cara*—"The capital has its order, the village its customs." In one part of Malaysia, someone asking how long it will take to get somewhere might be answered not in minutes but with "three rice cookings"—everyone simply knows how long it takes to cook rice—

the rice itself being a local variety. Obviously, state administrations had a practical-political reason to subsume all those pesky village measurements and time reckonings, or else face the inscrutability of "irreducibly local" measurements. The same was true for communication. Either local language was going to be hegemonic and inscrutable to the administration, or state language was going to be hegemonic and inscrutable to the village. To speak state language was to be comprehensible to the state; to be comprehensible was to survive in a land increasingly dominated by the state.* Like the Taylorists who codified and re-mystified working practices in their scientific timetables, thus de-skilling manufacturing workers and making them run more completely on factory managers' time, anyone seeking to disempower a social body would logically go for its language(s) first.

Thus, the colonial efforts to eradicate indigenous cultures targeted both linguistic and temporal practices. But if conquest meant internalization, then those schemes clearly failed. It's hard to kill a language in one fell swoop, or at all. In *The Colonisation of Time,* Giordano Nanni quotes Richard Elphick's observation that "two systems of thought do not 'collide'; rather, real people negotiate their way through life, grasping, combining and opposing different elements." In Nanni's account of South African colonies, that negotiation was sometimes violent: One Xhosa group, after burning down the colonists' mission, broke the mission bell over a stone—silencing the European zeitgeber that emanated Sabbaths and regular workdays. But even so-called acceptance was often appropriation and adaptation. Christian time could be put to useful ends—as evidenced by a Xhosa group who refused to hear missionaries on a Monday because (they pointed out) it wasn't the Sabbath.

Just because a language is imposed, it doesn't mean it can be controlled; and just because it's spoken, it doesn't mean it's been internal-

---

* For European colonial powers, this process took place at home as well. Scott writes that as the French language was imposed outwardly on French colonies, a process of domestic colonization took place in which foreign provinces like Brittany and Occitanie were "linguistically subdued and culturally incorporated." The more people had to use the official language (French), the more "those at the periphery who lacked competence in French were rendered mute and marginal."

ized. Although Native reservations in the United States were overseen by white agents well into the twentieth century, with traditional dancing typically restricted, the Lakota figured out in the 1920s that they could hold extensive dances on the Fourth of July—if they did it under the banner of patriotism. When this tactic worked, it spread throughout the northern and southern plains, with petitions for dances on New Year's Day, Washington's and Lincoln's birthdays, Memorial Day, Flag Day, and Veterans Day. In his book *Indian Blues,* John Troutman quotes Severt Young Bear: "The agents thought those weren't dangerous occasions, so we got to dance." These supposedly nationalist celebrations appeared dangerous, Troutman writes, only "when [agents] realized that they could not contain the symbolism of the holiday that they wished the Lakotas to observe."

Like the Xhosa using the Sabbath to get their way, there is something deeply funny about this story. It has the element of an inside joke, meaningful only to the Lakota and not to those less-savvy agents who flattered themselves with the incredible idea that Indians would want to celebrate the Fourth of July. Under the temporal and spatial surveillance of Indian agents, the Lakota were able to find an interstitial hiding space within the vicissitudes of language. This kind of adaptation has appeared over and over again, illustrated more recently by the Chinese citizens who used a mixture of homonyms, images, and sarcasm to evade Chinese internet censors in the 2010s.* An inside joke makes a new inside, a new center. If the state relies on intelligibility, the inside joke is a way of becoming simultaneously unintelligible to the overseer and mutually intelligible within a group.†

---

* A common example is the evolution of the Mandarin phrase that translates to "grass-mud horse," which became associated with resistance to internet censorship.

† There is nothing about this tactic that is inherently good, of course. Racist and conservative groups frequently make use of a similar strategy in a way that we would now refer to as a dog whistle, often falling back on the "jokiness" of the inside joke when confronted with their racism. Advertisers are also experts in linguistic innovation: What is a brand but a new word that's supposed to mean something recognizable to certain people? My point here is simply to highlight language use as a tool of power; like all tools of power, it can be used to harm or to liberate. As something we each participate in, language is also an easy access point for thinking about the nexus of the individual

In chapter 1, I mention how the measurement of labor time developed under plantation slavery. At the same time, between the ledger lines, enslaved people created "insides" within which forms of time could be protected. In "Plotting the Black Commons," J. T. Roane uses the term *plotting* to refer to (1) the plots of land on nineteenth century American plantations that enslaved people were given to grow their own food and create medicines; (2) the burial plots where West African funerary customs were drawn upon and molded to a new context; and (3) the wider context of the river and "interstices" that allowed foraging, hiding, and covert communication. In all cases, enslaved people "used the plot as stolen time to engage in their own independent visions of self, family, and community." Plotters found a way to speak a forbidden language, like the Lakota dancing on the Fourth of July: "By hiding in plain sight and developing social-geographic grammars unintelligible even as outsiders watched, the enslaved rendered Black holes in the landscapes of ostensible total control, mastery, and surveillance, providing the very basis of the Black commons."

In Roane's account, plotting is striking not only because it happened in the midst of some of the most exploitative and surveilled circumstances imaginable, but also because it comes from people themselves who were seen as the objects of capitalism's subject-object relationship. What prevailed in the Black commons was nothing less than "notions of value and values that ran anathema to capitalist enclosure and mastery." Building the commons, Black people were living a cosmology incompatible with the outside, "defy[ing] their simple thing-ification."

This kind of inside contains its own center. In a 2004 study of the relationship between task orientation and wage labor on the Pine Ridge Reservation, Kathleen Pickering cautions us not to view Lakota temporal practices as mere resistance—a caution that could easily extend to the innovation of "Black holes" as well. She writes that "Lakota constructions of time are about more than [the Lakotas'] place in relation to

---

and the collective, the informal and the structural. On this note, see James C. Scott's contrast, in *Domination and the Arts of Resistance,* between "public transcript" and the "private transcript," where severe domination tends to produce "a hidden transcript of corresponding richness."

Euro-Americans, they are about Lakota society itself." For example, in the twentieth century, being used to a task-oriented society, some Lakota actually thought that the idea that "time is money" implied laziness "because it limited work to only eight hours per day, regardless of whether the work was done." Pickering quotes a Lakota elder: "Time was never a specific minute, but rather spaces of time, like early morning, just afternoon, or just before midnight. The real meaning of Indian time comes from . . . *nake nula waun yelo,* a phrase in traditional songs that means 'I'm ready for whatever, any place, always prepared.'" Far from internalizing the Euro-American work ethic, the Lakota on the Pine Ridge Reservation observe and engage in wage labor only to the extent that it is necessary. To not see "mere resistance" here requires the Western viewer to try to be something other than, as Fred Moten calls it, "the settler who brings the center with them, as them, everywhere they go." One helpful illustration of this comes at the end of *The Colonisation of Time.* Nanni writes that, in 1977, a local town council erected a giant electrically operated clock in the middle of a remote Australian town. The town's inhabitants were mostly Pitjantjatjara and had no need for clock time. Thus the tower went unheeded: "Ironically, a white community worker pointed out a decade later that the clock was simply 'a waste of time.' 'The fact is,' the same person explained, 'that nobody looks at [it]. The clock has not been working for months. No one knew that it was not working.'"

This brings me back to "Filipino time." From one angle, the term looks derogatory, given that it was coined by the Americans who took over the Philippines at the turn of the century and who found Filipino people to be less than punctual. Still, it's often invoked as a sort of inside joke or even a wry point of pride, at least among people I know. When a recent memorial service that my mom attended started late, my cousin said, "What do you expect? It's a Filipino church."

A Medium post by Brian Tan, a product designer in the Philippines, exemplifies the view that Filipino time has outlived its welcome. His reasoning has to do with how it looks to the outside world, with its modern zeitgebers and productivity ethos. There was the risk that lateness would become "the brand of our people and our country," a serious disadvantage. I get it. Zerubavel wrote that observance of time systems, like any language, is what allows us to participate in an "intersubjective world," and the broadest intersubjective world right now is a global, capitalist one. But if, just for a moment, we leave behind historically and culturally specific notions of clock-based punctuality and time as money, then Filipino time actually doesn't appear to be a problem at all. If you and everyone you know are on it, then it's just time.*

THE BOOKS AND periodicals we want to look at are piling up on the table next to other people's piles. One person is flipping through an enormous handmade book about development in the South of Market district, consisting of more than fifty years of newspaper articles that a local man collected and pasted onto pages between the front and back cover of a drawing pad. Another has a 1966 issue of Ebony Magazine *open to an ethnographic account of "The Typical White Suburbanite." And the library's artist in residence is going through issues of an old magazine called* Display World, *looking for images of "men in suits . . . solving the world's problems with their self-appointed power and prestige," to use in her artist book.† Inevitably, our piles get into conversations, and so do we. The bodybuilder in* Physical Culture *is talking to the men in suits, who are talking to the 1960s "slum clearance" in the newspaper articles, who are talking to the suburbanites.*

* Filipino time might fruitfully be considered alongside other non-Western time designations, both in its original connotation of lateness or laziness and in its capacity to be reappropriated as something that resists Western time. Meg Onli, curator of a 2019 exhibition called *Colored People Time: Mundane Futures, Quotidian Pasts, Banal Presents* at the Institute of Contemporary Art at the University of Pennsylvania, drew on Ronald Walcott's exploration into "colored people's time" (CPT) in Black literature. In her curatorial statement, Onli writes that she was "drawn to CPT as both a living and liberatory phrase" because it "provided a linguistic tool for black people to navigate their own temporality, within and against the construct of Western time." Likewise, Vernelda Grant, a member of the San Carlos Apache Tribe, told *Indian Country Today* that "Indian Time"—also interpreted from the outside as lateness—became identifiable only once "the white man came and told Natives to 'get things done in a certain amount of measured time [breakfast at 7 A.M. versus getting up to prepare breakfast just before the sun rises].'"

† This is a work currently in progress by the artist Sarah Tell, titled *No Reason to Worry.* It will be a publication of Tell's imprint, Distress Press (@distress_press on Instagram).

*   *   *

I'M DWELLING ON these examples because, without them, it's too easy to read history as a linear story of the encroachment of capitalist time into all locales and areas of life. While that story is true from a certain remove, it carries the same risks as those I describe with the Anthropocene, where history appears as a smooth, deterministic, devastating onslaught where anything else ("resistance") looks like a postponement of the inevitable rather than an opening onto another trajectory.

Speaking a language is a way of participating in the making, preservation, and evolving of worlds. Time familects, under-the-radar languages, Black holes, and the possibility of new and old zeitgebers remind me of Fred Moten's articulation of "study." In an interview at the end of *The Undercommons: Fugitive Planning and Black Study,* he defines study thus:

> Study is what you do with other people. It's talking and walking around with other people, working, dancing, suffering, some irreducible convergence of all three, held under the name of speculative practice. . . . The point of calling it "study" is to mark that the incessant and irreversible intellectuality of these activities is already present. These activities aren't ennobled by the fact that we now say, "oh, if you did these things in a certain way, you could be said to have been studying." To do these things is to be involved in a kind of common intellectual practice. What's important to recognize is that that has been the case—because that recognition allows you to access a whole, varied, alternative history of thought.

To appreciate this kind of interaction as study is not only to take a different view of history, but also to blur the boundaries between what can otherwise feel hopelessly separate: personal agency and structural change. Sociality (coming together, speaking to one another) is suggested in the very name of one of the best-known agents of change, the union. And, traditional union or no, most shifts in a formidable power imbalance begin with the simple truism that "people are gonna talk."

In 2019, I attended an event called the Gig Economy, AI Robotics, Workers, and Dystopia San Francisco, after happening upon a flyer at the studio of KPFA, a local radio station. The event was part of Labor-fest, which commemorates the anniversary of San Francisco's 1934 General Strike, and took place in a small International Longshore and Warehouse Union building on the same waterfront where the four-day strike had occurred. The talks that evening were threaded through with the question of how to organize in a world where work had fragmented, traditional labor organizing was weakened, and companies possessed new technological forms of surveillance.

Mehmet Bayram, an IT worker with the International Labor Media Network, described the "mental barrier" that white-collar workers have when it comes to seeing themselves as part of the working class, in part because they use computers. Recounting the neo-Taylorist practices affecting the experience of IT workers, he said, "The tool changes but the run for profit does not." Then he told a story about trying to encourage his co-worker to join him in organizing. When the co-worker replied that his mother had cleaned offices and that *she* was working class, not him, Bayram pointed out that their conversation was happening as they were working at nine P.M., hours past the usual quitting time and essentially for free. It was a simple example of two people, late at night in the office, talking about what kind of workers they were and what their time was worth.

THE MATERIALS ON the table have reminded me of one of my favorite periodicals, Processed World. *I pull one out and show you the cover: It's the Terminator handing a pink slip to a strung-out employee clutching a coffee and surrounded by cigarette stubs. We open the magazine up to an essay about having a bad attitude at work, in the middle of which is a cartoon: a 1950s-era-looking illustration of a businessman pointing at the reader and placed above an image of a satellite dish. "MEN," it says. "You've seen what technology can do. Are you gonna let nature stop you now?"*

IN THE EARLY 1980s, a group of alienated office temp workers in San Francisco used paper from local branches of Bank of America, Federal

Reserve, and Crocker Bank to start printing *Processed World,* a collection
of articles, poetry, fiction, comics, and other visual art by pseudonymous
authors. Assembled at home and in basements, copies of the magazine
were distributed by hand to passersby in the Financial District; they were
also sent through radical collectives across the world and mailed to anyone who requested one, including people in prison. *Processed World* reads
like an irreverent Marxist labor magazine with shades of *Daria* and *Office
Space.* It is both serious and laugh-out-loud funny, often at the same
time. Alongside musings on alienation and coverage of various white-
collar strikes, a reader would find inside jokes and satirical ads—like one
for "BFB: Brains for Bosses, Inc.," which uses "the latest advances in
science to bring you the smartest, yet most pliable, workers *ever.*" One
issue contains a doctored conference program that *Processed World* members, dressed up in costume with video display terminals on their heads,
had handed out at the 1982 Office Automation Conference. To the cover
image, they added the text "The International Conference for the Per-
petuation of a Vacuous Existence" and changed the program's icon of a
figure sitting at a computer to one in which that same figure takes a bat
to the monitor.

Sometimes *PW* would include an artifact from the corporate world
nearly without comment; in this context, it *was* the joke. In a 1982 article
about office sabotage, a writer going by "Gidgit Digit" wryly contributed the certificate for "Team Spirit" that Bank of America had given
her. The "Spirit," smiling in the center of four Bank of America logos, is

covered in a white sheet that makes it look somewhat like a cutesy cartoon Klan member. Another issue reproduced the full text from "a genuine typing test administered by Temps, Inc. in San Francisco"—which I will reproduce in full here:

Time is the one thing in which we are all equal. There are the same number of hours in the day, the same number of minutes in the hour, and the same number of seconds in the minute for you just as for me. We are not all equal in ability to produce, of course; but many men learn to achieve maximum production, while others never seem to realize that they have the same number of hours in their day to work and to improve their production.

The time has come when you should recognize that it is the duty of every worker to give a full day's work for a full day's pay. Too many who are on the job are job holders rather than workers. They are frequently willing to give a full hour's work for a full day's pay. That isn't the way a business operates if it expects to survive. All the employers have a right to expect each worker to produce more than he is paid to produce. That "more" constitutes the profit requisite for business survival.

The dictionary is the only place where success comes before work. The material things we want just can't come to us out of thin air; they must be produced by somebody, and that means somebody must work. The quickest way to achieve success is to work for it—the surest way to get the material things we want is to work for them. If work isn't a thing of magic, it produces results more excellent than magic ever produces. To realize all we are capable of achieving, we must learn to love to work.

Alluding to Marx's *Capital,* the editors took the sole liberty of adding the sarcastic title "Labor Theory of Value?"

The concerns of *Processed World* in many ways foreshadowed the concerns of contract work and the gig economy. Founders of the magazine were mostly in their twenties and had taken temp jobs in return

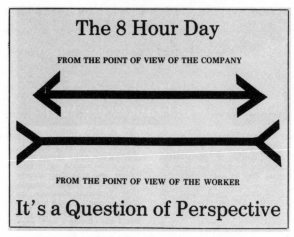

Image from the first issue of *Processed World,* in 1981

for some promise of free time, much as gig workers often cite schedule flexibility as a motivation for their choice. The temps also worked with computers and were subject to evolving forms of automation and surveillance of white-collar work. At a time when techno-enthusiasts and businesspeople were just as breathless about the future of work as they are now, *Processed World* maintained a skepticism reminiscent of that which manufacturing workers harbored toward Taylorism.

In her article on office sabotage, Gidgit Digit (winner of the "Team Spirit" certificate) correctly predicted that while computers might one day allow people to work from home, "it is unlikely that management will relinquish control over the work process." Incidentally describing a software like StaffCop, she foresaw that "rather than freeing clerks from the gaze of their supervisors, the management statistics programs that many new systems provide will allow the careful scrutiny of each worker's output regardless of where the work is done."

Processed World was, in a way, its own social media. The Letters section—often full of debate over the ethics of sabotage, the role of established unions, and whether the magazine was at risk of becoming a niche concern—often included isolated workers responding to other letters and expressing gratitude simply for the magazine's existence. "JUMPIN' JEHOSEPHAT! THERE'S INTELLIGENT LIFE OUT

THERE! WE ARE AT YOUR SERVICE," wrote two secretaries. "Nice to know that somebody out there breathes! . . . We would like to offer our services to The Noble Cause. We have limited copying capability with a high-resolution Minolta copier, if that will help." Another wrote that they'd received issues 4 and 5 from a friend "highly skilled in office sabotage," adding, "Christ on a bicycle, I don't think I've been this grateful since I was first taught to read!" One worker, toiling at a video desk terminal (a frequent target of *PW*), wrote:

> One day, at seven a.m. (much too early for the bosses to be there) I begin my shift and find a copy of *Processed World* that someone has left on my desk. Feeling dramatic and trying to be nonchalant, I slip it into my drawer, to later joyously suck in each page. In this partitioned, soundproof, PCB lined office jungle, truly the worst fate is to believe in your boss's dream, to strive for the company good. Thank you *Processed World,* for letting me know there are others who despise the purposes to which they are employed.

This kind of horizontal communication has become only more important as more people turn to part-time contracting and gig work, the work becomes more fragmented, and there is less guarantee of a shared physical space or time where conversation and solidarity might grow. Sometimes, new meeting places pop up incidentally; food delivery gig workers in Europe, for example, have naturally fallen into conversation at physical waiting points. But for the most part, the talk has moved to online forums. Geographically disparate people can trade information and stories, whether to commiserate or to try to demystify the algorithms that order their work and time.*

---

*In a 2020 study of "misbehavior" among Canadian Uber drivers, labor researchers highlighted the role of online forums, in particular the slogan "Don't take a poo!" The phrase referred to Uber's new UberPool service, which was unpopular among the drivers. UberPool rides not only were more complicated to execute, but actually paid less than regular Uber rides. In a Canadian city subgroup of UberPeople.net, drivers shared tips about how to avoid taking "poo" rides—for example by putting their phones on airplane mode intermittently. "Don't take a poo!" was an admonishment among drivers not to accept the rides so that Uber would have to change its ways.

Isolation is the harbinger of exploitation, and the forums give employees like gig workers the opportunity to compare notes and strategies. But even if some workers decide, for example, not to take batches below a certain pay level, other workers are often signing up for the first time in response to their own financial hardship, and "someone is always willing to take the run." Meanwhile, successful attempts to establish limits, and even get them enshrined into state law, can run aground of a global reality. Legally, traditional union organizing is bordered by space in ways that multinational companies are not. One Kenyan worker with GigOnline, when asked about the possibility of unionization, is clear-eyed on the matter: "They'll just take the job somewhere else if the unionised labour of freelancers in Nairobi don't want to do the work at a certain dollar amount. . . . They'll take it to Nigeria, they'll take it to Gabon, they'll take it to [the] Philippines. They'll take it to all kind of countries. . . . The unions will not have enough power, because I've seen what globalisation can do."

The situation demands new language and new channels of communication. In 2021, *The Nation* covered recent victories against exploitative gig work companies. In one instance, Uber workers in different countries studied financial news, anticipated Uber's IPO, and coordinated a walkout across twenty-five cities in order to get media attention at the most opportune time. This paved the way for the formation of a new, international labor organization called the International Alliance of App-based Transport Workers (IAATW), which managed to do its work using a "transnational network of resistance over forums, group chats, and video calls." In 2021, Deliveroo workers secured a victory similar to the Uber workers', under the Independent Workers' Union of Great Britain (IWGB).

Part of this language is the articulation of a global working class that extends way beyond traditional notions of blue collar and white collar. This communication is all the more remarkable given the many ways in which gig work atomizes and anonymizes its far-flung workers. Nicole, an IAATW member who helped coordinate the Uber protests, described this new map of resistance: "A worker in California is intimately connected to the Uber driver in Kenya, to the driver in India or Malaysia. . . . We all are suffering for a San Francisco billionaire's $40 million home."

The workings of the IAATW—and those of the more than century-old International Transport Workers' Federation that has pivoted to support them—exemplify what Oli Mould would call an actually creative activity, as distinct from "creativity" under capitalism. In *Against Creativity*, Mould observes that jobs of all sorts now encourage their employees to be "creative," which often translates to competitive flexibility, self-management, and individual assumption of risk. Meanwhile, even nominally anti-capitalist creative work, whether art, music, or slogans, is handily appropriated by the market. Mould writes that, in either case, creativity is not actually creative, because it merely "produces *more of the same* form of society." If it makes progress, it is the progress of capitalist logic into ever-more-minute corners of our daily lives, making what Braverman calls "the universal market" even more universal.*

It's an important distinction to make at a time when the Covid-19 pandemic has revivified conversations about work-life balance and the possibility of compressed workdays or weeks. What might look creative and emancipatory at first can turn out to be a re-entrenchment: Companies find that they can pay people less if they work for less time, while "the time they do spend at work is the most productive it can be." Already in the 1970s, Harry Braverman observed that corporations like IBM were "humanizing" work by changing the style of management rather than the position of the worker. Like Mr. Burns's funny hats, these strategies were nothing more than a "studied pretense of worker 'participation,' a gracious liberality in allowing the worker to adjust a machine, replace a light bulb, move from one fractional job to another, and to have the illusion of making decisions." It's the same insight as those 1950s companies with floodlit fields: Happy worker = higher output, and if the company can pay less, even better. The "criteria for success" have still not changed. On the one hand, you can't blame a business for speaking the language of the bottom line. On the other hand, you could also ask the old question: Why are individuals expected to be "resilient" when corporations are not?

---

* Specifically, the "universal market" refers to the market created when individual and community relations are replaced by transactions between consumers.

In that *Processed World* article on office sabotage, Gidgit Digit expresses similar ideas about what radical change actually means. Taking a dim view of the supposedly liberatory conveniences being offered in the 1980s, she writes:

> The individual "freedoms" that are created by the technological wonders of tele-shopping and home banking are illusory. At most they are conveniences that allow for the more efficient ordering of modern life. The basis of social life is not touched by this "revolution." As in the office it remains hierarchical. In fact, the power of those in control is enhanced because there is an illusion of increased freedom. The inhabitants of this electronic village may be allowed total autonomy within their personal "user ID's," but they are systematically excluded from taking part in "programming" the "operating" system.

While coming from very different contexts, Fred Moten's study, Oli Mould's creativity, and Gidgit Digit's (re)programming have something in common—a desire to be neither confined by nor supportive of the hierarchical forms of the market or the institution, but to dwell somewhere between the lines, somewhere interstitial and messy. Among other things, this can mean speaking a new or forbidden language (straightforwardly linguistic or no) to say what cannot currently be said. Inside the space of the inside joke, it's speaking crosswise rather than sucking up, using what Moten might call the "mutant grammar" of refusal.

As Carole McGranahan writes, "To refuse is to say no. But, no, it is not just that. To refuse can be generative and strategic, a deliberate move toward one thing, belief, practice, or community and away from another. Refusals illuminate limits and possibilities, especially but not only of the state and other institutions." Refusal may start in you but cannot end with you. It must be spoken, in messages, in magazines, on forums, and off-hours, in an ongoing "rehearsal." In summoning a world, it's the most creative thing you could possibly do.

*   *   *

ONE OF THE *librarians offers us some mint tea. While the water boils, we have time to look at some of the posters on the walls. One is printed on cloth and contains 198 nonviolent actions from Gene Sharp's* The Politics of Nonviolent Action. *Another is titled "Identifying and Dismantling White Supremacy in Archives: An Incomplete List of White Privileges in Archives and Action Items for Dismantling Them," from Michelle Caswell's Archives, Records, and Memory course at UCLA. Some of the privileges have to do directly with language. One reads, "When I look for materials from my community in archives, they will be described in the finding and catalog records using language we use to describe ourselves."*

IN EVERY LANGUAGE, certain things can appear or be spoken, and others cannot. That's an idea in the very title of Marilyn Waring's 1988 book, *If Women Counted: A New Feminist Economics.* Now known for lambasting the notion of the GDP as a criterion for success, Waring became a member of the New Zealand Parliament in 1975, at the age of twenty-three, where she was joined by just three other women. By the time she was appointed chair of the public expenditure committee in 1978, she was one of only two women. On the committee, whenever presented with economic jargon, Waring relied on the "art of the dumb question"—such as simply asking what a certain word meant—and soon found standards that rendered women's unpaid labor totally invisible, making sensible provisioning impossible.

Terre Nash's 1995 documentary on Waring's career shows her piecing together the relationship of personal, state, and international economics. She travels to countries in the Global South, talking to rural women about their unending work, producing visual time studies, and finding that things like inexpensive water pumps and new stoves would actually be the most "productive" interventions. Waring also visits the public accounts committees, treasury boards, and budget appropriation committees of other countries, gathering more information. Having thought that the "enormous paradoxes and pathologies" she's discovered in her own committee might be specific to New Zealand, Waring begins to realize that "it's nothing to do with New Zealand. These are the rules everywhere."

At a lecture in Montreal in the 1990s, she gives the example of Cathy, one of her constituents:

Cathy, a young, middle-class housewife, spends her days preparing food, setting the table, serving meals, clearing food and dishes from the table, washing dishes, ironing, keeping an eye on and playing with the children, dressing her children, disciplining children, taking the children to daycare or to school, disposing of garbage, dusting, gathering clothes for the washing, doing the laundry, going to the gas station and the supermarket, repairing household items, making beds, paying bills, sewing or mending or knitting, talking with the door-to-door salespeople, cutting the grass, weeding, answering the telephone, vacuuming, sweeping and washing the floors, shoveling snow, cleaning the bathroom and kitchen, and putting her children to bed.

Then the punch line: "Cathy has to face the fact that she fills her time in a totally unproductive manner. She too is economically inactive, and economists record her as unoccupied." The joke is about a mistranslation, a clash of languages, and the values assigned (or not) by those languages. Gloria Steinem, who appears briefly in the Waring documentary to say that most economists "seem to value their work in inverse ratio to their ability to be understood," praises Waring for reminding readers of what economics actually is: "the way we impute value to that which we consider valuable."

In *If Women Counted*, Waring recommends a change in imputation: altering official criteria to more accurately reflect what should be considered productive activity. The Wages for Housework movement, founded in the 1970s, represents a more expressly anti-capitalist body of thought based on similar observations.* The phrase "wages for housework" was first put forth by Selma James, who also coined the now-common term *unwaged labor* to refer to the housework, care work, and child-rearing that women were expected to do for free. She and others in the movement campaigned with mothers who were on income support in the United Kingdom and the National Welfare Rights Organization in the United States (led largely by Black mothers), which had a similar demand for a Guaranteed Adequate Income (GAI) and a recognition that "women's work is real work."

Wages for Housework drew on insights from Black welfare activists as well as from the Italian movement for workers' autonomy known as *operaismo* ("workerism"). From their perspective, women were slaves to wage slaves (men) and women's work upheld an overall system of exploitation that was hurting men and women. In 1975, James and the Italian autonomist Mariarosa Dalla Costa published *The Power of Women and Subversion of the Community*, in which they observe that, "where women are concerned, their labor appears to be a personal service outside of capital." Together with a study group, James had read the first volume of Marx's *Capital*, where she encountered the selling of labor power as a commodity and the division of labor. But she didn't see anyone talk-

---

*The 1970s saw the rise of several different but overlapping movements in the United States and the United Kingdom that were associated with "wages for housework," such as Black Women for Wages for Housework (co-founded by Margaret Prescod and Wilmette Brown), the Wages for Housework Committee (co-founded by Silvia Federici), and Wages Due Lesbians.

ing about *who made* labor power, or about where unwaged workers fell in that division. In her introduction to the book, James describes all the time that goes into making the kind of time that can be sold for a wage:

> The ability to labor resides only in a human being whose life is consumed in the process of producing. First it must be nine months in the womb, must be fed, clothed, and trained; then when it works its bed must be made, its floor swept, its lunchbox prepared, its sexuality not gratified but quietened, its dinner ready when it gets home, even if this is eight in the morning from the night shift. This is how labor power is produced when it is daily consumed in the factory or the office. *To describe its basic production and reproduction is to describe women's work.*

At the time, criticisms of Wages for Housework ranged from practical implausibility to the risk of cementing women in women's roles. But the movement has never been just about asking for wages for housework. First of all, the original demand was part of a set of others: a shorter workweek, reproductive freedom, wage parity, and guaranteed income for men and women. More important, it was a gesture: an attempt to imagine an option for women beyond taking on a second shift in a nuclear family or competing with men in what we'd now call *Lean In*–type feminism. By imputing value to women's work and thus to care, Wages for Housework sought a society in which care and collective liberation, not personal ambition and brutality, would be paramount— for everyone, and a benefit to all. In her discussion of the Wages for Housework movement, Kathi Weeks astutely emphasizes its use of the *demand:* not so much a plea for money as an unapologetic statement of power and an expression of desire. The demand is a total rejection of a situation that divides, per Braverman, "those whose time is infinitely valuable" from "those whose time is worth almost nothing," a fervent wish for a world where we don't keep dying in man-shaped cars.

In *The Problem with Work,* Weeks uses this kind of energy to lay out a demand for a universal basic income and a shortened workweek with no loss of pay.* Having spent the majority of her book probing the centrality and unquestioned goodness of work in modern life, Weeks sees something like universal basic income as operating on both a pragmatic and a moral-idealistic plane at the same time. On the one hand, UBI could provide relief to many people in the immediate term, addressing the fact that James and Dalla Costa put so simply in their book: "To 'have time' means to work less." But insofar as it shoehorns people out of total subservience of the wage, it is also creative, making space for further creativity. The demand for less work might be made "not so that we can have, do, or be what we already want, do, or are, but because it might allow us to consider and experiment with different kinds of lives, with wanting, doing, and being otherwise."

It is the "wanting" in Weeks's combination of "wanting, doing, and being" that stands out for me the most. Daily desires, intuition, and even quiet despair so often have the feeling of an undercurrent, an underlanguage, or an undercommons—under the zeitgebers of the workday, the workweek, the productivity spreadsheet, and the earnings report. At the very beginning of the documentary on Marilyn Waring, a journalist interviews people in the lobby who seem to have been brought to her lecture by a sly, questioning feeling. One young man tells the interviewer that he's attending due to "suspicions." About what? the interviewer asks. "Suspicions . . . that things are not as they seem, having come from a family where I was raised by a single mother and realizing there were great injustices," the man says.

Suspicion abides between the lines. Sometimes, it smolders. In 1980, a year before starting *Processed World,* its founders produced a sardonic leaflet called "Innervoice #1" for National Secretaries Day. A play on an invoice, it listed the costs of working as a secretary: six hours of nonstop typing caused "1 backache, 1 stiff neck," seventy hours a week of hypertension cost "1 sanity," and forty hours a week of tedious busy-

---

*It is worth noting here that, according to her 2020 piece for *The Independent,* Selma James does not support a universal basic income, favoring instead a care income that would go specifically to care workers. What James and Weeks would likely agree on, however, is that current models of remuneration reflect an unjust system that narrowly values particular types of working and being.

work cost "one imagination." Of course, to make Innervoice #1 itself
relied on whatever imagination was left over. The leaflet was a precursor
to *Processed World*'s sense of humor in general, which belied a painfully
recognizable rage about the scam of selling one's own life for the honor
of selling it some more.

Occasionally this showed through with surprising sincerity. Amid the
comic strips, fake ads, and cutting commentary, one two-page spread contained a melancholic collage of a face inside a computer terminal, hand-
cuffed hands, a phone, and a series of smaller heads, with the text "Another
Day at the Office: What Have We Lost?" In a letter, J.C. from Toronto
asked, "What does one do when one finds oneself marking time on the
job? One develops a lot of cynicism, apathy, and anger to which there is no
outlet." And J. Gulesian from San Francisco offered this meditation:

> *Dear PW,*
>
> *I would like to submit more observations on the daily life of a middle-*
> *aged secretary. It's all very hard, really, that daily life. It so often*
> *demands more than I can give and takes so much that my free time is*
> *spent trying to establish continuity between who I am and who I must*
> *be. Who I am means that I must establish and maintain human*
> *relationships. What I must be makes that dangerous and painful. You*
> *know how it is.*

Not everyone had time for this. In the very same issue, a Walter E.
Wallis of Wallis Engineering wrote to *PW* with the general tone of
someone yelling "Get a job!" at a bunch of hippies—or like Tommy
Anderberg, that taxpayer who didn't mind if physicists were chained to
their desks. After giving some tips on "making yourself worth more to
your company," Wallis ended by saying that "if the burden of apply-
ing yourself to your job so the customer is assured the best deal for the
money does not appeal to you, then fuck, snivel, whine, cheat, steal, and
bullshit your way through life, because you are nothing but a fucking,
sniveling, whining, cheating, thieving bullshitter." In response, a *PW* editor calmly dissected Wallis's arguments one by one, concluding with an
unabashed vision of a "new, freely cooperative and communal society

already latent within [this one]." It was Wallis who was poor, lacking in imagination: "Rather than contemplating such possibilities, he understandably prefers to give us vulgar and condescending advice on how to 'get ahead' in a world marching in lockstep toward the abyss."

The *PW* editor was trying to "contemplate the possibility" of a world in which all the assumptions of Wallis's world—one where you simply use personal ambition to outcompete others, blame yourself when you fail, and blame others for their failures—would be replaced with something else. For Wallis, you were responsible for your own time and your own time only within an unforgiving and inflexible structure; for the editors, time could truly mean something different insofar as the structure could be dissolved. For Wallis, the goal was individual power; for *PW,* I would argue, the goal was meaning and recognition.

In chapter 2, I suggest that an overworked achievement-subject should save herself by dialing down personal ambition. But just as ladder-climbing ambition is only one form of desire—one that exists on and reinforces a specific plane—there are many forms of frustration beyond what is trivially referred to as burnout. Some of those frustrations, whether you are advantaged or disadvantaged, include the following: having to sell your time to live, having to choose the lesser of two evils, having to say something while believing in another, having to build yourself up while starved of substantive connection, having to work while the sky is red outside, and having to ignore everything and everyone whom, in your heart of hearts, it is killing you to ignore. There is wanting more for yourself, and then there is simply *wanting more.*

Selma James is still active with the International Wages for Housework Campaign (now more commonly known as the Global Women's Strike). In 2012, she told journalist Amy Goodman about marching a year earlier in London with SlutWalk, a transnational anti-rape movement. James had felt vivified by the group's energy and anti-racism. "I didn't feel, walking with them, that I was surrounded by women who were ambitious," she said, referring by contrast to the part of the women's movement that had focused on climbing the ladder while welfare declined. "We really need to have another reason to be together, which is the real conditions of our lives, rather than an individual ambition." *Individual* is the operative word there. Just as Wallis was poor from *PW*'s perspective, an enthusiastic ladder climber is, in another sense, unambitious. It is far more "ambitious" to make the demand that James sensed at the march: "We want to have the freedom to live the lives as we like them, and we are together for that."

WE ASK TO see the artist books. They're up at the top of one of the shelves, so I wheel the big green ladder over and climb up there, where alphabetized gray boxes are waiting. Inside is everything from cloth-bound books to zines and sets of postcards, each in its own tan file folder carefully cut to its idiosyncratic dimensions. Many of these are by local artists, often representing projects done with research in this very library. The books and objects inside are like plants that grew in the archive, and now they give us seeds for our own projects.

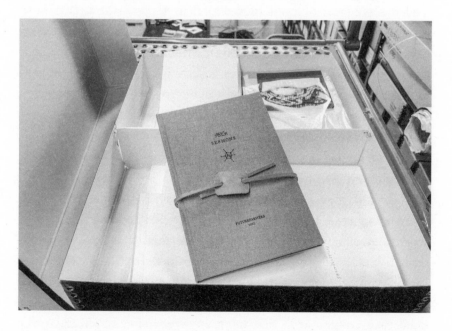

Inside a file folder in the E–F box, we find a small cloth-bound book with a leather strap made by the artist collective Futurefarmers. The title reads, "SOLE SERMONS," with the word SOUL printed over SOLE. You carefully undo the *strap and open the book. The text inside is letterpressed into the page but printed in a light ink, making it seem as though the words were both holding fast to the paper and in danger of evaporating. It's an essay about walking, by Rebecca Solnit. Her walking sounds like the opposite of "marching in lockstep":*

> Walking is made of steps, but a step is not a walk; a walk is made of perseverance, of continuing to step, and this process of repetition is not redundant but a form of inquiry. "Where are we going?" is the universal question, but the answer is just to go, to walk until your shoes wear out, and then to resole and keep walking. "I walked to the floor till I wore out my shoes / Lord they're killing me—I mean them lowdown blues," sang Hank Williams, and the walking keeps you alive. To keep walking is to keep living, to keep inquiring, and to keep hoping. Hoping and walking have preoccupied me the past dozen years, but I had to travel a long way down those two paths to realize that they were the same path whose rule is motion, whose reward is arrival in the unanticipated, and whose very nature is in contrast with the tenor of our time, a time preoccupied with the arrival and the quantifiable. Many love certainty so much more than possibility that they choose despair, itself a form of certainty that the future is notable and known. It is neither. To despair is to stop walking, and to stop walking is to fall into despair or those depressions that are both features of the landscape and states of mind—the hole deeper than a rut.

EARLIER THIS YEAR, I was in the garden of a septuagenarian friend while she was planting some beans. She told me that they were descended from beans she'd gotten twenty years before, from somewhere she couldn't quite remember—maybe Home Depot—and could never find again. At the time, she'd shared the beans with friends, all of whom loved them and couldn't find them anywhere else, either. But some friends let the bean pods mature and dry out, saving the beans and giving them back to her. She had no idea how many people had them by now and speculated that this line of beans could have spread across the entire country. As she planted them, we mused that although there had been a give-and-take between her and her friends, it was not exactly transactional—she wasn't taking back the things she had given them, though the two were certainly related.

She moved on to some lettuce beds, telling me I should take some lettuce. I thought she was just being polite, but she told me that she actually needed to get rid of the outside leaves so that the inside leaves would keep growing, before the plant reached maturity. She was constantly giving bags of lettuce to people, she said. This simple gesture, and the story of the beans, made me realize how broken my mental mechanisms were for thinking about anything beyond the transactional exchange. In part, this is because I've never lived anywhere I could garden. I'd sort of forgotten that a plant keeps growing, assuming that more lettuce leaves for me would mean fewer lettuce leaves for her.

But that wasn't the only thing I'd forgotten. Philosopher Ivan Illich worried in 1978 that "innumerable sets of infrastructures in which people coped, played, ate, made friends, and loved have been destroyed," leaving a barren social landscape of "huge zero-sum games, monolithic delivery systems in which every gain for one turns into a loss or burden for another, while true satisfaction is denied to both." In that moment, I felt not dissimilar to the young gig worker who told sociologists, in a study of why precarious workers didn't claim unemployment benefits during the pandemic in New York City, "You just sign up, say 'I don't have a job,' and the government gives you money? What is that about? If it was that easy, wouldn't everybody do it? I don't get it." Taking the lettuce was good for both me and my friend. I didn't get it.

A few months later, I was sitting in a different garden, this time a botanical garden with free admission. Two kids had taken over a lawn next to me and were playing a game of red light, green light. Their version of it, however, was considerably more complex than the one I played as a kid. "Red light" still meant stop, and "green light" still meant go, but for them, "purple light" meant to dance. "Light-blue light" meant to dance backward. "Golden light" meant to fall on the ground, and "green tree light" meant to moo and crawl at the same time. There were even more specific commands, like "throwing-shoes light" but also "go-back-to-shoe light." It was silly, but I was impressed that they never had to remind one another what any of the terms meant; they had created and memorized them together.

Time can have many rhythms, and rhythms can take on many meanings. Writing about the demoralization of work through processes like Taylorism, sociologist Richard Sennett has observed that "routine can demean, but it can also protect; routine can decompose labour, but it can also compose a life." It can be the construction of ritual, the way that the rabbi Abraham Joshua Heschel called the Sabbath "a palace that we build in time." Like the "lights" of the red light, green light game, the botanical garden had been composed and choreographed; different spots had different characters; things grew in different shapes and sizes and flowered at different times. The garden represented one set of gardeners' views about what kinds of arrangements would make up a harmonious whole, and visitors lingered in the parts they liked. While it wasn't a large space, the garden was dense—a space not just of biodiversity but of chronodiversity, inviting the human subject into conversations with different modes and speeds of life. Here, it was not only clear that time was not money, but that the category of "not money" could be infinitely elaborated.

Would it be possible not to save and spend time, but to garden it—by saving, inventing, and stewarding different rhythms of life? And wouldn't this simply be an acknowledgment and use of the chronodiversity that already exists for all of us on some level, individually or communally? The sociologist Barbara Adam, who has written about standardized, economic time, also knows that its dominance is as incomplete as it is unintuitive: "Tempo and intensity surround us at every level: we know that a birthday tomorrow can feel like an eternity to a little child whilst a birthday one year ago can seem like only yesterday to an old person. The dormant period of winter is followed by a burst of growth in spring. . . . 'Our' social time as it emerges from common usage is inseparable from the rhythms of the earth. Complexity reigns supreme."

If time can be gardened, then it's also possible to imagine its increase in ways other than individual hoarding. Before I left my friend's garden, she gave me some scarlet runner beans from a bean farm that no longer exists. They're now sitting on a metal shelf next to the store-bought beans that Joe and I, like many other people, started stocking up on during the pandemic. I've had so much time to look at and think about beans, but I had never considered what they actually were. I googled "can you plant store-bought beans," and the answer was yes. Those things in the bags—they weren't just commodities. Sure, you could eat them, but they weren't end-points and they weren't dead. At least some of them contained something: the possibility of future beans.

As I told more friends about this story, it became an inside joke, a new familect: *Time is not money. Time is beans.* It was as serious as many jokes are, which is to say about half. Saying it meant that you could take time and give time, but also that you could plant time and grow more of it and that there were different varieties of time. It meant that all your time grew out of someone else's time, maybe out of something someone planted long ago. It meant that time was not the currency of a zero-sum game and that, sometimes, the best way for me to get more time would be to give it to you, and the best way for you to get some would be to give it back to me. If time were not a commodity, then time, our time, would not be as scarce as it seemed just a moment ago. Together, we could have all the time in the world.

Chapter 7

Life Extension

THE COLUMBARIUM AND THE CEMETERY

Resonance, by contrast [with recognition] is always a dynamic event, the expression of a vibrant responsive relationship that can be seen perhaps most splendidly when a person's eyes light up. . . . [It] always refers to an occurrence between two or more subjects. I am recognized, but resonance is something that can only happen between us. Love as a resonant experience thus refers not to the fact of loving or being loved, but to the moment or moments of mutual, transformative, fluid, affecting encounter.

—HARTMUT ROSA, *Resonance: A Sociology of Our Relationship to the World*

We've crossed the Bay Bridge and are back in Oakland, facing east. From the sidewalk, you can hear "Signed, Sealed, Delivered" coming out of a Pilates studio with its door open on the ground-floor level of a luxury condo. "All right, y'all," the instructor says over the music, with pep and authority. "Let's push that left foot back in five, four, three, two, aaaaand one. You got this. We're very close to the end of class, y'all. Very close indeed."

Up ahead of us are the gates to a cemetery and a Spanish Revival–style building with a tower with small metal letters reading FUNERAL HOME—CREMATORY—COLUMBARIUM. *I push the heavy metal door open for you, and immediately we're enveloped by a sweet, funerary smell: living plants, wet rock, dust, ashes, incense. Muted sunshine filters through the skylights, huge tropical plants, and stone arches, while the only sound is the plaintive drip of a nearby fountain. The walls are dense grids of glass enclosures, giving the overall impression of a library. Unlike where we just were, though, the "books" behind the glass are urns modeled to look like books, with groups of different volumes containing the individuals of one family. Each book has a certain heaviness about it: a life's beginning and a life's end, the "covers" unopenable.*

WHEN I WAS very young, I encountered a terrifying story about time in a 1970s-era book my mom picked up from a garage sale, called *Magic Fairy Stories from Many Lands*. A boy who is impatient to grow up is wandering in a forest when a witch appears and gives him a ball with a golden thread sticking out of it. If he pulls the thread, she says, time will go faster. But he must use the device wisely, as the thread can no more easily be put back in than time can run backward. Predictably, the boy can't help himself: impatient to go home from school, he pulls the thread; impatient to marry his crush, he pulls the thread; impatient to have a child, he pulls the thread. All too soon, he finds himself at the end of his life without the sensation of having lived it.

The moral of the story is supposed to be about "living in the moment" and the folly of wanting to skip over the bad parts of life to get to the good ones. But when I read it, the thing I fixated on was the thread and the ball, simply as an illustration of the irreversibility of time. Even though it has a happy ending (the witch finds the old man and lets him live his life over again), I remembered this for a long time as a horror story.

Time management often trades on a form of this horror. Remember Kevin Kruse, the entrepreneur who put up the "1,440" poster in his office to remind himself how many minutes he had per day? In his book, shortly before introducing that poster, he asks you to hold your hand to your heart and become conscious of your breathing. Make no mistake: This is not a mindfulness exercise. Kruse says, "You will never get those beats back. You will never get those breaths back. In fact, I just took three beats away from your life. I just took two breaths from you." He immediately follows this with a pointed metaphor: You would never leave your wallet out in the open, and time is money. So, why are you letting people "steal" time from you?

Pursued to its logical conclusion, the idea of time as a personal and nonrenewable resource both evades and obsesses over mortality. After all, what is Kruse's "1,440" poster but a memento mori, as sobering as the skull in the corner of a seventeenth-century Dutch still life? Each day, every time you look at Kruse's poster, you have less than 1,440 minutes left. In "Why Time Management Is Ruining Our Lives," Oliver Burkeman observes that keeping a detailed log of your time use, in an effort to save time or spend it more wisely, ironically "heightens your awareness of the minutes ticking by, then lost forever." Whether on the level of minutes or of life stages and benchmarks, the more you stare at time, the more cruelly it seems to slip through your fingers.

There are many apps that purport to tell you how many years you have left to live. Out of timid curiosity, I recently downloaded one called When Will I Die? After answering a series of questions about my lifestyle and disposition and sitting for thirty seconds through an ad for a game called Wishbone ("Choose the cutest!" said some text above two random pictures of manicures), I saw a cartoon headstone whose inscription read, JENNY ODELL DIED AGED 0. The number then raced upward from zero, as though I were playing a casino game where the prize was life itself. I had a few moments to very palpably feel my desire not to die, for the number to be higher; eventually, it came to rest at ninety-five.

Silly and unscientific as the app was, one could imagine an ever-more-detailed app where literally every decision you made was logged

and incorporated into an algorithm that determined your time left to live. (This would not be dissimilar to the endgame pursued by some insurance companies.*) This represents one common response to the existential problem posed by time: to try to increase the overall amount of time you hold in your personal time bank. This version of the "logic of increase" explains the appeal of the resort I mention in chapter 3, where guests on Larry Ellison's island have their vitals and progression toward specific goals monitored—and is also probably the reason for productivity bros' power smoothies. A natural partner to time management, wellness is invoked both as a means to "perform" better and as a way to increase your overall years of life, as though you were a car or a watch.

Yet, just like productivity, this pursuit can easily surpass any reasonable goal of health in its search for a calculable optimum, another way of obsessively counting change. Numerical longevity and (a very specific

---

*Some automobile insurance companies use telematics and car trackers (similar to those mentioned in chapter 1) to collect data on driver behavior and determine rates accordingly. And Beam Dental, which uses a proprietary electronic toothbrush to collect brushing data from its users, promises a "lower rate based on group's participation in Beam® Perks wellness program and a group aggregate Beam score of 'A.'"

version of) wellness become the final metrics, avoiding the question of
what it is that we want to be well and live *for*, not to mention the irony
of a life consumed by the effort to make more of itself. This and other
problems appear succinctly in the subtitle of Barbara Ehrenreich's book
*Natural Causes: An Epidemic of Wellness, the Certainty of Dying, and Killing
Ourselves to Live Longer.* Ehrenreich subjects the wellness and anti-aging
industries to withering critique, questioning the monomaniacal project
of trying to become a lean, mean, living machine. For her, the product
offered by a capitalist version of wellness is "the means to remake oneself into an ever more perfect self-correcting machine capable of setting
goals and moving toward them with smooth determination." Citing a
long list of books on "successful aging," she observes a cruel dynamic
related to the bootstrapper ethos of individual time management:

> All the books in the successful-aging literature insist that a long
> and healthy life is within the reach of anyone who will submit to
> the required discipline. It's up to you and you alone, never mind
> what scars—from overexertion, genetic defects, or poverty—may
> be left from your prior existence. Nor is there much or any con-
> cern for the material factors that influence the health of an older
> person, such as personal wealth or access to transportation and
> social support. Except for your fitness trainer or successful-aging
> guru, you're on your own.

In this arena, things have not changed much from the days of *Physical Culture*, when being healthy and aging successfully meant not needing help from anyone and being able to stay ahead of the curve. This
answer leaves much to be desired. I remember it bothering me a lot in
high school, whenever I had a minute to think beyond the hypercom-
petitive atmosphere of accelerated classes and the gospel of PSAT prep.
My senior year, I developed a habit of cutting class and going to the park
to stare at the mallard ducks. One day, I stayed after art class to chat with
my friend Louise and our teacher, the painter William Rushton, who
made miracles happen with a public school's art budget. (We painted

with latex housepaint.) Judging by our conversation, I hadn't come up with any epiphanies among the ducks.

"I don't get it," I said. "You work hard in high school so you can get into a good college, then you work hard in college so you can get a good job, and then you work hard at your job so you can retire, and then you die. What's the point?"

Bill looked back at me with a mix of alarm and pity. "Jenny, that's just not how it works," he said.

YOU HAVE TO *be careful not to get lost in here. This labyrinthine building has three floors with dozens of rooms, and the logic of their layout is not at all apparent. After we hang a left from a giant atrium, the rooms widen and the displays become less formal. There are framed photographs and then more: eyeglasses (sometimes the same ones as in the accompanying photo); crosses; miniature cars; perfume; fishing tackle; a needlepoint of the date of someone's birthday; a midnight-blue bottle of Chinese liquor; a Wonder Woman PEZ dispenser; a glass frog on a ceramic plate; a watch that stopped at 7:10; a tiny set of green garden shears with a matching watering can and trowel; and two small jars of raspberry and boysenberry preserves with a first-place award from Santa Rosa, California, that looks to be from the 1980s.*

Since I was last here, this place has become more inhabited, and every time we see the number "2020," I have to wonder whether the person died of Covid-19—directly or of associated heartbreak and isolation. Sometimes, on the floor below the urns, there are little collections of things: sunflowers, roses, plates of citrus, incense in rice, bottled water, Vietnamese candied ginger. To see them there is like watching a wave of the desirous living washing up against a wall of death—which appears not so much like the end of a person but like a severing of ties.

WHEN I THINK back on that high school conversation, I know that Bill was trying to help me intuit that there was a different version of "the point" that I was not seeing. But imagining a different "point" requires more than bending the rules to play the same old game a different way. It requires grasping and playing a different game altogether, one where "winning" means something that previously may have been impossible to articulate.

In her book *What Can a Body Do? How We Meet the Built World*, the writer and design researcher Sara Hendren shows how useful a non-hegemonic perspective can be for imagining something outside the capitalist game. She provides a reading of the concept of "crip time," a term popularized by Irving Zola and Carol J. Gill and used to describe the tension between a disabled person's temporality and the clock-based, industrialized timetables of modern-day society. Alison Kafer has described crip time as an "awareness that disabled people might need more time to accomplish something or to arrive somewhere," something that ultimately "requires reimagining our notions of what could and should happen in time."

Crip time can apply to the short and the long term; Hendren adds that it can mean "bigger systemic fits and starts—the spiky, unpredictable time it might take a person to proceed through a fairly rigid K–12 education that's built on all kinds of normative chronologies." For Hendren, this latter understanding is personal. She not only teaches classes at the intersection of design and disability studies, but is the mother of Graham, who was diagnosed at birth with Down syndrome. Raising Graham has dislodged her family from the lockstep of the culture around them and its industrialized sense of time, something so deeply taken for granted that, for her family, "questions about time—about Graham and the diagnostics of his delays, but more urgently, the unknowns about his future," have proven more alienating than anything else about the experience.

But Hendren also says that Graham's "invitation to life on crip time" has been a gift to her, giving her the ability to view temporal norms from the outside. On the topic of the good life, whether among other parents, her students, or her own family, she notices a theme: "The economic tempo of the clock shapes our every conversation," while schools and workplaces assume "a form of able-bodied productivity, an ideal of speed and efficiency." Cast adrift from this time line, Hendren sees not a clock but an economic instrument appropriate to a world in which "economic productivity—a life performed in normative, regulated time—is still the unquestioned and overwhelmingly dominant metric for human worth." Meanwhile, Graham demonstrates both a different orientation toward time and a different way of being, thus showing how closely the two are intertwined. Noting that "the insistent, clock-driven measuring of childhood comes from others, not from him," Hendren is able to see something else entirely in and through her son:

> The neatness of the bell curves for milestones among typically developing children, how quick or how slow, even with some fuzziness around the edges—these were never generalizable or predictive for [Graham], not ever. Most timelines just didn't apply. And more important, his relative quickness or slowness compared to his peers and two younger siblings has thus far been of little interest to him as a source of self worth. His association with schooling and extracurricular activities like dance or sports has been primarily one of curiosity and friendship—not simple, mind you, just overall joyous, uncolored by comparative percentages and reading.

This gift, for Hendren, is analogous in some ways to the gift of disability studies in general, which is a questioning of what the good life means not just for disabled people, but for every person with a body that is not a machine and a soul that is more than a worker. A discussion of disability naturally brings up questions about what and whom we accommodate. "How long does it take, or should it take, for a body to move through the world, the forty-plus-hour work week, the demands of caregiving for ailing parents, the daily commute of the body with its changing needs over the life span—a pregnant body, an aging one, a body in recovery after a bad injury?" Hendren asks. "Is the clock of industrial time built for bodies at all?" In proposing a different kind of clock, crip time unsettles (as Sharma would say) *what time means*. Heterogeneous, nonstandard, and attentive to the body, it feels closer to the sundial than the clock.

Another eloquent articulation of the topography of crip time—and, really, all time when seen outside the clock, the grid, or the career ladder—comes at the very end of the documentary *FIXED: The Science/ Fiction of Human Enhancement* (the same one where Jamais Cascio worries about taking the wakefulness drug modafinil). The film pieces together debates about the good life among transhumanists, futurists, and disability scholars and activists. Echoing Ehrenreich's observations about the smoothly functioning machine, the activist and wheelchair user Patty Berne muses that the idea of human enhancement holds out the promise of always being "better than well," an ideal whose appeal she acknowledges. Anyone who is tired at the end of a workday, she says, could understandably think, "I want to be better than well. I'm tired. . . . I want to be excellent all the time." But Berne finds this idea to be, in a way, lifeless. Her conclusion is cut in over footage of her and another friend in a wheelchair riding through her neighborhood, going fast for the fun of it. "It's actually okay to be on a spectrum of reality. It means that there are times when it's juicier, there are times when it's drier, there's times when I'm gonna be tired, there's times when I'm going to have a lot of energy. It's actually part of being alive. It *is* being alive."

It's worth pausing here to note how different Berne's notion of "being alive" is from the cultural view that Hendren diagnoses, where being recognizably "alive" means producing and where producing means exhibiting a certain mastery of time. Berne's "being alive" is closer to the moral of that story about the thread, where the boy is supposed to learn that patterns of good times and bad times actually *comprise* the experience of life itself. To try to reduce the rich topography of experience to a means of maximal output is part of the same philosophy that would turn its back on the ocean or to one's inner landscape, where something new is always coming in on the tide.

Crip time abandons the rhetoric of mastery not only with regard to daily schedules and career tracks, but also to the future in general. Ed Yong observed in an April 2020 *Atlantic* piece that, with the Covid-19 pandemic, many abled people were thrust into a vexed relationship to time and proximity to mortality that sounded deeply familiar to those of the disability community. The scholar Ashley Shew described to Yong an experience of crip time that was not just about dissonance or inconvenience, but involved a different temporal center of gravity, one that must sit closer to the present: "Everything I enter in my calendar has an asterisk in my mind. . . . Maybe it'll happen, maybe it won't, depending on my next cancer scan or what's happening in my body. I already live in this world when I'm measuring in shorter increments, when my future has always been planned differently."

One thing this points to is the inhumanity of standard schedules and expectations for people with disabilities. But it also points to a truth about the human condition more generally. Steven Miller, a Seattle-based photographer, once shared a story with me about how he developed a habit of swimming out into the middle of a local lake after receiving a diagnosis for a rare cancer. The lake was hundreds of feet deep. Out there, he found himself contemplating the abyss, knowing that his paddling and the buoyancy of his body were the only things keeping him alive. It might seem like an unusual situation—not to have any idea how much time you have left—he said, but it was actually true for anyone. They, too, hovered over the same abyss.

Steven developed a fierce love for the lake and its depths, a lack of control that also meant a bracing experience of aliveness. In *The Burnout Society,* Byung-Chul Han finds something similar in a piece by Peter Handke called "Essay on Tiredness." Handke compares "divisive tiredness," the isolating exhaustion of burnout, with a more resigned "tiredness that trusts in the world" (or surrenders to a lake). Too worn out to grasp, and forced to sit back, the tired and resigned person finds that something else floods in: the world, in all its detail, its constantly acting and infinitely dispersed agents, and its minute-by-minute changes. Handke writes, "My tiredness articulated the muddle of crude perception . . . and with the help of rhythms endowed it with form—form as far as the eye could see." Like Josef Pieper's leisure, this "tiredness" is an inherently destabilizing experience, a loss of individual power that helps us find a home in something larger. Han adds to Handke's observation that "deep tiredness loosens the strictures of identity. Things flicker, twinkle, and vibrate at the edges."

I have been fortunate not to have experienced life-threatening illness up to this point in my life. But the "resigned tiredness" and its ensuing

opening do describe a life-changing moment I experienced at twenty-
seven. At the time, I had a full-time day job unrelated to the art I was
trying to make, and I had just pulled an all-nighter (something I can no
longer do) trying to finish an obsessively detailed project for an upcom-
ing exhibition. The following afternoon, so wired and tired I couldn't
nap, I splayed motionless on a couch in the apartment I shared with two
roommates, alone for the time being. In a state of open-mouthed passiv-
ity, I just happened to have my eyes pointed out the window and up at
the top of a redwood in the neighbor's backyard. At first, I felt I was hallucinating: The tree was growing little pears at the top, clustered closely
together. No: They were birds, at least thirty of them, all facing into the
sunset, an unearthly shade of lemon yellow.

Birds seen in 2013

At the time, I knew almost nothing about birds. But I couldn't get the
image out of my head, and in the coming months, I clumsily googled
things like "San Francisco yellow bird"—to no avail. It was only about
five years later, when I had put enough effort into getting to know the
local bird life, that I finally realized what they had been: cedar waxwings.

When I saw them, they would have been wintering in the Bay Area; at some point they would head north. But within that general pattern, cedar waxwings are nomadic, a mercurial expression of time. They follow the fruiting of berries in boisterous flocks, occasionally getting drunk on overripe ones. While many bird species are declining, cedar waxwings are actually increasing because they can eat the berries of popular suburban plants. Starting in the 1960s, some started to have bright orange tail bands instead of the usual yellow ones—they had been eating the berries from an exotic honeysuckle shrub found in suburban gardens, and the red pigment showed up in their feathers.

I remember that vision on the couch so clearly not just because it was the beginning of an abiding interest in birds and their territories. In a more general sense, I remember it as a kind of opening onto infinity. Through that opening, I saw something else—some*one* else, beckoning from a different version of time and space, where suburban gardens, far-flung wintering grounds, summer, and winter were all intertwined. From somewhere outside *me*. Similarly, Handke describes a certain kind of tiredness as enabling "more of less of me," the reality that expands when the ego recedes. Quoting Handke, Han writes, "The trusting tiredness 'opens' the I and 'makes room' for the world. . . . One sees, and one is seen. One touches, and one is touched. . . . Less I means more world: 'Now tiredness was my friend. I was back in the world again.'" Here lies a possible answer to my teenage complaint. Maybe "the point" isn't to live more, in the literal sense of a longer or more productive life, but rather, to be *more alive* in any given moment—a movement outward and across, rather than shooting forward on a narrow, lonely track.

BEHIND US, A *woman emerges from an elevator and heads to a small side room to fill a vase with water. There is an air of soft determination about her, as though she comes here often. Making our way along the glass cases, we begin to notice that the subjects of the framed photographs inside are not often alone. They hold children, lovers, pets. One person is scuba diving alongside a sea turtle; another smiles at something off-camera, her hair and coat dusted with snow. One sits in an old-growth redwood forest, tiny by comparison and looking up at a tree with an expression of complete serenity and appreciation. These are not just people who died; they are people who died on earth.*

BESIDES OFFERING A different set of values, crip time also offers an intuitive way to see time as a social fabric, in part because it flies in the face of dominant liberal concepts of independence, freedom, and dignity. Disability highlights something that is true for all of us: No matter how independent and fit we may feel, we are not simply *alive* but, rather, *kept alive*—against odds that some people are nonetheless privileged enough to ignore. In her book, Hendren cites the work of the philosopher Eva Feder Kittay, another mother of a disabled child, who observes that the relationship of dependency she has with her daughter is at once unique and commonplace. "People do not spring up from the soil like mushrooms," she writes. "People need to be cared for and nurtured throughout their lives by other people."*

If aliveness means touching and being touched—being *in* the world, being kept alive—then the scale between living and dead is inescapably social. In December 2020, responding to a year that had "awakened us to the fact that we die," B. J. Miller, a palliative care physician, wrote a *New York Times* op-ed in which he asked, "What Is Death?" The piece proceeded through different understandings, allowing that the answer might be different for everyone. Some people, he pointed out, might consider themselves "dead" if they are no longer able to have sex, read a book, or eat a pizza. Miller's own personal definition of being alive sounds remarkably similar to the photographer's relationship with the lake and to what Handke's "tiredness" allows: "For me, death is when I can no longer engage with the world around me; when I can no longer take anything in and, therefore, can no longer connect." Social distancing during the pandemic sometimes made him feel this way, he wrote, "but that's just me missing touching the people I care about. . . . Besides, I get to touch the planet all day long."

---

* Similarly, in *How We Show Up*, Mia Birdsong quotes Desmond Tutu's description of the South African idea of *Ubuntu:* "We say a person is a person through other persons. It is not I think therefore I am. It [is] rather: I am human because I belong, I participate, and I share."

Connection is a two-way street. If it's possible that we keep one another alive, it's also true that we're capable of keeping one another dead. I allude to this in chapter 4, in the case both of the "lesser minds" bias and the historical categorization of people "outside time." In the case of disability, a disabled person might be talked about as a lost cause or the static embodiment of a condition. Hendren, for example, writes of how her son's diagnosis created a painful dissonance between the way that she and the people in her close circle saw him and the way everyone else did. For them, Graham "became the diagnosis—forever described and understood and interpreted primarily by genetic status." There is a similar dissonance in Mel Baggs's video about language, where they note, "Ironically, the way that I move when responding to everything around me is described as 'being in a world of my own.'" Baggs and their environment are alive to each other, but neither appears fully alive to people from the outside.

WE EXIT THE *columbarium into a rush of daylight and wind, turn left, and pass through an iron gate and into a hillside cemetery. Amid the cedars and oaks, the parched expanse is dotted with small and not-so-small grave sites and dominated by an enormous hilltop monument—a building, really, with its own set of steps and a patch of lawn on what is called Millionaire's Row. It's the grave of Charles Crocker, one of the intercontinental railroad's Big Four, along with Leland Stanford. It had been Crocker's idea to hire Chinese workers, but industriousness was the only quality he was willing to ascribe to them. When the workers went on strike for shorter hours, Crocker was certain they couldn't be driving the protest. Opium dealers or a rival company must be to blame.*

Instead of visiting Crocker's grave, we take a right, passing a field of rectangular stone markers scarcely bigger than bricks, some completely hidden by encroaching grass, dandelions, and fallen sweetgum leaves. The next section after that doesn't look like a cemetery plot at all; there is almost nothing here but unwatered grass and a handful of redwoods and acacias. You would never know that this was the "strangers' plot," a place where, in the late nineteenth century, the city buried the indigent who had no one to speak for them. Some of the people buried here are Chinese workers who died in 1880 in an enormous dyna-

mite plant explosion in Berkeley—a plant whose product, which they called "the miner's friend," was also marketed for railroad construction. That accounts for twenty-two burials. But when a docent researched this plot in 2011, she found Chinese surnames in the hundreds.

THE TERM *SOCIAL DEATH* was first coined by Orlando Patterson in his 1982 survey of slavery throughout world history. It has since been taken up by scholars to describe myriad conditions in which an individual or a group is deprived of their status as human, existing in a liminal state between recognition and annihilation. In her book *Raising the Dead: Readings of Death and (Black) Subjectivity,* Sharon P. Holland suggests that death could be read not as an event but as "a figurative silencing or process of erasure." With the formal end of slavery in the United States, she writes, a certain dead-alive status persisted because "the transmutation from enslaved to freed subject never quite occurred at the level of the [white] imagination." Holland quotes bell hooks: "Reduced to the machinery of bodily physical labor, black people learned to appear before whites as though they were zombies, cultivating the habit of casting the gaze downward so as not to appear uppity. To look directly was an assertion of subjectivity, equality. Safety resided in the pretense of invisibility."

Social death is related to physical death in that the former puts one at greater risk for the latter, but social death also concerns a broader phenomenon of "death." The border between living and dead, for instance, becomes fuzzier when "some subjects *never* achieve, in the eyes of others, the status of the 'living.'" The socially dead take on a taboo quality that is at one with the American inability to think or talk about death, generally, or to reckon with its historical past.

In the United States, one of the most obvious examples of social death is mass incarceration, particularly at a time when, on any given day, one out of twelve Black men in their thirties is in prison or jail. Whereas earlier moments in the history of prisons—as evidenced by the very word *penitentiary*—cast imprisonment as rehabilitative, prisons had fewer such aspirations by the time Angela Y. Davis wrote *Are Prisons Obsolete?* in 2003. Davis pointed to the decline in prison educational programs, including a 1994 crime bill that disallowed Pell Grants for incarcerated students, thus removing decades-old programs that prisoners had fought to establish. (The ban was finally lifted in December 2020.) Davis describes a scene in the documentary *The Last Graduation* in which books are being taken out of the Green Haven Correctional Facility in Stormville, New York, after its program with Marist College was terminated: "The prisoner who for many years had served as a clerk for the college sadly reflected, as books were being moved, that there was nothing left to do in prison—except perhaps bodybuilding. 'But,' he asked, 'what's the use of building your body if you can't build your mind?' Ironically, not long after educational programs were disestablished, weights and bodybuilding equipment were removed from most U.S. prisons."

If prison is not rehabilitative, then what is it? For Davis and others who have defined the prison-industrial complex, it is part of a larger political-economic fabric that includes not only prisons but corporations, media, guards' unions, and court agendas. Prisoners may be "dead," but they and the prisons they fill still have economic value. On the level of public imagination, however, and particularly in the context of time, the prison simply becomes a black box: a screened-away place as unimaginable to the wider culture as death itself. In *Governing Through Crime,* Jonathan Simon calls this model the "toxic-waste-dump prison": "The distinctive new form and function of the prison today is a space of pure custody, a human warehouse or even a kind of social waste management facility, where adults and some juveniles distinctive only for their dangerousness by society are concentrated for purposes of protecting the wider community."*

This concept informs a chapter called "Project Exile," which Simon names after a 1990s criminal justice program that began in Richmond, Virginia, and gained broad popularity. Borrowing the name to describe strategies of total removal, Simon highlights an important temporal factor: an individual or group's "unchanging propensity" for crime, something that has proven a politically useful concept. The unchanging propensity is just one more way of seeing someone as outside time. Like disabled populations who are considered "a lost cause" or those groups slated for destruction under eugenics, those accused of crime are run through a system that projects them as indelibly marked, fundamentally being or containing a risk to society.

Over the last three decades, life sentences in the United States have grown at a rate that outpaces the general increase in prison population. According to the Sentencing Project, in 2020, one in seven people in prison was serving life with parole, life without parole, or virtual life (fifty years or more). In 2021, two-thirds of those serving life sentences were people of color. A life sentence represents one of the most extreme examples of rendering someone socially dead, by creating a person with no future. In her series of profiles on people given life sentences, Ashley Nellis recounts the denial of educational programming to one of her

---

* Rehabilitative programs do still exist in prisons and, in some cases, have increased. In the case of California, eight years after *Are Prisons Obsolete?* was published, the U.S. Supreme Court ruled that California's prisons were so overcrowded that they constituted cruel and unusual punishment. In response, the state increased funding for rehabilitative programs. When a 2019 study found that the results of the programs (measured in recidivism rates) had been disappointing, Lenore Anderson, executive director for Californians for Safety and Justice, told the *Los Angeles Times* that this was unsurprising given "multiple decades of a behemoth prison system devoid of a focus on rehabilitation." The report also found that the programs were more effective when coupled with community-based services for recently released inmates. With its inside-outside nature, this coupling could be understood as a way to break down the separation that Jonathan Simon describes in *Governing Through Crime*.

subjects, observing bluntly that "some administrations see it as a waste of money to provide programming to those who will never be released."

Without getting into the capriciousness of sentencing laws and parole regulations, it's possible to observe that "doing time" is more complicated than paying a certain number of years, if not one's entire life, to the state. Even after they've been disappeared from society into the "toxic waste dump," time still exists for an incarcerated person in the socially mediated and stretchy ways it does for all human beings. In one way, it gets slowed down, as the outside social world, with its changing customs and technologies, careens forward.* But, in another way, it speeds up: Studies have documented "accelerated aging" in incarcerated populations, with people numerically aged in their fifties exhibiting the health problems typical of people in their seventies.

This stretchiness extends to everyone connected to an incarcerated person. In her book *Carceral Capitalism,* Jackie Wang includes a melancholic interlude titled "Ripples in Time: An Update," in which she reflects on her brother's sentence of juvenile life without parole and how it has affected her life and her family. Asking herself, "What Is Prison?" she answers: "Immobility, yes, but also the manipulation of time as a form of psychic torture. The regimentation of time. The phenomenology of waiting. The agony of juridical limbo. The carceral ripple effect when any life is taken by the state, how it warps the temporalities of everyone in the orbit of the disappeared person."

Alongside Wang's personal recollections, Garrett Bradley's 2020 documentary film, *Time,* offers an evocative, visual sense of this "phenomenology of waiting." The film follows Sibil Fox Richardson, a formerly incarcerated mother of six, struggling for the release of her husband, Robert, who is serving a sixty-year sentence for robbery. Spliced into the

---

* One apt description of this comes up in *Facing Life,* Pendarvis Harshaw and Brandon Tauszik's series of video interviews with recently released prisoners who had been previously sentenced to life. Asked what the state could do to support people like her, Lynn Acosta says that those without friends or family to show the way are not provided with information on things like reestablishing credit, and that the evolution of technology presents an additional hurdle. "If you've been in for ten or more years, I've found, you're like a ghost in the machine," she says. "So you're basically starting from scratch."

narrative are clips from Richardson's video journal from decades earlier, entries of which are sometimes addressed directly to Robert. In them, she and their children speak to someone who is and is not there. The film, shot entirely in black-and-white, is full of images of waiting and of time: Richardson speaking the date in her videos, the date flashing across her car clock, billowing clouds passing slowly overhead, a two-minute-long shot of Richardson on hold with the court, and a giant drill rig hammering into the ground outside the window while she sits on the phone again, answering with smoldering politeness as she is told to try again later.

Garrett Bradley, *Time*, 2020

What *Time* makes palpable is the difference between abstract time and lived time, where the latter is an unstoppable procession that can never be recovered. Bradley cuts between Sibil Richardson's videos of the children goofing off and her own footage of them going about their business as grown men, and between Sibil as a defiant young mother and Sibil as an activist aged by twenty years of fighting. Later, Richardson gives a description that is about as close to the opposite of fungible time as you can get: "Time is when you look at pictures of your babies when they were smaller and then you look at them and you see that they have mustaches and beards, and that the biggest hope that you had was that before they turned into men, they would have a chance to be with their father."

In the meantime, emotional and financial exhaustion shapes the family's time line. One of Richardson's sons simply says that "this situation has just been a long time. A really long time." As Ismail Muhammad points out in his review of the film, the interior of the prison is never shown, nor is Richardson ever seen in a prison uniform. Instead, "the only images we see of the prison are shot from on high, giving us a bird's-eye view and emphasizing how it is occluded from the rest of society." This occlusion casts the prison as a kind of black hole, "warping" (as Wang would put it) the time outside it.

Once a person's "unchanging propensity" for crime has been declared, the time warp extends beyond formal release. In Joshua M. Price's *Prison and Social Death,* one man out of prison tells Price, "Never think or believe once you've paid your debt to society. There's no such thing. You're not a part of society no more. Never think you're part of society. You're an outcast." Diagnosing former incarceration as a "permanent condition," Price writes that, for those experiencing it, "time is oddly collapsed; the original criminal conviction still defines the person, years, even decades afterward," with a man convicted of a capital crime telling him, "I did my crime thirty years ago, but it might as well have been yesterday." As evidence of social death, Price gives an exhaustive list of rights denied to the formerly incarcerated, some standardized, some local, and some seemingly made up at the whim of parole officers. These examples, overlapping with spatial surveillance, often involve the control of one's time—curfews as early as seven P.M., having parole revoked for being late to a meeting, or being required to attend anger management classes or psychiatric therapy every day.

Noting how easily the lingering stigma of a sentence can provide a "useful cover or alibi" for racial discrimination and create second-class citizens—who, in the case of drug convictions, have had limited access to public assistance* on top of the everyday discrimination that a felony conviction brings—Price borrows Patricia Williams's term *spirit murder,* which Williams calls "the disregard for others whose lives qualitatively depend on our regard." Price, whose book reflects not only research and analysis but also social familiarity with incarcerated and formerly incar-

---

* In April 2022, the U.S. Department of Housing and Urban Development began exploring ways to lower barriers to public housing for people with criminal histories. The Consolidated Appropriations Act of 2021, the same act that removed the ban on Pell Grants for incarcerated people, also extended eligibility for federal student aid (FAFSA) to applicants with drug convictions. But in some states, people with felony drug convictions still face challenges in accessing the Supplemental Nutrition Assistance Program (SNAP, formerly known as food stamps), and in South Carolina, they are barred from it for life.

cerated subjects, suggests that the cost of social death for the people he knows is borne not just by them, but by everyone else, too. "The hidden cost of spirit murder may be in missing the rich realities among us," he writes, "replacing the curiosity one might have in the interior life of contemporaries and compatriots with phantasms that inspire hostility and loathing." Those who traffic in social death, in other words, imagine a world of zombies.

Price's field notes illuminate, in contrast, subjects who are obviously alive with hopes and desires and oriented toward a future:

> December 2008. I am talking to a group in protective housing at the jail about pursuing their education. We are trying to put together a pilot program at the jail since we are only minutes from a state university. Many of them tell me they would like to go back to school after they get out. Two men sitting off from the main circle call out that they would like to understand opera better. A young man says a bit shyly that he would like to learn ancient Greek and a few people chuckle. Another man says he draws and would like to learn how to put together a graphic novel. After the session breaks up, he brings over a few of his drawings to show me.

People touched by incarceration are perhaps uniquely suited to perceive second chances and aliveness where others don't. A 2019 article about the closing of Rikers Island and its garden—planned and staffed by prisoners and run by the Horticultural Society of New York, where some former prisoners work as paid interns—included an observation of some guinea fowl pecking around the feet of an inmate. Among the flock, a gift from a prison farm on Long Island, one bird stands out: Limpy, injured when he flew into a barbed wire fence. Hilda Krus, the director of the garden, reports that the prisoners have special affection for Limpy, saying, "This bird is like me. I'm also injured, they might want to get rid of me, but they won't succeed." Krus adds that they take the same attitude toward damaged or unattractive plants. "The students tell me, 'We don't want to get rid of things that are imperfect.' They do everything that they can to save them."

This story, among others from Price's book, demonstrates the ways that the socially dead create a social life, often through the connections to others that incarceration seeks to destroy.* Describing the growth of mutual and self-regard within a brutal space of disregard, Price calls this *grace*. Grace persists in spite of rather than because of incarceration, Price notes, as "violence is neither necessary nor desirable to achieve grace." To me, grace is linked to the need for self-transcendence as described by Viktor Frankl, the author of *Man's Search for Meaning*. In "Self Transcendence as a Human Phenomenon," he describes something that sounds like the opposite of an "unchanging propensity": "It is a constitutive characteristic of being human that it always points, and is directed, to something other than itself. It is, therefore, a severe and grave misinterpretation of man to deal with him as if he were a closed system. Actually, being human profoundly means to be open to the world, a world, that is, which is replete with other beings to encounter and with meanings to fulfill."

Incarceration is the logical response to the fantasy of closed-system people. At the same time, as an extreme form of institutionally codified social violence—something that, as Price says, turns social death into "an unequivocal social fact"—it exists on a spectrum with subtler shades that are no less significant. In the Comments section of a 2021 *Washington Post* article about a Sentencing Project report, one person perfectly demonstrated the relationship between casual racism and social death when they asked, "Does the report say anything about non-whites' possibly significant predilection to committing crimes worthy of life imprisonment?" Echoing eugenics language, this person suggested that nonwhite people could somehow *contain* such predilections rather than existing as individuals in complex and intergenerational networks of risk, harm, and trauma. For anyone who can so readily imagine "closed-system people," the concept of restorative justice is neither possible nor desirable.

In chapter 2, I bring up Ta-Nehisi Coates's quote about "the inescapable robbery of time." When he writes this to his son, Coates is not

---

*One example that Price gives is All of Us or None, an advocacy group of formerly incarcerated people.

talking about something as cut-and-dried as years thrown into the social waste dump of a prison. Instead, he is describing something granular and closer to home, a kind of spirit murder that happens at the level of identity and everyday interaction in a white-dominated world. Like Garnette Cadogan's "cop-proof wardrobe" and minute movements in the street, this robbery represents an attrition: "an unmeasured expenditure of energy, the slow siphoning of the essence" that "contribute[s] to the fast breakdown of our bodies." It is the cost in time and experience of being told to be "twice as good" and to accept "half as much":

> It struck me that perhaps the defining feature of being drafted into the black race was the inescapable robbery of time, because the moments we spent readying the mask, or readying ourselves to accept half as much, could not be recovered. The robbery of time is not measured in lifespans but in moments. It is the last bottle of wine that you have just uncorked but do not have time to drink. It is the kiss that you do not have time to share, before she walks out of your life. It is the raft of second chances for them, and twenty-three-hour days for us.

Codified or not, forms of disregard can be felt in any social hierarchy: race, gender, ability, class. And shifts between the two can happen in a relative blink of an eye (recall Cadogan's shock in New Orleans, followed by ease when he revisits Jamaica). Marc Galanter, a psychiatrist who conducted more than a decade's worth of studies of various cults and charismatic groups, once described a surreal moment in which he very quickly went from "in group" to "out group" and back again. Galanter and a colleague were visiting a national festival held by the Divine Light Mission on the outskirts of Orlando, Florida, and were treated warmly because a respected member had vouched for them. But when a suspicious member asked them if their project had been approved by more senior figures, they were unable to answer definitively. When a request for confirmation was sent up the hierarchy and came back negative, Galanter recalls, "I soon felt myself to be a nonperson, treated civilly but coolly, having become an outsider as rapidly as I had been made an insider. The very people who had hovered around us to help with our plans now found making conversation uncomfortable. People seemed to be looking *through* my colleague and me rather than at us." Then, when the decision from on high was reversed, their status changed back: "As if automatically triggered, a renewed air of intimacy suffused our exchanges." Galanter and his colleague were real, three-dimensional people once again—in effect, socially dead and then resurrected.

I began this chapter by talking about the impulse to increase one's life by lengthening one's numerical life span. When it becomes pathological in the way that Ehrenreich describes—where life is an imagined reserve of fungible time in a zero-sum game—I'm reminded of Donald Trump's logic for not exercising. Seeing the human body as being like a battery with only so much energy, Trump could imagine exercise only as a permanent subtraction from his energy bank. In contrast to this hoarding mentality, I want to suggest another way of "increasing" life, one that has to do with the regard missing from social death. This would be a kind of life extension that reaches outward instead of forward, an increase in aliveness *for everyone* that begins with mutual regard—a world with living beings in it, not zombies.

What I don't want to suggest is that those privileged in any social hierarchy could bring beings back from the dead by suddenly affording them attention in a way that leaves those hierarchies intact. Again, connection is a two-way street. As Price implies when he writes about "the hidden cost of spirit murder," people who move through a dead world are themselves less alive than they could be. People and things are alive when we *become alive to one another*. To regard someone is a balancing of power, an agreement not just to shift one's center of gravity, but to admit to two centers. Hendren suggests a similar unsettling when she imagines a world in which Graham would be allowed the fullest measure of humanity. That world would have to do more than translate him more nicely into current economic notions of personhood, and it would have consequences for everyone: "My son doesn't need a gentle and pacifying form of 'inclusion.' Inclusion is necessary, but it will never be sufficient. He needs a world with a robust countervailing understanding of personhood and contribution and community in it, human values that are alive and operational outside the logic of the market and its insistent clock. He needs it, and so do the rest of us."

There is nothing abstract about the regard we have for one another; it makes and takes lives every day. Just as incarceration crystallizes social death as a "social fact," encoding it into specific policies and suggesting atemporal nonpersons in the process, there is much to be gained by trying to move in the opposite direction. Toward the end of his autobiography, Albert Woodfox, a prison activist and former Black Panther who spent forty-three years in solitary before being released from prison in 2016, on his sixty-ninth birthday, writes, "I have hope for humankind. It is my hope that a new human being will evolve so that needless pain and suffering, poverty, exploitation, racism, and injustice will be things of the past." Woodfox implores the reader not to turn away from the incarcerated, listing organizations and campaigns working to abolish solitary confinement and dismantle the prison-industrial complex.* When he quotes Frantz Fanon—"Superiority? Inferiority? Why not simply try to touch the other, feel the other, discover the other?"—it is a reminder of how much is possible. Working to end carceral logic makes way for a beautiful discovery: a world more alive to itself, full of spirit life instead of spirit murder. If it's true that time is simply aliveness, then this is the surest way to make time.

LIFE UNFOLDS FURTHER *as we pass a couple of ponds, where creek water halts on its way down to the bay. Football-shaped night herons haunt the dense branches on the ponds' edges, unmoving, watching the water for fish. A trio of people on the other side are laughing about something. Geese are talking, too, wandering across the grass. The wind makes a brittle sound in the oak trees, and attached sideways to a cedar trunk is—you just caught it—a brown creeper.*

We've reached the outer bounds of the cemetery. When we turn around, we can see the bay, its water blinding white against the hills and sky. There it all

---

*Woodfox cites the Black Lives Matter movement, the Safe Alternatives to Segregation Initiative, Stop Solitary (a campaign of VOTE [Voice of the Experienced]), Solitary Watch, *Prison Legal News*, Critical Resistance, and the Malcolm X Grassroots Movement.

is: the cranes of the Oakland port, sliding their containers; the freeway, choked with the slow stream of cars; the Santa Cruz Mountains, holding back a layer of fog; the South of Market, hiding our library; the rooftop of the columbarium, letting in the light. Our whole day is spread out in front of us, in space. Most of my life has taken place within our current view, and my childhood is just out of sight to the south. I could point to different things on this rolling tapestry and tell you everything that I remember. Maybe if we sat here long enough, and I did a good enough job, then you could really know me: who I've been, who I am, and who I wish to become.

TOWARD THE END of their report on the lesser minds problem I mention in chapter 4, the authors include a surprising insight: A dehumanizing bias can also exist on an "intrapersonal" level. That is, we're capable of viewing not just others but our future and past selves, too, as having lesser minds and as being less alive. Furthermore, we seem to do it for the same reasons: A lack of "direct access" to the mental states of those selves makes us less prone to see them as having evolving inner lives.

I have kept journals since I was very young. I tend to revisit them when my relationship to time feels especially punitive, when I've been castigating myself for what I have not yet become or achieved. In these journals, instead of snapshots of a closed-system person, I find a live self who is always questioning, always "trying to get it together," always writing toward a future and refiguring the past.

This past year, when I was writing chapter 4, I visited my parents' house and dug through their garage to find the high school journals where I'd mentioned "it." I took one back with me, unthinkingly plopping it onto my desk at home next to my current journal. It was surreal to see them side by side like that, with nearly three decades intervening—and yet they were written with the same hand. When I was younger, I used to think my impulse to write in journals was a kind of bid for immortality, a jealous snatching and pressing of time, as if moments were butterfly specimens. But now I value the process for the way it dispels the myth of a finished self. Looking at the two journals, I thought, *I'm thirty-five and still looking for "it."* For a moment, I overflowed my temporal container: I inhabited a moment that was not exactly linear, but closer to harmonic.

At the time, I had just finished watching the British documentary series *Up*. Starting in 1964, the series sought to give "a picture of the United Kingdom in the year 2000" by choosing a handful of British seven-year-olds from different backgrounds to interview about their opinions and dreams. The conceit of the film is that basic aspects of one's personality are in place by the time one is seven years old. After 1964, Michael Apted took over the project as director, revisiting those people every seven years as they proceeded through different schooling, jobs, marriages, divorces, and children and grandchildren of their own, all the way through to *63 Up* in 2019.

The most logical way to watch this series at any given point would have been to watch the most recent one, as each contains at least some clips from previous segments, catching the viewer up on the subjects' lives thus far. Instead, Joe and I watched every single episode starting from the one in 1964, which meant that by the time we reached the last one, we'd seen certain clips of the subjects as children so many times that we'd nearly memorized their answers (such as Nicholas Hitchon, later to become a physicist, responding to a question about what he'd like to be when he grows up: "I'd like to find out all about the moon and all that").

No documentary—nor any representation, for that matter—is capable of portraying a person or a place completely, and the *Up* series is no exception. Indeed, many of the participants complain at various points about being pictured inaccurately, especially with the films' initial hyper focus on the effects of class background. Nevertheless, seeing the past clips pile up gives an undeniable depth to subsequent segments, with each one appearing like the lighter-colored new growth on a plant each spring. The layered editing process seems even to have influenced the director himself. In the final segments, Apted begins to shift from interviewer to interlocutor, inviting conversations with the films' subjects about how the framing of his questions had affected them. Familiarity and concern show in his voice, no longer that of an objective "observer." One gets the sense that he sees them increasingly as people, not test subjects. In turn, some of the subjects who were critical of Apted seem to soften up on him, as they find themselves together journeying toward the end of life. In *63 Up*, one subject has died and the physicist has been diagnosed with cancer; Apted himself passed away in 2021.

It's perhaps for this reason that Joe remarked, upon finishing *63 Up*, on how well the series worked as an "empathy machine," Roger Ebert's term for what film is capable of doing. While the thesis of the documentary had been that the gist of someone's character is set at age seven, the series nonetheless denied any tendency to see an individual as a bounded point in space or time. On the one hand, there are definitely personality traits that seem present in some of *Up*'s subjects from the beginning. On the other hand, nothing about the events of their lives, or their responses to those events, proves predictable. That both could be true speaks to the way that anything that lives in this world exists as an expression of time.

I thought of how different these people's identities felt from those in arenas like social media—pictured as game players who've appeared fully formed, sui generis, and instantly identifiable. There, our icons interact as if in a Newtonian billiard ball universe—ageless, knocking into one another in an abstract space, and unchanged by the impact. In contrast, the quiet grandeur of *63 Up* is like that of Garrett Bradley's *Time*, in that it comes from a dimensionality that involves not only people, but time in the sense of development, decay, and experience. Like the old moss that Robin Wall Kimmerer knows money can't buy, it takes no fewer than fifty-six years to make a film series chronicling fifty-six years of change. Nor can there be a sixty-three-year-old person without the sixty-two-year-old, the sixty-one-year-old, the sixty-year-old, and so on, backward past even their birth and into the histories of their ancestors.

It's perhaps unsurprising that Peter Handke, that same writer who described an ego-softening "tiredness," has a poem that perfectly describes a self that is more of a harmony than a single changing note. Translated as "Song of Childhood," the poem contains stanzas each starting with "When the child was a child."* Toward the beginning of the poem is a list of sorrowful contrasts: When the child was a child, "it

---

*Handke wrote "Song of Childhood (Lied Vom Kindsein)" for Wim Wenders's *Wings of Desire* (1987); the poem is read at various points. The translation and line breaks here are based on the English subtitles.

saw many beautiful people, while today that's a stroke of luck"; where the child once "could imagine paradise very clearly," it now "can only guess at it"; and where the child once "played with enthusiasm," it now "can only muster it up / when it's about work." So far, this sounds like a linear trajectory. But the final part of the poem concerns something that remains open:

> When the child was a child,
> Berries filled its hand as only berries do,
> and it's still the same today.
> Fresh walnuts made its tongue raw
> and they still do now.
> Atop every mountain it longed
> for yet a higher mountain.
> And in every city it longed
> for a larger city.
> And it still does.
> It reached up into the treetop for cherries
> as excitedly as it still does today.
> It was shy around strangers
> and it still is.
> It waited for the first snow
> and it still waits that way.
> When the child was a child
> it hurled a stick at a tree like a lance,
> and it still quivers there today.

The poem illustrates Frankl's notion that to be human is to be directed to something other than oneself and Patty Berne's observation that the experience of tension within time *is* life itself. It also explains why, in those moments of true encounter that unsettle the boundary between myself and something or someone else—when time seems to stop, then expand—I sometimes notice a strange side effect. Like an oceanic upwelling, long-buried memories come to the surface: images and states of mind I remember from childhood, from college, from my early adult life. These memories are often of similar moments of encounter, as though under the grid of calendar years and career milestones there were another dimension, one where all these encounters spilled into one another. Bergson might identify this as the dimension of "the deep-seated self," arguing that the truest, most willful actions are the ones that come from it. When we say we are "moved," I think it is this self, not just today's self, that is moved.

This open-endedness suggests the final reason I want to think about life extension as an outward movement rather than a forward one, particularly when it comes to death. As I suggest in chapter 2, denying the logic of increase means allowing the idea of limits, including the limit of one's life. No matter how optimized, healthy, and productive I am, I simply will not become more or better forever, which means there are things I will never do and never be. Just like this book, which could have been anything when I started it, my life will take some paths and not others—and then it will end, the thread pulled out of the ball, with no witch to indulge me by taking it back. Realizing that I cannot be everything is in one sense incredibly freeing: It means I am not responsible for being everything. Yet the fact that life ends, for anyone who enjoys being alive and in the world, is also inherently sorrowful.

The time-honored insight of religions and cultures through history has been to respond to this condition by dissolving the bounds of the individual, seeing death as a sacred reinsertion into the world. The deceased are placed into the earth; they are burned and their ashes scattered over water and hillsides; they are covered in bark and sealed off in the hollows of trees; they are left at high elevations for the birds to eat; they are cast off to sea. In "What Is Death?" Miller makes a similar acknowledgment of boundlessness, noting from a physical standpoint that the atoms of your body and the energy that animated them don't simply disappear, any more than they could appear out of nowhere. As earthly beings, we have someplace to return, a substrate in which this energy transmutes into something else.

I think this physical description can be transposed into the social as well. As the civil rights activist Yuri Kochiyama said, life is not yours alone but also "the input of everyone who touched your life and every experience that entered it." This is true both now and after you've gone. I think of my step-grandmother, whom I have looked up to my entire life and whom I lost during the pandemic. The last time I saw her was just weeks before the lockdown, when my boyfriend, my parents, and I met her for lunch. At the table, she squeezed my hand warmly and congratulated me on *How to Do Nothing*. I'll forever see her walking away from us in the parking lot, looking spry and flashing a winning smile as she waved over her shoulder. Now that she is gone, tiny things about myself will unexpectedly remind me of her: a certain laugh, a certain posture, or even the way I position a pin in my hair. Though this haunting is no substitute for her presence and still carries the sharpness of loss, I welcome it nonetheless. Her life has extended into mine.

When Ehrenreich emphasizes nonhuman agency in *Natural Causes,* this porousness of identity is part of what she is interested in. Whether on the cellular or the social level, the bounded self is an illusion, the confederation of "me" potentially anarchic. For thirty-six Gregorian calendar years, an identifiable pattern of stuff and influences has persisted recognizably in my person, animated by I know not what. After "me," they will continue on to do, and be, something else. From this perspective, the prospect of one's own death looks somewhat less lonely. Ehrenreich, who was in her seventies when she wrote *Natural Causes,* wryly joking that she was "old enough to die," offers this reflection at the end:

> It is one thing to die into a dead world and, metaphorically speaking, leave one's bones to bleach on a desert lit only by a dying star. It is another thing to die into the actual world, which seethes with life, with agency other than our own, and, at the very least, with endless possibility. For those of us, which is probably most of us, who—with or without drugs or religion—have caught glimpses of this animate universe, death is not a terrifying leap into the abyss, but more like an embrace of ongoing life.

That embrace, too, can be ongoing. When we think of loved ones who are no longer with us, we surely wish we had embraced them more, literally and figuratively. Older people looking back on their lives sometimes say that, if they had the chance to do it over, they would have embraced it more fully. Like Miller's definition of aliveness as "touch[ing] the planet," Handke's description of touching and being touched, or Hartmut Rosa's "resonance" in the epigraph for this chapter, my definition of being alive is simply that: the embrace. I feel alive if I'm not alone in the air, but embraced by it. I feel alive when someone's eyes light up, and mine do too. I feel alive if I can look at a deer *and* see it looking back at me; if, when geese speak, it sounds like language; if, when I walk on the ground, I feel it pushing back against me. I'm alive to the extent that I can be moved.

But for this to happen, for there to be "more of less of me," the forward-leaning ego that grasps at time has to die—at least for that moment. This death can feel like a trust fall into time and mortality itself. The philosopher Jiddu Krishnamurti writes that, in a state of complete attention, "the thinker, the center, the 'me' comes to an end." This supposed emptiness makes way for so much more, as "it is only a mind that looks at a tree or the stars or the sparkling waters of a river with complete self-abandonment that knows what beauty is, and when we are actually seeing[,] we are in a state of love." That state, he says, has "no yesterday and no tomorrow." This is easier said than done, of course, and my whole life so far seems to consist of forgetting and remembering this wisdom. But whenever I do remember, I forgive myself for forgetting. I've come to see the truly alive, ego-dissolving state less as a goal to arrive at and closer to something like rain. It comes and goes, and when it comes, you make use of it and give thanks.

Strangely enough, it has "rained" even in my sleep. About once a month, a lucid dream will pop up in the middle of one of my usual stress dreams: rushing through an airport, late for a bus or to my own class, or being unprepared to give a speech. At first, nothing changes, except that I've suddenly stopped and become suspicious that I'm in fact sleeping. The settings and the props remain, but they feel neutralized, cast adrift from the anxiety-producing script that created them. In turn, they come to the foreground as objects of fascination, unfrozen in time. I, too, get unfrozen: I find that I can move with agency, as though I've gained control of my arms and legs for the first time.

Lucid dreams are truly a liminal state between sleeping and waking; I know in the dream what day it is, what I'm wearing, and what I've done in other lucid dreams. I am also aware that the dream will probably end in a few minutes, so the question becomes what to do in that time. But this is a very different kind of "point" from the one I would have been propelled by just moments ago, given that, for the most part, I would sum up my purpose in a lucid dream as "just looking around." I know that I will wake up soon, and I want to prolong the dream. But I'm not afraid of waking up. I have only gratitude for this ephemeral fluke, making the most of my time by perceiving and testing the environment around me. When I reach out my hand, I often become aware of a gripping sensation, but it is not the iron grip of fear. Instead, it feels like holding fast, like "touching the planet" before I inevitably drift away.

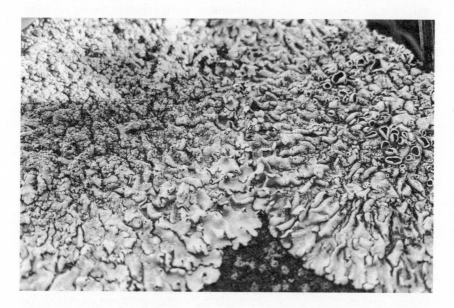

ATTACHED TO OUR hill, we are blanketed with sound: cars, bluebirds, people, maintenance vehicles, and air; the wind in our ears, the rustle of tiny coyote brush leaves a few feet away and of the trees below us, among the graves. Next to us is a boulder of greenstone, a metamorphic rock that once bubbled as lava into an ancient sea. Now lichens are growing on it, a tiny civilization you can feel with your finger. A bumblebee—one of those harmless ones that really bumbles—gets closer and farther away. Bee activity.

The sun is eventually going to set behind the skyline. But, in the meantime, if you cast your eyes upward, there's a different view I wanted to show you. In high school, my art teacher gave me another piece of advice: To get an accurate California blue sky, the trick is to add a near-imperceptible bit of alizarin crimson. There, between us and outer space, is a blue full of crimson and everything else—the circling hawk, the turkey vulture now headed west, the restless flock of tiny swallows zipping around unpredictably over our heads, like air molecules. Though we can't see it yet, the earth is ever so slowly rotating our view, altering the blue, lengthening our shadows. It is holding us fast, turning us toward tomorrow.

Conclusion

Halving Time

Scientists are saying the future is going to be far more futuristic
than they originally predicted.
—*Southland Tales* (2006)

No one is responsible for an emergence; no one can glory in it,
since it always occurs in the interstice.
—MICHEL FOUCAULT, "Nietzsche, Genealogy, History"

In the winter of 2010, a partnership of California state agencies and non-profits launched a citizen science initiative called the California King Tides Project, with the tagline "Snap the Shore, See the Future!" Residents were asked to visit recommended areas along the shoreline to take photos during the year's king tide, a regularly occurring phenomenon in which the positions of the sun and moon align a certain way and increase the tide by a few feet. The fact that this natural, temporary rise happened to correspond with the anthropogenic sea level rise expected over the next few decades was something that the California Coastal Commission sought to exploit. They suggested that while observing the king tide, the viewer should "imagine seeing these tides (and the associated flooded streets, beaches, and wetlands) almost daily." This imaginative exercise would render the future sea level rise more palpable and, ideally, "motivate us to stop burning fossil fuels." Like someone from a time machine warning people in the past, the king tide appears like an explosion of the future into a present that otherwise can't access it.

Eleven years later, the California King Tides Project was still going. On its site, I navigated through the dutifully contributed photos of the 2020 king tide, clicking blue dots on a satellite map of California. Familiar places looked unfamiliar: Near my studio, a set of steps at Jack London Square where I often see people hanging out was completely

submerged, a set of handrails disappearing into the water. At Middle
Harbor Shoreline Park, signs about not swimming or wading were up
to their necks in bay water. In San Francisco, there was significantly less
beach at Baker Beach. In fact, clicking on any beach area produced a
strange temporal dissonance, as the sand in the photo was only a sliver
of what showed on the satellite map.

The photo that struck me the most intensely was by Alan Grinberg,
a man who lives along the bluffs in Pacifica. Titled "The Most Expensive
Photograph I Have Ever Taken," it shows the entirely white background
of a wave crashing behind a Buddha statue in his backyard. On Grinberg's
Flickr account, I saw the next five photos he took and the reason for the
title: The wave gets closer, overtakes the Buddha, overtakes the yard, and
then overtakes his camera. The initial image was striking not just because
of its backstory, but because of its contrasts: the statue with its eyes closed
and hands folded calmly as the future arrives in all its violence.

Sitting like that in the midst of a chaotic meantime, the Buddha
reminded me of an anecdote from Ajahn Chah, a Thai meditation master: "You see this goblet? I love this glass. It holds the water admirably.
When the sun shines in, it reflects the light beautifully. When I tap it, it

has a lovely ring. Yet for me this glass is already broken. When the wind knocks it over, my elbow knocks it off the shelf, and it falls to the ground and shatters, I say, 'Of course.' But when I understand that this glass is already broken, every minute with it is precious."

Opening and closing the photos on the Coastal Conservancy map, I, too, thought, *Of course,* but I felt incapable of such equanimity. Instead, I would look at the ocean on a nice day and feel like I might implode from the pressure between now and not-now. In 2020, the king tides receded as they always have. But in the photos, they left something permanent: a memory of the future that hangs over the present like a pall, like the wave suspended by the camera shutter.

THE MEANTIME IMPLIES waiting, a less important area between two specified times. In the case of dread or, really, any overemphasis on some future point, the meantime also appears empty: With nothing but distance between you and your destination, it may as well have already taken place. It's as though you had an amazing set of binoculars that let you see something far away in such detail that you didn't actually need to go there. *Let's just get it over with,* says the heartbroken subject, unable to enjoy her already-broken cup.

Allowing that a certain amount of change is locked in, I can still hear Bergson's complaint about this attitude toward the meantime. You're turning time into space, he would say. You're imagining empty blocks of time stretching out in front of you, mentally crossing that distance toward the thing you think has already happened, instead of admitting the creative aspect of time that is ever evolving and shifting, each second heaving the world—and you—through the crust of the present and into the future.* And yet, remember that this "distance" would be like the space in the abstract grid of a cartographer, as opposed to Bjornerud's physical "timefulness." While recognizing the limits of abstract space as a metaphor for time, I think there is a different kind of spatial under-

---

*In *Time and Free Will,* Bergson writes: "It will no longer do to shorten future duration in order to picture its parts beforehand; one is bound to *live* this duration whilst it is unfolding."

standing (or, at least, one that initially seems spatial) that can help the Western subject grasp something concrete in the meantime.

In *How to Do Nothing,* I draw on the concept of bioregionalism, a sense of familiarity with and responsibility to a particular place that informs one's identity. Although the term was popularized in the 1970s, its core concepts aren't new; at its best, bioregionalism reflects the way that indigenous people have related to land, exhibiting care for and acknowledgment of a network of life forms, waterways, and other agents specific to each place. Bioregions are different, but their boundaries are porous, and as networks, they connect to both the large (a weather system or an ocean current) and the small (micro climates and symbiotic complexes of species). Previously, I used bioregionalism as a model for identity, as it offers a study of flows, interdependence, and a form of difference without boundaries that I'd found especially useful as a biracial person.

As it turns out, bioregionalism can be a useful way of thinking about time as well. I allude to this in chapter 6 with the idea of chronodiversity, "gardening" time, and Barbara Adam's observation that, in the psychological experience of time, "complexity reigns supreme." My friend and mentor the poet John Shoptaw has a poem called "Timepiece" with topographical language I often think of, lines about "a steep night, a tangled week, an August that shelves / down toward a swift dream." Could it be that the opposite of looking assuredly through binoculars at a flat space would be the perspective you get when rounding a mountain trail—one where, even though you know where you are, things look different at every turn?

Bioregionalism is useful here both as metaphor and as concrete demonstration, in that its timescales overlap and sometimes lie outside the human perspective. Expressed simply as change, ecological and geological time are full of difference: Things happen both quickly and slowly, at both tiny and inconceivably epic scales. Rocks like sandstone form gradually, whereas volcanic rocks like obsidian form in violent contact. Different mountain ranges rise at different rates, with some speculated to have pushed up (relatively speaking) "like a Popsicle." I wrote this conclusion while in sight of Mount Rainier (Tacobud), whose massive mudslide around 5,700 years ago reduced its summit by half a mile and may be preserved in the Nisqually Native American oral tradition.* In the next few hundred million years, geologists predict, the continent I am on will crash into Asia. In the meantime, we have earthquakes whose rupture can happen at ten times the speed of sound in dry air.

The year I wrote this, Brood X, a group of cicadas that emerge in seventeen-year periods, swept over the East Coast and Midwest, at one point clogging the auxiliary power unit of the plane used for Joe Biden's first international trip as president. An arborist in Royal Oak, Michigan, received the usual calls from people worried about their trees suddenly dropping huge numbers of acorns and had to inform them about mast years, a temporal phenomenon in which trees coordinate to drop their fruit all at once. (Describing masting in pecan trees, Robin Wall Kimmerer points to studies that suggest the trees may be using underground mycorrhizal networks—in other words, talking to one another—in order to enact such a "unity of purpose.") West of the Sierras, a bristlecone pine tree that started growing five thousand years ago continued photosynthesizing in soil made white by ancient limestone. In Tillamook County, Oregon, people kept visiting the Neskowin Ghost Forest, a graveyard of the stumps of Sitka spruce killed when an earthquake in 1700 suddenly inundated them with mud, and discovering that you could see them only at low tide.

I am purposely mixing together what would seem to be biological and geological examples in part to emphasize the overlapping quality of different cycles, but also because, in reality, rocks are hard to separate from what we (today) typically consider alive. Carboniferous limestone is made of the shells and hard parts of marine organisms. In the Santa Cruz Mountains, there is a recognizable plant community that tends to pop up wherever the soil contains serpentinite, a kind of iron- and magnesium-rich mantle rock that altered as the Pacific Plate slid under the North American Plate. Iain Stewart, in the documentary series *Rise*

---

*The Nisqually oral tradition includes a story of a time when Mount Rainier was a monster that devoured everything in its path, until the Changer appeared in the shape of a fox, causing the mountain to burst a blood vessel. Vine Deloria, Jr., notes that this story is repeated with few differences by four separate tribes of the region. Some have speculated that the burst blood vessel refers to the massive mudslide.

of the Continents, points out another such echo: Because skyscrapers tend to be built where hard rock is close to the surface, the shape of the Manhattan skyline can be read as a translation of the underground presence of Manhattan schist. Like serpentinite, schist has a composition inseparable from its history: The reason it's so hard is that it was compressed over three hundred million years ago under a mountain range with heights similar to those of the Himalayas today. That range formed when two landmasses collided in the formation of Pangaea.

Schist outcrop on Bennett Ave. in New York City

The Manhattan skyline, as well as the schist that can presently be found sticking up out of the ground in Central Park, are both examples of the fuzzy boundary between past and present. Other fuzzy boundaries have to do with what counts as an individual and what counts as a life span or event—questions that end up being deeply related. An enormous fungal network that occupies thousands of acres of soil in Oregon's Blue Mountains and could be anywhere from 2,400 to 8,650 years old has "rekindled the debate of what constitutes an individual organism," according to an article in *Scientific American.* One scientist suggested that an organism is "one set of genetically identical cells that are in communication with one another that have a sort of common purpose or at least can coordinate themselves to do something." In the case of Pando, a clonal colony of quaking aspen trees in Utah—always referred to in the singular and named after the Latin for "I spread"— the life span of its individual trees is a little more than a hundred years, but the connected root system is thousands of years old. One image of Pando shows a boundary line drawn on top of a satellite image of what might otherwise appear to be simply a tree-covered hillside.

Outline of Pando

The trees of Pando and the visible mushrooms of a fungal network are both examples of bodies that are also totally embedded in some other kind of body. Events can take on a similar ambiguity. When John McPhee writes about the San Gabriel mountain debris flow filling up a house in six minutes, it's difficult to see this event in isolation from its preconditions: for example, an earthquake that broke up the rocks, or a fire on the mountain during an earlier summer. Indeed, McPhee describes a summer fire in 1977 that prompted Hidden Springs officials to warn residents of the possible resulting debris flow in the coming winter (to no avail, even though they ended up being right). Did the debris flow start when the rocks moved? Or did it start when the chaparral caught fire?

This tension is implied in another *Scientific American* article, whose title ("The Longest Known Earthquake Lasted 32 Years") seems at odds

with its subtitle ("The 'Slow Slip' Event Preceded a Devastating 1861
Quake of at Least Magnitude 8.5 in Sumatra"). The opening paragraph
reveals an event within an event, like a slow-ripening fruit falling suddenly off the branch: "A devastating earthquake that rocked the Indone-
sian island of Sumatra in 1861 was long thought to be a sudden rupture
on a previously quiescent fault. But new research finds that the tectonic
plates below the island had been slowly and quietly rumbling against
each other for 32 years before the cataclysmic event."*

To a non-Western perspective uninterested in certain types of boundaries or models of subjecthood, separating things from their context probably isn't as much of a problem. It wasn't for Bergson, either. In *Creative
Evolution,* where duration is a process of becoming and where states are
always breaking through to other states, he had to view individuality not
as an absolute category but as existing on a spectrum. "For the individuality to be perfect," he wrote, "it would be necessary that no detached
part of the organism could live separately. But then reproduction would
be impossible. For what is reproduction, but the building up of a new
organism with a detached fragment of the old?" All living things contain
the means for overrunning their own boundaries. In that sense, Bergson
notes, individuality actually "harbors its enemy at home."

Trying to draw a line around myself, I am forced to ask, *Am I Jenny or
am I my mother's daughter, my grandmother's granddaughter?* and so on. If I
am an event, when did I start? Thirty-five years ago? Hundreds of years
ago? Thousands? Am "I" not like the visible mushroom growing out of
a substrate outside which I would be incomprehensible, even impossible? Though my episodic memory goes back only so far, my existence
is explained by older things: my mother's immigration, a war whose
exigencies threw my grandparents together, and the fish swimming off
the coast of Estancia, on the eastern tip of Iloilo. The people who fished

---

*In *Time and Free Will,* Bergson invokes a similar dynamic when describing the process
of personal deliberation and arriving at a choice. While we typically think of deliberation as "an oscillation in space" (between two or more outcomes), for Bergson, the
deliberative process is instead "a dynamic progress in which the self and its motives, like
real living beings, are in a constant state of becoming." It is this progress that "the free
action drops . . . like an over-ripe fruit."

there have something to do with me, just as I continue to have something to do with them.

In *Sand Talk,* Yunkaporta complains that "it is hard to write in English when you've been talking to your great-grandmother on the phone, but she is also your niece, and in her language there are no separate words for time and space." He explains that in his great-grandmother/niece's kinship system, there is a reset every three generations where your grandparents' parents get classified as your children, because "the granny's mother goes back to the center and becomes the child." Furthermore, a question that translates in English as "What place?" actually means "What time?" According to the paradigm his great-grandmother/niece uses, these two features are naturally intertwined: "Kinship moves in cycles, the land moves in seasonal cycles, the sky moves in stellar cycles, and time is so bound up in those things that it is not even a separate concept from space. We experience time in a very different way from people immersed in flat schedules and story-less surfaces. In our spheres of existence, time does not go in a straight line, and it is as tangible as the ground we stand on."

Note how different "the ground we stand on" is from abstract space. Yunkaporta's "ground" is not a metaphor. It is referring to real ground, every bit as concrete as the Newtonian, imagined grid of space is empty, abstract, and "flat."

What we think time is, how we think it is shaped, affects how we are able to move through it. Flat time offers only so many options. Thinking about the relationship between seeing and moving, I'm reminded of a scene from the 1986 Jim Henson movie, *Labyrinth,* that I still remember thirty years after a babysitter named Liz brought the VHS tape to my house. Sarah (played by a young Jennifer Connelly) has just started out in a formidable labyrinth, at the center of which is a castle where the Goblin King (a fantastically coiffed David Bowie) waits. Finding herself in a long, unbroken section that stretches only forward and back, Sarah complains, "What do they mean 'labyrinth'? There aren't any turns or corners or anything. This just goes on and on." Momentarily suspicious that it might actually be something different, she starts running, but she soon gets tired, pummeling the bricks with her fists before slumping to the ground.

Although she doesn't notice, some mosslike plants with eyes at their tips turn to look at her. Then, on one of the protruding bricks, a small worm with blue hair and a red scarf yells out, "Ello!" Once she gets over her shock, Sarah asks the worm if he knows his way through the labyrinth. He says he doesn't and invites her to "come inside and meet the missus" instead. She has to solve the labyrinth, she says, reiterating her complaint that it has no corners or openings. "Well, you ain't look-ing right," says the worm. "It's full of openings. It's just you ain't seeing 'em." Then he points to what appears to be a brick wall. Sarah walks up to it, looking back at the worm with skepticism. She doesn't see it. "Things are not always what they seem in this place," the worm says. "You can't take anything for granted."

The moment that made such an impression on me as a kid was when, with great hesitation, Sarah puts up her hands and miraculously walks into and through a wall that is actually an optical illusion. Thanks to 1980s special effects, she begins to disappear to the left. "Don't go that way!" says the worm. "Never go that way." After she changes course and disappears to the right, the worm delivers the punch line: "If she'd have kept on going down that way, she'd have gone straight to that castle." The last-minute twist reminds me of what Yunkaporta says right after his complaint that the word *nonlinear* casts linearity as the default. He mentions a man who "tried going in a straight line many thousands of years ago and was called *wamba* (crazy) and pun-ished by being thrown up into the sky," adding that "this is a very old story, one of many stories that tell us how we must travel and think in free-ranging patterns, warning us against charging ahead in crazy ways."

Sarah's disappearance into the wall also feels like an enactment of the difference between *chronos* and *kairos*. As I observe at the outset of this book, *chronos* is homogenous while *kairos* is more heterogeneous, suggesting a critical moment for action. In Astra Taylor's "Out of Time: Listening to the Climate Clock," an essay that fundamentally influenced my entire line of questioning in this book, she notes that *kairos* in modern Greek now means "weather" and goes on to describe its usefulness in ecological terms: "Perhaps the opportune time to intervene is fleeting, like a passing thunderstorm or the peak of spring, and we risk a mismatch by striking too late." It occurs to me, reading this, that the phrase isn't "seize time." It's "seize *the* time."

Compared to *chronos, kairos* sounds like the domain of those wayfarers who knew that time is inseparable from space and that every place-moment demands close attention, lest you miss your opportunity. It's not that you can't plan, but that the time in the plan doesn't appear flat, dead, inert. Instead, in the "meantime," you wait with your ear to the ground for patterns of vibration that will never repeat themselves. Faced with flatness, you look for an opening. When it comes, you take it, and you don't look back.

I WROTE THIS conclusion at a residency on Maury Island, which sits connected to Vashon Island in Puget Sound and is accessible only by ferry. The house I stayed in was on a wide, sleepy road that passed along Quartermaster Harbor, a large inlet. I took to walking on that road after I was done with work for the day, which meant that it was often dusk. One evening, I saw a strange-shaped figure far ahead of me, moving slowly, stopping, and then moving again. In the fast-gathering dark, there was a moment when I genuinely could not tell if it was an animal or a person. I could sense my brain vacillating, not knowing what to do with this shape, and thus having to wait and watch it much more closely than I ordinarily would have. Eventually the figure turned out to be a person wearing a large cloak, now disappearing around a shrubby corner.

In that brief pause, I had experienced doubt, and doubt had increased my sensitivity to everything. The etymology of the word *doubt* contains the Proto-Indo-European root *dwo,* for "two," with the later Latin *dubius* meaning "of two minds, undecided between two things." Though the temporarily unknowable figure had stopped me in my tracks, preventing forward movement, doubt is not stillness. Something grows in that cut, even if it's collapsed in the very next moment.

In *Time and Free Will,* Bergson was willing to grant that much of our thinking and activity was constrained by habits that we simply let run, like automated local processes. Over time, he wrote, those habits would form a "thick crust," rendering our true agency unrecognizable to us. But the crust wouldn't always hold. He gave the example of some problem you might be working through, soliciting advice from friends who all made very reasonable recommendations. You might be well on your way to drawing the logical conclusion from this array of advice when something entirely different takes hold:

> Then, at the very minute when the act is going to be performed, *something* may revolt against it. It is the deep-seated self, rushing up to the surface. It is the outer crust bursting, suddenly giving way to an irresistible thrust. Hence in the depths of the self, below this most reasonable pondering over most reasonable pieces of advice, something else was going on—a gradual heating and a sudden boiling over of feelings and ideas, not unperceived, but rather unnoticed. (emphasis in original)

Such a "bursting" might later come to look like cooled lava or ossified history, letting us forget the contingency of the moment in which it happened. You can make the same mistake in the opposite direction, forgetting that the future will contain many such moments of doubt—or even neglecting to notice when you're in one.

My friend, the artist Sofía Córdova, told me that when she became pregnant, she decided to undertake the Latin American tradition of the *cuarentena,* a period following birth where the mother stays at home with the baby for forty days.* Through her work as an artist, Sofía had long thought about "time outside history," or "women's time, queer time, Black, Indigenous time. . . . The time that doesn't get recorded in the 'great archives' of our species by the white men who write them, who

---

* A similar practice is observed in other countries, including China, Korea, India, Iran, and Israel. The Chinese American writer Fei Lu has also written in *Atmos* about adapting *zuo yuezi* ("sitting/doing the month/moon"), ordinarily a postpartum practice, to the process of recovering from gender affirmation surgery.

write of progress." But the *cuarantena* took her from thinking about this time to experiencing it. Birth was, of course, "just one possible portal or escape hatch from the flow of our shared concept of time"; one need not give birth to experience it. This is how she described what lay on the other side of the portal:

> Since what we—baby and I—had just gone through was so inherently in and of the body, the scale of my thinking and experience was diminished (not meant pejoratively) to the space of the body and at farthest, the edges of our home. . . . The whole universe was here and within the walls of my body that universe continued on. In other words, my insides and limited outsides felt as one, and moreover, the only ways of telling time. The physical experience of reassembling your body, having a steep drop-off of hormones (which have been running the show and which for me were akin to a constant low dose of psychedelics), sharing your body with an infant, being vigilant in a new way, even in deep sleep, sleep itself becoming reshaped, all these things keep you tightly anchored to the experience of the time and tasks of healing, feeding, sleeping at all hours. It's very cellular.
>
> The cyclical nature of this pattern as you and this new being are rapidly moving to a new understanding of self and world, coupled with the specificity, disorientation, and uncanny nature of this fleeting moment, make for some pretty hard edges when it comes to defining this time against everything that has happened before or since. There are of course other glimpses into this type of time, but for me they've been shorter, more fleeting (seeing the glint of the sun off the surface of the sea, entering a body of water, singing or playing music together with people you love).

Sofía's experience of the *cuarentena* was something we talked about the first time we had seen each other in person since the start of the pandemic. By then, it was September 2021, and the entire world, in myriad ways and to different degrees, had experienced a disruption of its usual temporalities. Like Tricia Hersey (of the Nap Ministry), Sofía was one of the many people who did not want to see things go "back to normal." Wasn't there a lesson to be learned from the experience of quarantine, of interruption? Hadn't something grown in this moment of doubt, if only grown unsettled?

Sofía Córdova, *Underwater Moonlight (days of blood + milk)*, 2019

If your priority is speed and a need to remain ahead of the curve, doubt can look only like a cost, like those "delays" and "interruptions" of the inevitable and taken-for-granted that Daniel Hartley identified in "Anthropocene, Capitalocene, and the Problem of Culture." But for anyone for whom progress just feels like the road to certain death, doubt is a lifeline, a little space for agency bursting through Bergson's "crust," a strangely un-collapsible *kairos*. Simply as an opening, it can contain the seed of the "non-time" that Hannah Arendt identifies in *Between Past and Future:*

> It may well be the region of the spirit or, rather, the path paved by thinking, this small track of non-time which the activity of thought beats within the time-space of mortal men and into which the trains of thought, of remembrance and anticipation, save whatever they touch from the ruin of historical and biographical time. This small non-time-space in the very heart of time, unlike the world and the culture into which we are born, can only be indicated, but cannot be inherited and handed down from the past; each new generation, indeed every new human being as he inserts himself between an infinite past and an infinite future, must discover and ploddingly pave it anew.

This passage occurs in the preface, where Arendt describes the moment after the unexpected fall of France to the Nazis in 1940. European writers and intellectuals—"they who as a matter of course had never participated in the official business of the Third Republic"—were suddenly "sucked into politics as though with the force of a vacuum," into a world where word and deed were inseparable. Arendt argues that this created a public intellectual realm that, only a handful of years later, would collapse as they all went back to their private careers. Yet those who participated remembered a "treasure," one where "he who 'joined the Resistance, *found* himself.'" The treasure was nothing less than a sense of traction interrupting a life of petty careerism, a time in which one's actions meant something different—meant something at all. Arendt writes that, in those years, these writers and intellectuals were "visited for the first time in their lives by an apparition of freedom . . . because they had become 'challengers,' had taken the initiative upon themselves and therefore, without knowing or even noticing it, had begun to create that public space between themselves where freedom could appear."

Arendt distinguished the "activity of thought" that can happen in this non-time from more programmatic ways of thinking that are closer to habit, deductive and inductive reasoning where "rules . . . can be learned once and for all and then need only to be applied." Instead, what Arendt had in mind was like Oli Mould's creativity in that it made something *new* through the dialogue of free agents. It is also similar to Selma James and Mariarosa Dalla Costa's description of what happened when formerly isolated housewives began to talk, study, and organize. "In the sociality of struggle," they wrote, "women discover and exercise a power that effectively gives them a new identity." Self-described activists or not, many people know the feeling of going "off script" in any number of ways, where you feel you are truly making something new with others. Even if in seemingly small ways and only very briefly, you may find that you are bringing into existence a new territory of ideas, language, and actions that could never have been predicted, even by you. Exhilarating as those moments can be, they are also full of the discomfort of leaving behind the familiar. Full of doubt.

In this context, doubt is actually the valuable thing, the thing we want to seize. But encountering freshness and agency in this way, Arendt wrote, required someone to hold their ground "between the clashing waves of past and future." Otherwise, you'd get flattened by certainty: The past would crush you with tradition, and the future would crush you with determinism. Hence the importance and fragility of the "gap" (another term for "non-time") in the title of Arendt's preface, "The Gap between Past and Future."

To live in the gap between past and future is quite simply the human condition, even if culturally dominant and politically convenient views of time, history, and the future obscure it from us. Looking mournfully to the future in which something new can never happen, we can't see ourselves standing in the gap, the only place where anything new is capable of happening. It makes me wonder if one meaning of "having time" is to *halve time*—to make a cut in *chronos* and hold the past and the future apart as much as hope will allow.*

EVERY PIECE OF writing is a time capsule. It assembles fragments of its own world and sends them onward to a reader who exists in a different one, not just in space but also in time. Even writing privately in a journal presupposes a future self who will be reading it—and a future at all. In the case of this book, I cannot know what has happened between the time I am writing this and the time in which you are encountering it. But I can tell you that I am living in a moment of doubt. Perhaps you are, too.

That evening when I saw the indistinguishable figure, I had been headed to the place where the road ends, a designated "natural area" called Raab's Lagoon. There, after the pavement turns to grass, you pass under alder and fir trees and come to a bench dedicated to a man who died in 2016. If you keep going, the pathway juts out into the water, part of an artificial barrier between the waters of Quartermaster Harbor and the smaller lagoon. Across a small breach through which the harbor water flows, the barrier continues on until it hits the other side of the lagoon. The first time I visited, the water in the breach didn't seem to be moving in any particular direction. It was high tide, though I didn't know that at the time; having just arrived, I thought the area always looked like that.

Over the course of a few weeks, I inevitably became familiar with the tides, because Quartermaster Harbor was right outside the door of my room. When the tide was high, you could hear the water plopping and the plastic canoe docks banging against the wooden posts, something I started to call "the song of the dock." When the tide was lower, white-winged scoters, migratory diving ducks with a surreal flourish of a white feather under their eyes, would appear in loose flocks and dive for the mussels at the bottom. When the tide was all the way out, the

---

* Likewise, Vine Deloria, Jr., compared changes in culture and civilization to a mosaic in which you can recognize neither the old pattern nor the new one coming into being. For him, too, this was a fragile state; if we couldn't navigate a "dreadful middle ground . . . in which we substitute the same meaningless pieces over and over again," we risked "lapsing into a new and more sophisticated barbarism."

mussel shells were maximally revealed, and both people and glaucous-winged gulls would walk by on the exposed rocky beach.

I realized I actually knew nothing about tides. Some of my days of googling resembled an elementary school science module, where I learned that there were "high high tides" and "low high tides," and that the highest tides—the king tides that I describe at the beginning of this conclusion—happened during a new or full moon when the moon was in perigee (closest to Earth) and Earth was in perihelion (closest to the sun).* I learned that there were such things as "earth tides," where those same forces actually moved solid land a little bit. I found out that while the moon pulls on our water, the water pulls back, speeding the orbit of the moon and causing it to drift away from us. I studied the local tide chart, where the curves had a periodicity and logic of their own but existed out of sync with the calendar boxes and hour markings of the grids in which they were shown. Several nights, the moon came out full and bright and clear, a reminder.

One day, I happened to go to Raab's Lagoon at a medium-low tide that was rising. By that time, I'd found out the reason for the barrier: It had been part of a road whose route I'd unwittingly been walking and that formed the entire shape of the tiny park. The road had originally gone to a nearby sawmill, and it had passed over the lagoon on a causeway with a bridge in the middle. In that same lagoon where I now saw only geese and the occasional heron, they used to store logs. Then, in the 1950s, the bridge was burned to allow boats in from the harbor, at least at high tide. With that, the old road was interrupted; now I was standing on the grassy banks, looking down into non-road.

Except at high tide, the leftover narrows exaggerated the motion of the water, becoming an indicator that acted like an arrow. When the tide was low, water flowed southward, out of the lagoon and back into the harbor. It stayed that way until the tide came back up to six feet, when—at some exact moment I often tried to catch but never could—

---

* For a very helpful and intuitive animation of this, see the Exploratorium's video, "King Tides | Full Spectrum Science—Shorts | Ron Hipschman" on YouTube. Incidentally, the video contains yet another great Pacifica photo from Alan Grinberg.

the water reversed direction and began to flow from the harbor into the lagoon. It would start to rush, something locals called the "Vashon rapids." Then this rushing would slow as the water levels became even, and the whole process would begin again when the tide went back down. Being from the Bay Area, I hear rushing water and think of rainfall, as it comes down a mountain through creeks and springs. But here in the harbor, the rushing back and forth was a sign of gravity, a message about the position of things in outer space.

That day in the breach, the increase of water was flowing into the lagoon through a series of tide pools right in the middle. When a tiny jet of water shot up from the exposed ground nearby, I clambered down over some concrete and wooden debris to investigate. I crouched down only to have one of the "pebbles," like one of the eyeball-bearing plants in *Labyrinth*, open up wide and shoot water into my unsuspecting face.

It was the siphon portion of a rough piddock clam that was buried in the mud. Between these little jets, the whole ground seethed and hissed as air bubbles escaped the former and soon-to-be lagoon floor.

Under the northward-flowing water and tucked partway under a rock was an improbably large, purple lump of something studded with white, as though it'd been dusted with confectioner's sugar. It was a purple ochre sea star, one of the many species of sea star threatened by sea star wasting syndrome. Since 2013, this nightmarish disease had been causing sea stars to disintegrate all along the western Pacific coasts and even in aquarium tanks; in the photos, they just sort of fall apart and melt away. Within what has been called the largest disease epidemic ever observed in wild marine animals, ochre sea stars have been one of the hardest-hit species. The disease was even documented at this lagoon in the past. And yet, the sea star I saw seemed healthy. It was busy circulating seawater through its body and eating mussels, as the water level very slowly rose around it.

Sea star wasting syndrome is still not totally understood, though the virus that causes it has been present in sea stars for a long time. Its new destructiveness seems to have something to do with an increase in the temperature of the water, which might stress the sea stars and make them more vulnerable. There's no mystery, of course, to what's causing that warming. About the healthy sea stars in Raab's Lagoon, the director of the island's nature center speculated that, for now, the cooler, moving waters of the lagoon might be favoring them. Or perhaps the sea star I was looking at had inherited the resistance to the syndrome that scientists documented in a 2018 study. Inspiring as it was, one of the study's authors cautioned that resistance was "a small, distant bright light on a pretty stormy sea." And more outbreaks had been observed in Puget Sound a month before I arrived.

Knowing all this made the three-dimensional, bodily presence of the ochre sea star feel like a small miracle, even more of a miracle than a sea star already is. I could not view it apart from its potential disappearance. Writing from this gap between the past and future, I have to acknowledge the very real possibility that this animal—like so many things—is rare or gone in your world. At the same time, I can't take that outcome for granted, because if I do, there is less of a chance that you will ever see one.

That's the irony of determinism: It involves something of a choice. In another of Ted Chiang's stories, a narrator from the future decides to warn the past of a coming technology called the Predictor, a device that uses a "negative time delay" to show a flash right as you move to hit the button to cause the flash. The Predictor can't be outwitted. As a demonstration that free will does not exist, it eventually causes its own kind of wasting syndrome, in which people lose all motivation and live in a "waking coma." When doctors try to argue with them, pointing out that "no action you took last month was any more freely chosen than one you take today," the patients reply, " 'But now I know.' And some of them never say anything again."

Throughout the story, the narrator—who, it turns out, is using a negative time delay to send the message—says he knows that free will does not exist. But in the actual message he is trying to relay, he contradicts himself: "Pretend that you have free will," he exhorts the denizens of the past. "It's essential that you behave as if your decisions matter, even though you know that they don't. The reality isn't important: what's important is your belief, and believing the lie is the only way to avoid a waking coma. Civilization now depends on self-deception. Perhaps it always has." He recognizes, too, that in some ways his message makes no sense: "There's nothing anyone can do about it—you can't choose the effect the Predictor has on you. Some of you will succumb and some of you won't, and my sending this warning won't alter those proportions. So why did I do it?" His answer is a paradox: "Because I had no choice."

Chiang's story shows the inextricability of time, will, aliveness, and desire. The narrator's "having no choice" at the end is ambiguous, but one way to read it is as Bergson's "deep-seated self rushing up to the surface," which happens against all logic and odds. To want something at all, to love something and fear its disappearance, is to dwell in that gap between past and future, allowing that "sudden boiling over of feelings and ideas." That day, in the gap in the old road, I saw a living ochre sea star that was not wasting away. It made me desperate to imagine a future with ochre sea stars in it.

By saying this, I am talking about much more than an individual animal. When the zoologist Robert T. Paine introduced the concept of a "keystone species" in 1969, it was based on observations of the effects of ochre sea stars in an intertidal environment. Their role in eating mussels, clearing a certain amount of space at a certain height on the rocks, was so important that the biodiversity of that entire intertidal ecosystem could collapse without them, the effects radiating outward to other ecosystems. The years since sea star wasting syndrome took hold have been an uninvited extension of Paine's original experiment, where he removed the ochre sea stars to see what would happen. As a recognition of interconnectedness, the keystone species concept represents the life-or-(collective)-death stakes of how we view the bounds of a person, place, or time.

Like me, like a rock, sea stars bear the traces of things that happened in time, both near and far. For a long time, scientists were unsure of how sea stars evolved their arms, because they seemed to arrive "fully formed" in the fossil record. Only in 2003 did a Moroccan research team find a possible missing link in the Fezouata Formation, a "Pompeii of paleontology" where even the soft body of a sea star might be preserved. This ancestor was called *Cantabrigiaster fezouataensis,* and it was the oldest starfish-like animal in the fossil record. Then, this past year, researchers from Harvard University and the University of Cambridge noticed that *Cantabrigiaster* also had features in common with sea lilies, flower-shaped filter feeders whose "stems" attach to the seafloor and whose "petals" catch plankton particles from the water.

At some point around the great Ordovician Biodiversification Event, a window in which conditions in certain places caused an explosion of diversity, *Cantabrigiaster* may have done something weird and unexpected: It switched its orientation. "The five arms of starfish are a relic left over from these [sea lily] ancestors," one of the researchers wrote. "In the case of *Cantabrigiaster,* and its starfish descendants, it evolved by flipping upside-down so its arms are face down on the sediment to feed." Now, the sea star in front of me was hugging its rock, either right side up or upside down, depending on which temporal frame you took. Not that the sea star would much care. It was in sea star time-space, perhaps sensing a dark blob (me) with the compound eyes on the ends of its arms.

When I looked up, the tide had come in farther, the water edging closer to the ground I was standing on. Soon I would climb back up to the old road, from which I could see the rest of the harbor, ringed by fiery yellow leaves. The local newspaper had reported that the Coast Salish S'Homamish had named this area Tutcila'wi after their fellow residents: those trees now called bigleaf maple trees, much more frequent here than where I'm from. Since I'd arrived, the trees had changed their color at a rate whose difference was noticeable day by day and that seemed to be speeding up.

Down below, the water had begun to rush over the rocks. Though I was standing still, I felt I was rushing, too, parts of me dying and other parts coming to life. There were annual growth rings on the clam shells; they reminded me of the lines on my forehead that appeared during the pandemic. Apparently, it was a time when many people aged faster, a collective compression of our biological clocks.

As the water rose, I knew I could not stand there and watch the sea star forever, but I did for as long as I could. What I felt in that meantime was not exactly joy, but neither was it despair. It was something tidal, an oscillation back and forth, impossible to pin down and yet possibly legible to those around me: to the ducks, who would migrate again; to the trees, which would turn green again; to the mussels, who would be submerged again; to the water, which would flow back out again. Nor did my body misunderstand. In the center of me a muscle was beating, a series of creation events ongoing for now that I hadn't started and wouldn't stop. Under the rush of the water, I felt my heartbeats as words. They were saying what they always had: *Again. Again. Again.*

Acknowledgments

This book was written on unceded Lisjan (Ohlone) territory. I encourage readers to learn about the Sogorea Te' Land Trust (sogoreate-landtrust .org) and to seek out iterations of the Landback movement in their own surroundings.

This project has been bolstered by the care of others from the very beginning. I knew it would be safe in the hands of my agent, Caroline Eisenmann, even when it was just a fragile outline. The Vashon Artist Residency and Little Joshua Tree granted me the time and space to think about time and space. My editor, Hilary Redmon, clearly shared my enthusiasm for the big questions and was willing to follow them wherever they went. I appreciate her patience, precision, and ability to render my language more accessible. Thanks to everyone at Random House for taking the time, to Jia Tolentino for the advice, and to Dan Greene for the Herculean task of fact-checking.

I am deeply indebted to Rick and Megan Prelinger of the Prelinger Library (which we visit in chapter 6), not just for guidance but for the warmest encouragement one could ask for. Prelinger librarian Devin Smith kept my project in mind and dug out treasures like Laird's *Increasing Personal Efficiency*. If what I believe is true—that the past is trying to speak to the present in the form of unanswered questions and unredeemed desires—then archival projects like the Prelinger Library are some of the only conduits through which this can happen. The library is truly a gift that keeps on giving, a place full of seeds (beans?) waiting to change the future.

I presented some very early thoughts on this book at Stanford's Digital Aesthetics Workshop in 2019. Thanks to Shane Denson for inviting me, and to the workshop members for generously opening up new paths of inquiry. While writing this, I also had the good luck to be in conversation with Rebecca Solnit, Jess Nordell, Helen Macdonald, and Angela Garbes about their new books. And in the spaces in between, I was kept afloat by an epic thread of emails with R. O. Kwon, Indira Allegra, Bahar Behbahani, Ingrid Rojas Contreras, Rachel Khong, Raven Leilani, Antoinette Nwandu, and Camille Rankine.

I thank the artists and writers who have allowed me to include their work here, and the folks who shared their experiences and expertise with me, an interdisciplinary interloper. Some of the biggest contributions, though, can't be contained in the notes. When I look at this book, I see the jeweler's loupe that my friend Joshua Batson gave me to look at leaves, and the East Bay trail that we went loupe-ing on. I see the patch of redwoods where I sat with my friend and teacher for life, John Shoptaw, who knows how to use poetry to dissolve the clock. And I see the migratory birds in Lake Merritt that my fellow "birdnoticer" Joe Winer was also noticing on his side of the bay. Reflected here are so many other exchanges with friends and thinkers in common: Helen Shewolfe Tseng, Laura Hyunjhee Kim, Raenelle Tauro, Cara Rose DeFabio, Neeraj Bhatia, Christina Corfield, Cat Ferguson, Gary Mao, and Ansh Shukla.

Above all, I see the hundreds of conversations I had with my partner, Joe Veix, on our daily pandemic walks, when time was strange and nothing seemed given. Joe lived and thought through every question with me, turning it over and inside out until something new emerged. And if writing on hard topics ever felt like "drinking poison," it was his ability to make me laugh under any circumstance that was so often the antidote.

To my parents: Thank you for the gift of trust, unexamined time, and a safe haven for curiosity growing up. As a respite care provider for foster families, my mom continues to show me the relationship between love and listening. I also thank my dad for encouraging the philosophical questions that have continued into this book (even if fourteen was too young to be given Plato's *Republic*).

Lastly, so much of what I have tried to synthesize here came directly from teachers, human and nonhuman, in my neighborhood and in the Santa Cruz Mountains. Thanks to Devora for the lettuce, Tom for the "soft eyes," the bay trees for encoding my memories, the rocks for making time real, and the birds for modeling hope when I couldn't see it. I'm grateful for the enlivenment.

Notes

INTRODUCTION: A MESSAGE FOR THE MEANTIME

xi **Compared to vascular plants** Robin Wall Kimmerer, *Gathering Moss: A Natural and Cultural History of Mosses* (Corvallis: Oregon State University Press, 2003), 97.

xi **On Antarctica** Wynne Parry, "Antarctic Mosses Record Conditions on the Icy Continent," Live Science, December 30, 2011, livescience.com/17686-antarctic -mosses-climate-change.html.

xii **disagreement about when a moss** J. M. Glime, *Bryophyte Ecology* (Houghton: Michigan Technological University, 2022), vol. 1, chap. 5-2, 4. Glime writes that "there is no general agreement on the definition of spore germination," and that furthermore, some species exhibit an intermediate phase between swelling and distension, called the protrusion phase.

xii **the earliest mosses evolved** Kimmerer, *Gathering Moss*, 23.

xii **hundreds of millions of years** In *Gathering Moss*, written in 2003, Kimmerer puts this development at 350 million years ago (23). A 2018 study of new evidence suggested that land plants may have arisen 400 to 500 million years ago. Elizabeth Pennisi, "Land plants arose earlier than thought—and may have had a bigger impact on the evolution of animals," *Science*, February 19, 2018, science .org/content/article/land-plants-arose-earlier-thought-and-may-have-had -bigger-impact-evolution-animals.

xii **a moss that was frozen** Becky Oskin, "1,500-Year-Old Antarctic Moss Brought Back to Life," *Scientific American*, March 17, 2014, scientificamerican.com/article /1500-year-old-antarctic-moss-brought-back-to-life/.

xii **ability of some species** Glime, *Bryophyte Ecology*, vol. 1, chap. 7-3, 2–3. Glime cites a 1982 observation of an *Anoectangium compactum* that revived after nineteen years. See also Kimmerer, *Gathering Moss*, 35–43.

xii **It was this very quality** Janice Lee, "An Interview with Robin Wall Kimmerer," *The Believer*, November 3, 2020, culture.org/an-interview-with-robin -wall-kimmerer/.

xiii **live in that spot** Svante Björk et al., "Stratigraphic and Paleoclimatic Studies of a 5500-Year-Old Moss Bank on Elephant Island, Antarctica," *Arctic and Alpine Research* 23, no. 4 (November 1991): 361.

xiii **time is horizontal** Josef Pieper, *Leisure, the Basis of Culture* (San Francisco: Ignatius, 2015), 49–50. This book was first published in German in 1948, with an English translation in 1952.

xiv **employer and employee could both benefit** Carl Honoré, *In Praise of Slowness: Challenging the Cult of Speed* (New York: HarperOne, 2009), chap. 8.

xiv **Marge gets a job** *The Simpsons,* season 4, episode 7, "Marge Gets a Job," directed by Jeff Lynch, aired November 5, 1992, on Fox.

xvi **"Woman Waiting for Evidence"** Justine Jung, "Woman Waiting for Evidence That World Will Still Exist in 2050 Before She Starts Working Toward Goals," Reductress, March 23, 2022, reductress.com/post/woman-waiting-for-evidence-that-world-will-still-exist-in-2050-before-she-starts-working-toward-goals/.

xvi **"the clock can tell me"** Michelle Bastian, "Fatally Confused: Telling the Time in the Midst of Ecological Crises," *Environmental Philosophy* 9, no. 1 (2012): 25.

xvi **In the body** I learned this from my friend Joshua Batson as he was completing a course on anatomy and massage.

xvii **Obviously self-care was necessary** Minna Salami, at Is There Time for Self-Care in a Climate Emergency? (online event), Climate Emergence—Emotional and Ecological Wellbeing Strategies, July 12, 2021, climateemergence.co.uk /time-for-selfcare-event-recording.

xviii **"marching in lockstep toward the abyss"** Louis Michaelson, letter from the editors, *Processed World* 5 (July 1982): 8.

xviii **"The smallest act"** Hannah Arendt, *The Human Condition* (Chicago: University of Chicago Press, 1998), 190.

xviii **this distinction relegated colonized people** Giordano Nanni, *The Colonisation of Time: Ritual, Routine and Resistance in the British Empire* (Manchester, UK: Manchester University Press, 2012), 10.

xix **In 2021, Tropicfeel** Jack Morris, "Say Yes to Climbing an Active Volcano!" YouTube video, July 1, 2021, youtube.com/watch?v=OpUb_k_LP98.

xx **What Morris has seen** Megan Lane, "Sulphur Mining in an Active Volcano," BBC News, February 9, 2011, bbc.com/news/world-asia-pacific-12301421. Lane refers to the filming of sulfur mining for the BBC series *Human Planet:* "During filming, the BBC crew was enveloped in a toxic cloud 40 times the UK's safe breathing level (no limits exist for the miners). Corrosive air-borne particles ate into the cameras, which promptly broke down." Mari LeGagnoux, "Sulfur Mining in Indonesia," *Borgen Magazine,* July 7, 2014, borgenmagazine.com /sulfur-mining-indonesia/; Martha Henriques, "The Men Who Mine the 'Devil's Gold,'" BBC Future, February 21, 2019, bbc.com/future/article/20190109 -sulphur-mining-at-kawah-ijen-volcano-in-indonesia; Coburn Dukehart, "The Struggle and Strain of Mining 'Devil's Gold,'" *National Geographic,* November 16, 2015, nationalgeographic.com/photography/article/the-struggle-and -strain-of-mining-devils-gold; *Where Heaven Meets Hell,* directed by Sasha Friedlander (ITVS, 2013), pbs.org/video/global-voices-where-heaven-meets-hell/; Andrew Nunes, "Stark Photos Document the Dangerous Lives of Ijen's Sulfur Miners," *Vice,* May 21, 2017, vice.com/en/article/vb48y4/stark-photos -document-the-dangerous-lives-of-ijens-sulfur-miners.

xxi **The man trying to sleep** Abby Narishkin and Mark Adam Miller, "VIDEO: Why Miners in Indonesia Risk Their Lives to Get Sulfur from Inside an Active Volcano," *Business Insider,* January 21, 2022, businessinsider.com/sulfur-miners -active-volcano-indonesia-dangerous-jobs-2022-1. This video depicts a day's work for Mistar, a sulfur miner at Ijen, and also shows the mining company's associated sulfur-refining plant.

xxi **the tools that we would now** Caitlin Rosenthal, *Accounting for Slavery: Masters and Management* (Cambridge, Mass.: Harvard University Press, 2018).

xxi **Most of the sulfur** Ivan Watson et al., "Volcano Mining: The Toughest Job in the World?" CNN, July 7, 2016.

xxii **This phenomenon, in which one** Allen C. Bluedorn, *The Human Organization of Time: Temporal Realities and Experience* (Stanford, Calif.: Stanford University Press, 2002), 147–49; Sarah Sharma, "Speed Traps and the Temporal: Of Taxis, Truck Stops and TaskRabbits," in *The Sociology of Speed: Digital, Organizational, and Social Temporalities,* eds. Judy Wajcman and Nigel Dodd (Oxford, UK: Oxford University Press, 2017), 133.

xxii ***experience economy*** B. Joseph Pine II and James H. Gilmore, "Welcome to the Experience Economy," *Harvard Business Review* (July–August 1998), 97–105.

xxii **On the Tropicfeel website** Jack Morris, "Volcano Adventure in Ijen," Tropic-feel, tropicfeel.com/journeys/volcano-adventure-in-ijen/.

xxii **Its story began** Corentin Caudron et al., "Kawah Ijen Volcanic Activity: A Review," *Bulletin of Volcanology* 77, no. 16 (2015): 15–16, link.springer.com/article/10.1007/s00445-014-0885-8; H. K. Handley et al., "Constraining Fluid and Sediment Contributions to Subduction-Related Magmatism in Indonesia: Ijen Volcanic Complex," *Journal of Petrology* 48, no. 6 (2007): 1155, academic.oup.com/petrology/article/48/6/1155/1564285; Hobart M. King, "Kawah Ijen Volcano," Geology.com, geology.com/volcanoes/kawah-ijen/; Brian Clark Howard, "Stunning Electric-Blue Flames Erupt from Volcanoes," *National Geographic,* January 30, 2014, nationalgeographic.com/science/article/140130-kawah-ijen-blue-flame-volcanoes-sulfur-indonesia-pictures.

xxiii **"I believe that we are"** William McKibben, "The End of Nature: The Rise of Greenhouse Gases and Our Warming Earth," *The New Yorker,* September 11, 1989, newyorker.com/magazine/1989/09/11/the-end-of-nature.

xxiii **"How can I translate"** Natalie Diaz, "The First Water Is the Body," in *New Poets of Native Nations,* ed. Heid E. Erdrich (Minneapolis, Minn.: Graywolf Press, 2018), 101.

xxiv **a track by Daniel Deuschle** "Daniel Deuschle," Musicbed.com, musicbed.com/artists/daniel-deuschle/43856.

xxiv **When Instagram was in its early** Charlotte Cowles, "'We Built a House with Our Influencer Money,'" *The Cut,* November 15, 2019, thecut.com/2019/11/travel-influencers-built-a-house-with-their-instagram-money.html; "Travel Blogger Couple Lauren Bullen and Jack Morris Split," News.com.au, April 8, 2021, news.com.au/travel/travel-updates/travel-stories/travel-blogger-couple-lauren-bullen-and-jack-morris-split/news-story/e08140753d5e89612887ca742d442e4a.

xxiv **"For more than a year"** Jack Morris (@jackmorris), "Touch down in Egypt!" Instagram, April 7, 2021, instagram.com/p/CNXG7bjhFi5/.

xxv **"acquisitive mood"** Susan Sontag, *On Photography* (New York: Farrar, Straus and Giroux, 2011), 4.

xxvi **Kamaʻehuakanaloa was formerly known** Bobby Camara, "A Change of Name," *Ka Wai Ola,* October 1, 2021, kawaiola.news/aina/a-change-of-name/.

xxvii **explored by social theorists** Some social theorists who have explored the relationship of agency to structure include Anthony Giddens (in *The Constitution of Society: Outline of the Theory of Structuration*) and Margaret Archer (in *Realist*

Social Theory: The Morphogenetic Approach). For example, Archer writes in *Realist Social Theory*, "Society is that which nobody wants, in the form in which they encounter it, for it is an unintended consequence. Its constitution could be expressed as a riddle: what is it that depends on human intentionality but never conforms to their intentions?" (165).

xxvii **individual and institutional biases** Jessica Nordell, *The End of Bias: A Beginning* (New York: Metropolitan, 2021), 111, 250.

xxviii **importance of work-time flexibility** Robert E. Goodin et al., *Discretionary Time: A New Measure of Freedom* (Cambridge, UK: Cambridge University Press, 2008), 390–93.

xxviii **the American Dream exploits our fears** Mia Birdsong, *How We Show Up: Reclaiming Family, Friendship, and Community* (New York: Hachette, 2020), chap. 1.

xxix **You can see this work** As an example of this tension, Annie Lowrey recognized in *The Atlantic* that mutual aid during the Covid-19 pandemic might "point to a better way of envisioning community," while Joanna Wuest worried in *The Nation* that reliance on and romanticization of mutual aid would mean an abandonment of efforts for more structural change. This would simply uphold what Wuest called a "libertarian fantasy, with only atomized acts of compassion for those left out." Annie Lowrey, "The Americans Who Knitted Their Own Safety Net," *The Atlantic*, March 24, 2021, theatlantic.com /ideas/archive/2021/03/americans-who-knitted-their-own-safety-net/618377/; Joanna Wuest, "Mutual Aid Can't Do It Alone," *The Nation*, December 16, 2020, thenation.com/article/society/mutual-aid-pandemic-covid/.

xxix **"a panoramic assault"** David Hockney, *That's the Way I See It* (San Francisco: Chronicle Books, 1993), 112.

xxix **her own fear regarding the future** Naomi Klein, *This Changes Everything: Capitalism vs. the Climate* (New York: Simon and Schuster, 2014), 465.

xxx **"the slow time of COVID"** Herman Gray, "The Fire This Time," at Race at Boiling Point (online event), University of California Humanities Research Institute, June 5, 2020, youtube.com/watch?v=3I22E2Sezi8.

xxx **the pandemic had invited some amount** Mia Birdsong, interview with Carrie Fox and Natalie S. Burke, *Mission Forward*, podcast audio, July 6, 2021, trustory .fm/mission-forward/mf307/.

CHAPTER 1: WHOSE TIME, WHOSE MONEY?

3 **"Time to me is about life-span"** Barbara Adam, *Timewatch: The Social Analysis of Time* (Cambridge, UK: Polity, 1995), coda.

3 **"Moments are the elements of profit"** Karl Marx, *Capital* (New York: Penguin, 1990), 1:352.

4 **In July 1998** Quoted in Mario Macrì et al., "Clocking and Scientific Research: The Opinion of the Scientific Community," November 13, 1998, openaccessrepository .it/record/21217?ln=en.

6 **The very first image** *Modern Times*, directed by Charlie Chaplin (United Artists, 1936), criterionchannel.com/modern-times.

7 **"Time costs you money"** Advertisement for International Time Recording Company of New York, *Factory: The Magazine of Management* 16–17 (February 1916): 194.

7 **"You pay them CASH!"** Advertisement for Calculagraph, *Industrial Management*, August 1927, 65.

7 **deductions from time paid** Quoted in E. P. Thompson, "Time, Work-Discipline, and Industrial Capitalism," *Past and Present* 38 (December 1967): 81.

8 **"Working in an Amazon warehouse"** Emily Guendelsberger, *On the Clock: What Low-Wage Work Did to Me and How It Drives America Insane* (New York: Little, Brown, 2019), 11.

8 **"My scanner gun"** Guendelsberger, *On the Clock,* 79.

9 **Installed on workers' computers** Sara Morrison, "Just Because You're Working from Home Doesn't Mean Your Boss Isn't Watching You," *Vox,* April 2, 2020, vox.com/recode/2020/4/2/21195584/coronavirus-remote-work-from-home-employee-monitoring; Aaron Holmes, "'Bossware' Companies That Track Workers Say the Tech Is Booming and Here to Stay—but Employees and Privacy Advocates Are Ringing Alarm Bells," *Business Insider,* June 20, 2021.

9 **"Get the Most out of Your Employees' Time"** Workpuls, "Employee Monitoring Software," Insightful (Workpuls), workpuls.com/employee-monitoring.

9 **a contractor for a translation agency** Quoted in Morrison, "Just Because You're Working from Home."

9 **the systems' features promote productivity** Gadjo Sevilla, "The Best Employee Tracking Software for 2020," *PC Mag,* October 29, 2020 (since updated). See original at web.archive.org/web/20201103091025/pcmag.com/picks/the-best-employee-monitoring-software.

9 **"Just because the employees know"** Workpuls, "Employee Monitoring Software."

9 **StaffCop shows the employer** "Filters," Staffcop Enterprise, 4.10 User Manual, docs.staffcop.ru/en/work_with_data/filters.html; StaffCop, Employee Monitoring Software, "StaffCop Enterprise: Time Tracking Report," YouTube video, February 20, 2019, youtube.com/watch?v=2uh7-wO3D_k&t=61s.

10 **After unveiling the Productivity Score** Jared Spataro, "Our commitment to privacy in Microsoft Productivity Score," Microsoft 365, December 1, 2020, microsoft.com/en-us/microsoft-365/blog/2020/12/01/our-commitment-to-privacy-in-microsoft-productivity-score/.

10 **"The Shitty Tech Adoption Curve"** Cory Doctorow, Twitter thread, November 25,2020, twitter.com/doctorow/status/1331633102762831873.

10 *the containers are all the same* Barnaby Lewis, "Boxing Clever—How Standardization Built a Global Economy," International Organization for Standardization, September 11, 2017, iso.org/news/ref2215.html; Craig Martin, "The Shipping Container," *The Atlantic,* December 2, 2013, theatlantic.com/technology/archive/2013/12/the-shipping-container/281888/. For a history of containerization specifically in the Oakland context, see episode 1 of Alexis Madrigal's 2017 podcast series, *Containers,* podcasts.apple.com/us/podcast/containers/id1209559177.

11 **fungible time** Bluedorn, *The Human Organization of Time,* 28. Bluedorn contrasts fungible time with epochal time: unique events or durations.

11 **"nothing more than personified labour-time"** Marx, *Capital,* 1:352–53.

12 **Understanding the invention** David Landes, *Revolution in Time: Clocks and the Making of the Modern World* (Cambridge, Mass.: Belknap Press/Harvard University Press, 2000), 13.

12 **"simply intricate oddities"** Landes, *Revolution in Time,* 39–40. While Landes goes on to say that this is "clearly a defensive statement" in what was seen as

a game of technological one-upmanship—the Europeans similarly disdained Chinese clocks—it is still true that the clock may have appeared uniquely irrelevant. Before this statement, Landes quotes Carlo Cipolla: "Foreign machinery could not be properly appreciated because it was not the expression of a Chinese response to the problems set by a Chinese environment" (39).

12 **a crucial deviation** Landes, *Revolution in Time,* 54–56.

12 **The Rule, which subsequently spread** Columba Stewart, "Prayer Among the Benedictines," in *A History of Prayer: The First to the Fifteenth Century,* ed. Roy Hammerling (Leiden, Netherlands: Brill, 2008), 210.

12 **"idleness is the enemy"** Saint Benedict of Nursia, *Saint Benedict's Rule for Monasteries, or, Rule of Saint Benedict* (Collegeville, Minn: The Liturgical Press, 1948), chap. 48, Project Gutenberg eBook, gutenberg.org/files/50040/50040 -h/50040-h.htm.

12 **punishments include being made to stand** Saint Benedict of Nursia, *Saint Benedict's Rule for Monasteries, or, Rule of Saint Benedict,* chap. 43.

12 **Five centuries later, Cistercian monks** Landes, *Revolution in Time,* 58, 73.

13 **Canonical hours are not equal hours** Landes, *Revolution in Time,* 74–82. Technically, the ancient Greeks used twenty-four equal hours for their theoretical calculations as early as the second century B.C., but laypeople continued to use variable hours. Michael A. Lombardi, "Why Is a Minute Divided into 60 Seconds, an Hour into 60 Minutes, Yet There Are Only 24 Hours in a Day?" *Scientific American,* March 5, 2007, scientificamerican.com/article/experts-time-division -days-hours-minutes.

13 **"an unintended consequence"** Landes, *Revolution in Time,* 82.

13 **"o'clock" means "of the clock"** John Durham Peters, *The Marvelous Clouds: Toward a Philosophy of Elemental Media* (Chicago: University of Chicago Press, 2015), 220.

13 **the marine chronometer** Landes, *Revolution in Time,* 155–66. Landes also relates further developments in chronometer design to competition between the colonial powers of England and France, at a time when colonial trade was growing quickly (167–68).

14 **there was no need to coordinate** Eviatar Zerubavel, "The Standardization of Time: A Sociohistorical Perspective," *American Journal of Sociology* 88, no. 1 (July 1982): 6.

14 **"master clocks"** Nanni, *The Colonisation of Time,* 51–52; see also Jay Griffiths, "The Tyranny of Clocks and Calendars," *The Guardian,* August 28, 1999, theguardian.com/comment/story/0,,266761,00.html.

14 **"There is no 'Standard Railroad Time'"** *Travelers Official Railway Guide of the United States and Canada, June 1868: 100th Anniversary Facsimile Edition* (New York: National Railway Publication Company, 1968), 13.

14 **Sandford Fleming imagined the exact opposite** Sandford Fleming, "Time-Reckoning for the Twentieth Century," in the *Annual Report of the Board of Regents of the Smithsonian Institution, Showing the Operations, Expenditures, and Condition of the Institution for the Year Ending June 30, 1886* (Washington, D.C.: Smithsonian Institution Press, 1889), 350–57.

15 **"The committee is aware"** Fleming, "Time-Reckoning for the Twentieth Century," 355n.

15 **twenty-four international time zones** W. Ellis, "The Prime Meridian Conference," *Nature* 31 (1884): 7–10, nature.com/articles/031007c0.

16 **a standardized approach to time** In addition to the examples given in this chapter, something similar happened closer to home, on Spanish missions in California: "The regulation of the day by the padres into a rigid schedule announced by Church bells and reinforced by punishments for breaking the rules, ran completely counter to the traditions of the Native Americans." Fred Glass, *From Mission to Microchip: A History of the California Labor Movement* (Berkeley: University of California Press, 2016), 42.

16 **"the project to incorporate the globe"** Nanni, *The Colonisation of Time,* 2.

16 **"You must know that today"** Quoted in Nanni, *The Colonisation of Time,* 25.

16 **In some southern African towns** Nanni, *The Colonisation of Time,* 162–65.

16 **in the Philippines and Mexico** Peters, *The Marvelous Clouds,* 228.

16 **A faltering exchange in Coranderrk** Nanni, *The Colonisation of Time,* 101.

17 **colonial missions tried "to induce people"** Nanni, *The Colonisation of Time,* 96.

17 **Puritanism entered a "marriage of convenience"** Thompson, "Time, Work-Discipline, and Industrial Capitalism," 95.

17 **a passage "decidedly lacking in subtlety"** Nanni, *The Colonisation of Time,* 198.

18 *In the 1860s, Chinese railroad workers* Gordon H. Chang, "Op-Ed: Remember the Chinese Immigrants Who Built America's First Transcontinental Railroad," *Los Angeles Times,* May 10, 2019, latimes.com/opinion/op-ed/la-oe-chang-transcontinental-railroad-anniversary-chinese-workers-20190510-story.html.

18 *the brutal winter of 1866/67* "Tunneling in the Sierra Nevada," *American Experience,* PBS, pbs.org/wgbh/americanexperience/features/tcrr-tunneling-sierra-nevada/; Abby Stevens, "Dynamite, Snow Storms, and a Ticking Clock," *Moonshine Ink,* January 14, 2017, moonshineink.com/tahoe-news/dynamite-snow-storms-and-a-ticking-clock/.

19 *charging them for room and board* "Workers of the Central and Union Pacific Railroad," *American Experience,* PBS, pbs.org/wgbh/americanexperience/features/tcrr-workers-central-union-pacific-railroad/. Irish workers were provided with free room and board.

19 *In June 1867* Chang, "Op-Ed: Remember the Chinese Immigrants."

19 *cutting off their food supply* "The Chinese Workers' Strike," *American Experience,* PBS, pbs.org/wgbh/americanexperience/features/tcrr-chinese-workers-strike/; Nadja Sayej, "'Forgotten by Society': How Chinese Migrants Built the Transcontinental Railroad," *The Guardian,* July 18, 2019, theguardian.com/artanddesign/2019/jul/18/forgotten-by-society-how-chinese-migrants-built-the-transcontinental-railroad.

19 *it did later quietly raise* "Railroad—Chinese Labor Strike, June 24th, 1867," Museum of Chinese in America (MOCA), June 24, 2019, mocanyc.org/collections/stories/railroad-chinese-labor-strike-june-24th-1867/.

19 **Capitalist practices also had roots** Harry Braverman, *Labor and Monopoly Capital: The Degradation of Work in the Twentieth Century* (New York: Monthly Review, 1998), 44–45. See the footnote in which Braverman quotes a letter from Marx to Engels: "In general, the army is important for economic development. For instance, it was in the army that the ancients first fully developed a wage system. . . . The division of labour *within* one branch was also first carried out in the armies."

19 **"Before inventors created engines"** Lewis Mumford, *Technics and Civilization* (London: G. Routledge, 1934), 41.

19 **While the systematic management** Caitlin Rosenthal, *Accounting for Slavery: Masters and Management* (Cambridge, Mass.: Harvard University Press, 2018), introduction.

20 **the Barbadian Society** Justin Roberts, *Slavery and the Enlightenment in the British Atlantic, 1750–1807* (Cambridge, UK: Cambridge University Press, 2013), 74.

20 **slaves should "[do] as much"** Letter from George Washington to John Fairfax, January 1, 1789, founders.archives.gov/documents/Washington/05-01-02 -0160.

20 **"Four good fellows"** Quoted in Mark M. Smith, *Mastered by the Clock: Time, Slavery, and Freedom in the American South* (Chapel Hill: University of North Carolina Press, 2000), chap. 3 endnotes.

20 **Plantation accounting systems** Roberts, *Slavery and the Enlightenment*, 76.

20 **some West Indian sugar planters** Roberts, *Slavery and the Enlightenment*, 125–26.

20 **When clocks arrived on the plantation** Smith, *Mastered by the Clock*, introduction.

21 **"[enslaved people] could not quit"** Rosenthal, *Accounting for Slavery*, introduction.

21 **In early-nineteenth-century America** Amy Dru Stanley, *From Bondage to Contract: Wage Labor, Marriage, and the Market in the Age of Slave Emancipation* (Cambridge, UK: Cambridge University Press, 1998), 62.

21 **it was compared to prostitution** Lawrence B. Glickman, *A Living Wage: American Workers and the Making of Consumer Society* (Ithaca, N.Y.: Cornell University Press, 2015), 15. "Nineteenth-century workers frequently spoke of 'wage slavery' or 'prostitution,' invoking the most degraded states the race-conscious, patriarchal white male American workers could imagine."

21 **"none of us who toil"** Quoted in Glickman, *A Living Wage*, 22.

21 **"In What does slavery consist?"** Quoted in Glickman, *A Living Wage*, 43.

21 **Wage labor, or "the unfettered ability"** Glickman, *A Living Wage*, 44, 49. In a chapter of *From Bondage to Contract* called "The Labor Question and the Sale of Self," Amy Dru Stanley describes the debate over whether the buying of labor as a commodity was comparable to enslavement. Some leaders even thought the labor question might prove to be the "'logical sequence of the slavery question'" (61). About this moral debate, Stanley suggests that "although in the agrarian South the question of labor and self entitlement remained yoked to the issue of landowning, in the industrial North this question became principally about the sale of time" (62).

21 **the waged workplace contained** Braverman, *Labor and Monopoly Capital*, 45–46.

22 **When workers did organize** Alex Vitale, *The End of Policing* (London: Verso, 2017), 77–80.

22 **"Americans have come to think"** Philip Dray, *There Is Power in a Union: The Epic Story of Labor in America* (New York: Anchor, 2011), 119.

22 **agreeing to institute certain policies** Dray, *There Is Power in a Union*, 49.

22 **blacklist employees across the board** Dray, *There Is Power in a Union*, 43, 46.

22 **This kind of behavior** Joey La Neve DeFrancesco, "Pawtucket, America's First Factory Strike," *Jacobin*, June 2018, jacobinmag.com/2018/06/factory-workers -strike-textile-mill-women.

22 **"I object to the constant hurry"** Quoted in Dray, *There Is Power in a Union*, 26.

23 **"If [workers] arrived only two"** John Brown, *A Memoir of Robert Blincoe, an Orphan Boy, Sent from the Workhouse of St. Pancras, London, at Seven Years of Age, to Endure the Horrors of a Cotton Mill, Through His Infancy and Youth, with a Minute Detail of His Sufferings, Being the First Memoir of the Kind Published* (Manchester, UK: J. Doherty, 1832), 59.

23 **"The Superintendent shall again ring"** Thompson, "Time, Work-Discipline, and Industrial Capitalism," 85.

23 **"wholesome discipline of factory life"** Dray, *There Is Power in a Union*, 54.

24 **"5th. You must apply yourself industriously"** Quoted in Dario Melossi and Massimo Pavarini, *The Prison and the Factory: Origins of the Penitentiary System*, trans. Glynis Cousin (Totowa, N.J. : Barnes and Noble Books, 1981), 153.

24 **Jeremy Bentham's designs for the panopticon** Janet Semple, *Bentham's Prison: A Study of the Panopticon Penitentiary* (Oxford, UK: Clarendon, 1993), 123–24.

25 **"When we leave this sphere"** Marx, *Capital*, 1:280.

25 **It was only through protracted efforts** Marx, *Capital*, 1:389–416.

25 **" 'petty pilfering of minutes' "** Marx, *Capital*, 1:352.

25 **Sometimes employers practiced outright deception** Thompson, "Time, Work-Discipline, and Industrial Capitalism," 86.

25 **"closer filling-up of the pores"** Marx, *Capital*, 1:534.

26 **But unlike the Ancient Greeks** William James Booth, "Economies of Time: On the Idea of Time in Marx's Political Economy," *Political Theory* 19, no. 1 (February 1991): 16.

26 *headlines about the supply chain* Paul Berger, "Why Container Ships Can't Sail Around the California Ports Bottleneck," *The Wall Street Journal*, September 21, 2021, wsj.com/articles/why-container-ships-cant-sail-around-the -california-ports-bottleneck-11632216603; Lisa M. Krieger, "As Cargo Waits and Costs Climb, Port of Oakland Seeks Shipping Solutions," *Mercury News*, July 25, 2021, mercurynews.com/2021/07/25/as-cargo-waits-and-costs-climb -port-of-oakland-seeks-shipping-solutions/.

27 *beaks more than half the length* Hugh Jennings, "Bird of the Month: Long-billed Curlew," Eastside Audubon, April 3, 2018, eastsideaudubon.org/corvid -crier/2019/8/26/long-billled-curlew.

27 *They are back for the time* Megan Prelinger, email to author, July 14, 2022; Lia Keener, "Catching Up to Curlews," *Bay Nature*, January 18, 2022. baynature .org/article/catching-up-to-curlews/.

27 **such societies exhibited their own** Michael O'Malley, "Time, Work, and Task Orientation: A Critique of American Historiography," *Time and Society* 1, no. 3 (1992): 346.

27 **the "ambiguous position" of American clocks** O'Malley, "Time, Work, and Task Orientation," 351.

28 **time-saving systems** Catharine Beecher, *Treatise on Domestic Economy* (Boston, Mass.: Thomas H. Webb, 1843), 258; Ivan Paris, "Between Efficiency and Comfort: The Organization of Domestic Work and Space from Home Economics to Scientific Management, 1841–1913," *History and Technology* 35, no. 1 (2019): 81–104.

29 **those people are simply acting** Marx, *Capital*, 1:381. "Under free competition, the immanent laws of capitalist production confront the individual capitalist as a coercive force external to him."

29 **"[He] remarked: 'All they lack'"** Braverman, *Labor and Monopoly Capital*, 232.

29 **"The way to effect savings"** C. Bertrand Thompson, "The Stop Watch as Inventor," *Factory: The Magazine of Management* 16–17 (February 1916): 224.

29 **As the foremost time-study man** Braverman, *Labor and Monopoly Capital*, 63–68.

31 **"Every step in the labor process"** Braverman, *Labor and Monopoly Capital*, 58.

31 **"metric time first gave us"** D. T. Nguyen, "The Spatialization of Metric Time: The Conquest of Land and Labour in Europe and the United States," *Time and Society* 1, no. 1 (1992): 46.

32 **when domestic work did become waged** Nina Banks, "Black Women's Labor Market History Reveals Deep-Seated Race and Gender Discrimination," Working Economics Blog, Economic Policy Institute, February 19, 2019, epi.org /blog/black-womens-labor-market-history-reveals-deep-seated-race-and-gender -discrimination/.

32 **"Since housework does not generate profit"** Angela Y. Davis, *Women, Race, and Class* (New York: Vintage, 1983), 228; Barbara Adam, *Timewatch* (Cambridge, UK: Polity, 1995), chap. 4.

33 **Well into the twentieth century** Jacqueline Jones, "Black Workers Remember," *The American Prospect*, November 30, 2000, prospect.org/features/black-workers -remember/.

33 **One kilogirl** Claire L. Evans, *Broad Band: The Untold Story of the Women Who Made the Internet* (New York: Portfolio/Penguin, 2018), 24.

33 **In 2014, when Amazon released data** Jay Greene, "Amazon Far More Diverse at Warehouses Than in Professional Ranks," *The Seattle Times*, August 14, 2015, seattletimes.com/business/amazon/amazon-more-diverse-at-its-warehouses -than-among-white-collar-ranks/. (The 2015 article refers to data released in 2014.) Katherine Anne Long, "Amazon's Workforce Split Sharply Along the Lines of Race and Gender, New Data Indicates," *The Seattle Times*, September 22, 2021, seattletimes.com/business/amazon/amazons-workforce-split-sharply-along -the-lines-of-race-gender-and-pay-new-data-indicates/.

33 **an "automaton"** Karl Marx, *Grundrisse: Foundations of the Critique of Political Economy* (London: Penguin, 1993), 739. Marx writes of changes in labor "whose culmination is the *machine,* or rather, *an automatic system of machinery* . . . set in motion by an automaton, a moving power that moves itself; this automaton consisting of numerous mechanical and intellectual organs, so that the workers themselves are cast merely as its conscious linkages."

33 **"your movements are controlled"** Quoted in Emily Reid-Musson, Ellen MacEachen, and Emma Bartel, "'Don't Take a Poo!': Worker Misbehaviour in On-Demand Ride-Hail Carpooling," *New Technology, Work, and Employment* 35, no. 2 (July 2020): 151.

33 **"[The sensors] reported"** Jessica Bruder, "These Workers Have a New Demand: Stop Watching Us," *The Nation*, May 27, 2015, thenation.com/article/these -workers-have-new-demand-stop-watching-us/.

33 **Data collected through telematics** Joseph Carino, "Uber Driver Tracking and Telematics," Geotab, January 15, 2018, geotab.com/blog/uber-driver-tracking/.

34 **an Amazon warehouse** *Secrets of the Superfactories,* season 1, episode 2, directed by Paul O'Connor, aired October 14, 2019, on Channel 4.

34 **"lights-out manufacturing"** Andrew Wheeler, "Lights-Out Manufacturing:

Future Fantasy or Good Business?" Redshift by Autodesk, December 3, 2015. Accessed December 3, 2020. Archived: web.archive.org/web/20210616205134 /redshift.autodesk.com/lights-out-manufacturing/.

34 **humans "increasingly have to compete"** Guendelsberger, *On the Clock,* 54.

34 **the graph of growth versus wages** Guendelsberger, *On the Clock,* 77–78.

34 **"above the API"** Peter Reinhardt, "Replacing Middle Management with APIs," personal blog, February 3, 2015, rein.pk/replacing-middle-management-with -apis. Thanks to Nick Pinkston for making me aware of this term.

35 **Laura Morales, an employee at OutPLEX** "How Bots Will Automate Call Center Jobs," *Bloomberg,* August 15, 2019, bloomberg.com/news/videos/2019-08-15 /how-bots-will-automate-call-center-jobs-video.

36 **reading books like** *Scientific Office Management* Braverman, *Labor and Monopoly Capital,* 223.

36 **"The work is still performed"** Braverman, *Labor and Monopoly Capital,* 220.

37 **a "chronophage," something that "eats time"** Richard Seymour, *The Twittering Machine* (London: Verso, 2020), 195.

37 **"What is the average time"** This post has since been "restricted by the owner" on Askwonder.com, but it can still be accessed via the Internet Archive's Wayback Machine. Ashley N. and Carrie S., "What Is the Average Time Someone Spends Looking at an Instagram Post?" Askwonder.com, April 14, 2017, web .archive.org/web/20200814110334/https://askwonder.com/research/average -time-someone-spends-looking-instagram-post-o1oyu31rb.

37 **"I love Wonder"** Danielle Narveson, review of Wonder Research, Askwonder .com, askwonder.com/.

37 **a Wonder Research worker in 2018** "Good in theory, awful in execution" [employee review of Wonder Research], Glassdoor, April 23, 2018, glass door.com/Reviews/Employee-Review-Wonder-Research-RVW20272240 .htm; "Very Difficult but Rewarding" [employee review of Wonder Research], Glassdoor, May 20, 2019, glassdoor.com/Reviews/Employee-Review-Wonder -Research-RVW26220924.htm.

38 **"Potemkin AI"** Quoted in Gavin Mueller, *Breaking Things at Work: The Luddites Were Right About Why You Hate Your Job* (London: Verso, 2021), 118.

38 **Moderators were required** Casey Newton, "Bodies in Seats," The Verge, June 19, 2019, theverge.com/2019/6/19/18681845/facebook-moderator-interviews -video-trauma-ptsd-cognizant-tampa.

39 **"even humans have a hard time"** Quoted in Drew Harwell, "AI Will Solve Facebook's Most Vexing Problems, Mark Zuckerberg Says. Just Don't Ask When or How," *The Washington Post,* April 11, 2018, washingtonpost.com/news /the-switch/wp/2018/04/11/ai-will-solve-facebooks-most-vexing-problems -mark-zuckerberg-says-just-dont-ask-when-or-how/.

39 **"If I wanted to work"** Mueller, *Breaking Things at Work,* 114.

39 **"Welcome to Spinify!"** "Spinify," spinify.com. The website has since changed; the closest version to the one referenced here is the Internet Archive Wayback Machine's June 5, 2021, snapshot: web.archive.org/web/20210605141235/spinify .com/.

40 **Here, a nameless woman** *Merger,* directed by Keiichi Matsuda (2019), km.cx /projects/merger.

CHAPTER 2: SELF TIMER

43 **"More important than any"** P. K. Thomajan, "Annual Report to Yourself," *Good Business*, January 1966, 12. *Good Business* was one of several publications by the Unity School of Christianity (now known as Unity). In a 1936 book called *Prosperity* (republished in 2008 as *Prosperity: The Pioneering Guide to Unlocking Your Mental Power*), Unity co-founder Charles Fillmore included a "revised" Twenty-third Psalm:

> *The Lord is my banker; my credit is good.*
> *He maketh me lie down in the consciousness of omnipresent abundance;*
> *He giveth me the key to His strongbox.*
> *He restoreth my faith in His riches;*
> *He guideth me in the paths of prosperity for His name's sake.*
> *Yea, though I walk in the very shadow of debt,*
> *I shall fear no evil, for Thou art with me;*
> *Thy silver and Thy gold, they secure me.*
> *Thou preparest a way for me in the presence of the collector;*
> *Thou fillest my wallet with plenty; my measure runneth over.*
> *Surely goodness and plenty will follow me all the days of my life,*
> *And I shall do business in the name of the Lord forever.*

Charles Fillmore, *Prosperity: The Pioneering Guide to Unlocking your Mental Power* (New York: TarcherPerigree, 2008), 77–78.

43 **"Just because you're going forward"** Billy Bragg, "To Have and Have Not," on *Life's a Riot with Spy vs Spy (30th Anniversary Edition)*, Cooking Vinyl, 2013, streaming audio, open.spotify.com/track/5OL8fXk5wyeB7g2eg5B9Xh?si=5ce1b2fdbdf647fa.

44 **"we must constantly demonstrate our usefulness"** Oliver Burkeman, "Why Time Management Is Ruining Our Lives," *The Guardian*, December 22, 2016, theguardian.com/technology/2016/dec/22/why-time-management-is-ruining-our-lives; see also Burkeman, *4,000 Weeks: Time Management for Mortals* (New York: Farrar, Straus and Giroux, 2021).

44 **Hedonismbot** *Futurama*, season 4, episode 18, "The Devil's Hands Are Idle Playthings," directed by Bret Halaand, aired August 10, 2003, on Fox; *Futurama*, season 6, episode 12, "The Mutants Are Revolting," directed by Raymie Muzquiz, aired September 2, 2010, on Comedy Central. *Futurama*, season 4, episode 8, "Crimes of the Hot," directed by Peter Avanzino, aired November 10, 2002, on Fox.

45 **Protestantism arose** Jon D. Wisman and Matthew E. Davis, "Degraded Work, Declining Community, Rising Inequality, and the Transformation of the Protestant Ethic in America: 1870–1930," *The American Journal of Economics and Sociology* 72, no. 5 (November 2013): 1078–79. Citing David Landes, the authors suggest that the close relationship between the Protestant work ethic and the commercial bourgeoisie is further evidenced by a "strong Calvin-like work ethic" that also attended the rise of merchants in seventeenth-century Japan.

45 **As a form of Protestantism** Charles Taylor, *Sources of the Self: The Making of the Modern Identity* (Cambridge, Mass.: Harvard University Press, 1989), 184.

45 **"fashion[ing] himself as both preacher"** Margo Todd, "Puritan Self-Fashioning: The Diary of Samuel Ward," *Journal of British Studies* 31, no. 3 (July 1992): 260.

45 **assembly-line jobs offered little room** Robert Eisenberger, *Blue Monday: The Loss of the Work Ethic in America* (New York: Paragon House, 1989), 10.

45 **"it is to the greater productivity"** Frederick Winslow Taylor, *Principles of Scientific Management* (Norwood, Mass.: Plimpton, 1911), 141.

46 **"a practical and detailed manual"** Donald Laird, *Increasing Personal Efficiency* (New York: Harper and Brothers, 1925), front cover.

46 **"Engineers have improved this world remarkably"** Laird, *Increasing Personal Efficiency*, 179.

46 **"a personal question"** Laird, *Increasing Personal Efficiency*, 6.

46 **"avoid excessive eye movements"** Laird, *Increasing Personal Efficiency*, 67.

47 **"Effective Thought"** Laird, *Increasing Personal Efficiency*, 123.

48 **"turning toward hard work"** Samuel Haber, *Efficiency and Uplift: Scientific Management in the Progressive Era, 1890–1920* (Chicago: University of Chicago Press, 1964), ix.

48 **"Do I spend more time"** Laird, *Increasing Personal Efficiency*, 136.

49 **"Your own time"** Roy Alexander and Michael S. Dobson, *Real-World Time Management* (New York: American Management Association, 2009), 4.

49 **"I encourage you to try it"** Kevin Kruse, *15 Secrets Successful People Know About Time Management: The Productivity Habits of 7 Billionaires, 13 Olympic Athletes, 29 Straight-A Students, and 293 Entrepreneurs* (Philadelphia: Kruse Group, 2015), chap. 1.

50 **"you can't make more time"** Kruse, *15 Secrets Successful People Know About Time Management*, chap. 16.

51 **"pulling oneself up by one's bootstraps"** "Bootstrap (n.)," Online Etymology Dictionary, etymonline.com/word/bootstrap; J. Dorman Steele, *Popular Physics* (New York: American Book Company, 1888), 37.

51 **To do so, she must invest** Freelance and gig work are probably one of the more concrete examples of bootstrapping. As evidenced by the fight over California AB-5, a law requiring companies like Uber to classify gig workers as employees, there is some disagreement in the United States over how exploitative this type of work is, what moves people to do it, and how best to go about improving its conditions. Yet what both sides seem to be able to agree on is that freelance work is more precarious than regular full-time work in that it asks the individual to assume risk (financial and otherwise). A survey by freelancing platform Contently that focused on freelancers' opposition to AB-5 also mentioned that survey participants would in fact welcome government assistance on healthcare, affordable insurance, assistance with unpaid wages, and lower self-employment taxes. Likewise, in a Rest of World survey of 4,900 gig workers in fifteen countries, 60 percent of gig workers reported being financially satisfied, but 62 percent also said "they were frequently anxious and scared on the job, afraid of accidents, assaults, illness, or simply not making enough money to cover their costs." Meanwhile, any meaningful gains in autonomy have often sidestepped or eroded labor protections, particularly in places like the Philippines, which has the sixth-fastest-growing freelance economy. (In 2021, the House of Representatives of the Philippines responded by passing the Freelance Workers Protection Act.) Ruth Berins Collier, Veena Dubal, and Christopher Lee Carter, "Labor Platforms and Gig Work: The Failure to Regulate," IRLE Working Paper No. 106-17, September 2017, irle.berkeley.edu/files/2017/Labor-Platforms-and

-Gig-Work.pdf; Philip Garrity, "We Polled 573 Freelancers About AB5. They're Not Happy," The Freelance Creative, January 30, 2020, contently.net/2020/01/30 /resources/we-polled-573-freelancers-about-ab5-theyre-not-happy/; Peter Guest, "'We're All Fighting the Giant': Gig Workers Around the World Are Finally Organizing," Rest of World, September 21, 2021, restofworld.org/2021/gig-workers -around-the-world-are-finally-organizing/; Seha Yatim, "Unique Gig Economic Situation in PH Calls for Nuanced Approach," The Manila Times, May 30, 2021, manilatimes.net/2021/05/30/opinion/unique-gig-economic-situation-in-ph -calls-for-nuanced-approach/1801152.

51 **62 percent of U.S. respondents** "The American–Western European Values Gap," Pew Research Center, November 17, 2011, updated February 29, 2012, pewresearch.org/global/2011/11/17/the-american-western-european-values -gap/.

51 **a median of 53 percent** Laura Silver, "Where Americans and Europeans Agree—and Differ—in the Values They See as Important," Pew Research Center, October 16, 2019, pewresearch.org/fact-tank/2019/10/16/where-americans -and-europeans-agree-and-differ-in-the-values-they-see-as-important/.

51 **compared to Democrats** Samantha Smith, "Why People Are Rich and Poor: Republicans and Democrats Have Very Different Views," Pew Research Center, May 2, 2017, pewresearch.org/fact-tank/2017/05/02/why-people-are-rich-and -poor-republicans-and-democrats-have-very-different-views/.

52 **"forces outside our control"** Pierre Bourdieu, Practical Reason: On the Theory of Action (Stanford, Calif.: Stanford University Press, 1998); Harry Frankfurt, "Freedom of the Will and the Concept of a Person," The Journal of Philosophy 68, no. 1 (January 1971): 5–20.

52 **"President," "Scum," or "Capitalism"** John McLeod, "President," pagat.com /climbing/president.html. This is the personal website of McLeod, a noted British card game researcher whom David Parlett cites when describing the game in The Penguin Book of Card Games. Games similar to this one exist in Japan, Australia, and some European countries.

52 **"the Swap Game"** Mario D. Molina et al., "It's Not Just How the Game Is Played, It's Whether You Win or Lose," Science Advances 5, no. 7 (July 17, 2019).

53 **"If you're tired of spending 90%"** "About Entrepreneurs on Fire," eofire.com /about/.

53 **In 2016, Dumas launched a Kickstarter** "'The Freedom Journal': Accomplish Your #1 Goal in 100 Days," Kickstarter, kickstarter.com/projects/eofire /the-freedom-journal-accomplish-your-1-goal-in-100/posts/1459370; John Lee Dumas, The Freedom Journal (self-published, 2016).

53 **"Success Pack for 2017"** "The Mastery Journal by John Lee Dumas," Kickstarter, kickstarter.com/projects/eofire/the-mastery-journal-master-productivity -discipline; "Master Productivity!" themasteryjournal.com/.

53 **"the world's most disciplined man"** Craig Ballantyne, "Time Management Hacks From the World's Most Disciplined Man," YouTube video, July 18, 2022, youtube.com/watch?v=jnCdIYvbcEg; Craig Ballantyne, "This Morning Routine Will Increase Your Productivity and Income," YouTube video, June 3, 2020, youtube.com/watch?v=114Qgvn_wD8&ab_channel=TurbulenceTraining.

54 **crushing or dominating the following** Steve Costello, "How to Overcome Entrepreneurial Anxiety, Banish Stress, and Crush Your Goals: 'Unstoppable'

author Craig Ballantyne outlines 12 habits that will help you beat anxiety, refo-
cus, and excel," *Entrepreneur,* April 16, 2019, entrepreneur.com/article/332049;
JayWongTV, "Dominate Your Competition, Be Unstoppable & Live Your Best
Life | Craig Ballantyne Interview," YouTube video, January 24, 2019, youtube
.com/watch?v=FYN6dAsY774; Craig Ballantyne, "5 Skills for Crushing It in
Sales," YouTube video, April 15, 2020, youtube.com/watch?v=O_HRks1xD6I;
JasonCapital, "High Status Summit: Craig Ballantyne Teaches You How to
Overcome Shyness & Crush It on Social Media," YouTube video, April 23,
2020, youtube.com/watch?v=x-3ZB2r9GTY; Craig Ballantyne, "How to Dominate Your Life with 4-Quadrants," YouTube video, June 18, 2018, youtube
.com/watch?v=hjVut7yzQq4; Optimal Living Daily, "441: Follow Your Own
Rules to Crush Life by Craig Ballantyne with Roman Fitness Systems (The
Jack LaLanne Story—A Hero)," YouTube video, March 26, 2018, youtube.com
/watch?v=g3VY9U-_yD8.

54 **"modern methods like the 80/20 rule"** Ari Meisel, *The Art of Less Doing: One
Entrepreneur's Formula for a Beautiful Life* (Austin, Tex.: Lioncrest, 2016), back cover.

54 **"making money from our fleet"** At some point in early 2022, screwtheninetofive
.com began redirecting to wealthycoursecreator.com, run by the same people but
now offering a thirty-seven dollar online marketing crash course. Screw the Nine
to Five, web.archive.org/web/20220128034955/www.screwtheninetofive.com/;
Wealthy Course Creator, wealthycoursecreator.com/.

55 *a company accused* XPO Global Union Family, *XPO: Delivering Injustice,* February
2021, xpoexposed.org/the-report.

56 **In the search for affordable housing** Richard Schenin, "Bay Area Commuting
Nightmares: Jobs in City, Affordable Homes in Exurbia," *Mercury News,* September 30, 2015, mercurynews.com/2015/09/30/bay-area-commuting-nightmares
-jobs-in-city-affordable-homes-in-exurbia/. The article cites as examples the following drives: "Manteca to Mountain View (140 miles round trip); Los Banos to
San Francisco (240 miles); American Canyon to Santa Clara (150 miles); Discovery Bay to South San Jose (130 miles); Patterson to Palo Alto (170 miles); Tracy to
Walnut Creek (90 miles); Modesto to Campbell (170 miles); Hollister to Mountain View (120 miles); Newman to downtown San Jose (190 miles)."

56 **"You wake up in the morning"** Arnold Bennett, *How to Live on 24 Hours a Day*
(Garden City, N.Y.: Doubleday, Doran & Company, 1933), 23.

56 **Henry Ford gave five hundred copies** John Adair, *Effective Time Management*
(London: Pan, 1988), 9.

56 **she had given up** May Anderson in discussion with the author, August 11, 2021.

56 **a "cruel joke"** Robert E. Goodin, "Freeing Up Time," *Law, Ethics, and Philosophy* 5 (2017): 37.

57 **"The computer chip didn't free us"** Marc Mancini, *Time Management,* The Business Skills Express Series (New York: Business One Irwin/Mirror, 1994), v.

57 **"You tell me to work"** Alexander and Dobson, *Real-World Time Management,* 5.

57 *zeitgeber* Allen C. Bluedorn, *The Human Organization of Time: Temporal Realities
and Experience* (Stanford, Calif.: Stanford University Press, 2002), 150.

58 **Sarah Sharma illustrates this negotiation** Sarah Sharma, "Speed Traps and the
Temporal: Of Taxis, Truck Stops and TaskRabbits," in *The Sociology of Speed: Dig-
ital, Organizational, and Social Temporalities,* eds. Judy Wajcman and Nigel Dodd
(Oxford, UK: Oxford University Press, 2017), 132.

58 **"structuring relation of power"** Sharma, "Speed Traps and the Temporal," 133.

59 **"Say I simplify"** Sarah K.'s review of *Do Less: The Unexpected Strategy for Women to Get More of What They Want in Work and Life,* Goodreads, November 15, 2019, goodreads.com/review/show/3049904518. Similarly, Elizabeth Spiers has observed that as work became more casual, with remote working practices blurring work and home life, younger remote workers were expected to be more available. At the same time, however, those workers found that their bosses became less, not more, available—evidence that "24/7 availability only works top-down." Elizabeth Spiers, "What We Lose When Work Gets Too Casual," *The New York Times,* February 7, 2022, nytimes.com/2022/02/07/opinion/culture /casual-workplace-remote-office.html.

59 *both men and women* **expect women** Linda Babcock et al., "Gender Differences in Accepting and Receiving Requests for Tasks with Low Promotability," *American Economic Review* 107, no. 3 (2017): 724.

59 **"We're all socialized"** Charlotte Palermino, "For Years I Said Yes to Everything. Saying 'No' Finally Got Me Ahead at Work," *Elle,* February 28, 2018, elle.com /culture/career-politics/a18754342/saying-no-at-work/. This story also quotes Katharine O'Brien, who conducted a study with similar findings: "Women typically are regarded as nurturers and helpers, so saying 'no' runs against the grain of what might be expected of them." Furthermore, a group of studies showed that not accepting work hurts women's work evaluations but not men's. M. E. Heilman and J. J. Chen, "Same Behavior, Different Consequences: Reactions to Men's and Women's Altruistic Citizenship Behavior," *Journal of Applied Psychology* 90, no. 3 (2005): 431–41.

59 **"If I don't accept"** Ruchika Tulshyan, "Women of Color Get Asked to Do More 'Office Housework.' Here's How They Can Say No," *Harvard Business Review,* April 6, 2018, hbr.org/2018/04/women-of-color-get-asked-to-do -more-office-housework-heres-how-they-can-say-no.

59 **Car crash dummies** Specifically, Hybrid III, the current standard crash test dummy, was based on the fiftieth percentile of men in the 1970s. The National Highway Traffic Safety Administration began using female-style dummies in 2003, but they are still basically scaled-down versions of the male dummies, and are sometimes tested only in the passenger seat. Riley Beggin, "Female Crash Dummies Need to Be Updated for Accuracy, Rep. Lawrence Tells Feds," *The Detroit News,* February 15, 2022, detroitnews.com/story/business/autos/2022/02/15 /female-crash-dummies-need-updated-rep-lawrence-tells-feds/6797643001/; Caroline Criado Perez, "The Deadly Truth About a World Built for Men—from Stab Vests to Car Crashes," *The Guardian,* February 23, 2019, theguardian.com /lifeandstyle/2019/feb/23/truth-world-built-for-men-car-crashes; Alisha Haridasani Gupta, "Crash Test Dummies Made Cars Safer (for Average-Size Men)," *The New York Times,* December 27, 2021, nytimes.com/2021/12/27/business/car -safety-women.html.

60 **finding your dream job** Laura Vanderkam, *168 Hours: You Have More Time Than You Think* (New York: Portfolio/Penguin, 2011), chap. 3.

60 **outsourcing tasks** Vanderkam, *168 Hours,* chap. 6.

60 **"core competencies"** Vanderkam, *168 Hours,* chap. 2.

60 **"pummeling the life out of life"** Review of Laura Vanderkam's book *168*

Hours: You Have More Time Than You Think, Publishers Weekly, March 29, 2010, publishersweekly.com/978-1-59184-331-3.

61 **"cognitive tools that think"** Kevin K. Birth, *Time Blind: Problems in Perceiving Other Temporalities* (Cham, Switzerland: Palgrave Macmillan, 2017), 99.

61 **"It is an intoxicating concern"** Sharma, "Speed Traps and the Temporal," 134.

61 **"child care should be socialized"** Angela Y. Davis, *Women, Race, and Class* (New York: Vintage, 1983), 232.

62 **"fair workweek laws"** Stephanie Wykstra, "The Movement to Make Workers' Schedules More Humane," *Vox,* November 5, 2019, vox.com/future -perfect/2019/10/15/20910297/fair-workweek-laws-unpredictable-scheduling -retail-restaurants.

62 **One pilot UBI program** "Health and Well-Being," Stockton Economic Empowerment Demonstration (SEED), stocktondemonstration.org/health-and -wellbeing; "Front and Center: For Tia, Guaranteed Income Provided 'a Little Push,'" *Ms.* magazine, April 15, 2021, msmagazine.com/2021/04/15/front-and -center-1-tia-guaranteed-income-black-mothers-women-ms-magazine-magnolia -mothers-trust/.

62 **"time tax"** Annie Lowrey, "The Time Tax," *The Atlantic,* July 27, 2021, theatlantic .com/politics/archive/2021/07/how-government-learned-waste-your-time -tax/619568/.

63 **"white people own time"** Brittney Cooper, "The Racial Politics of Time," TEDWomen 2016, October 2, 2016, ted.com/talks/brittney_cooper_the_racial _politics_of_time?language=en.

65 **describing a hypothetical character named Linda** Hartmut Rosa, "De-synchronization, Dynamic Stabilization, Dispositional Squeeze," *The Sociology of Speed,* eds. Wajcman and Dodd, 27.

65 **Linda does not have access to** *Feierabend* Rosa, "De-synchronization, Dynamic Stabilization, Dispositional Squeeze," 29.

65 **"yuppie kvetching"** Elizabeth Kolbert, "No Time," *The New Yorker,* May 19, 2014, newyorker.com/magazine/2014/05/26/no-time.

65 **He compares Linda's situation** Rosa, "De-synchronization, Dynamic Stabilization, Dispositional Squeeze," 30.

66 **"discretionary time"** R. E. Goodin et al., "The Time-Pressure Illusion: Discretionary Time vs. Free Time," *Social Indicators Research* 73, no. 1 (2005): 45.

66 **one quarter of adjuncts** American Federation of Teachers, *An Army of Temps: AFT 2020 Adjunct Faculty Quality of Work/Life Report,* 2020, aft.org/sites/default /files/adjuncts_qualityworklife2020.pdf.

66 **drugs that slow people down** Rosa, "De-synchronization, Dynamic Stabilization, Dispositional Squeeze," 39.

67 **Given a legitimate prescription** Quoted in *Fixed: The Science/Fiction of Human Enhancement,* directed by Regan Brashear (New Day Films, 2014), fixed.vhx.tv/.

67 **"This obsession is the only way"** Vanderkam, *168 Hours,* chap. 3.

67 **"the drive to maximize production"** Byung-Chul Han, *The Burnout Society* (Stanford, Calif.: Stanford University Press, 2015), 9.

67 **achievement-subjects are "entrepreneurs of themselves"** Han, *The Burnout Society,* 8.

67 **"wears down in a rat race"** Han, *The Burnout Society,* 42.

68 **"disappearance of domination"** Han, *The Burnout Society,* 11.

68 **"positivity of *Can*"** Han, *The Burnout Society,* 9.

68 **"auto-aggression" of the master** Han, *The Burnout Society,* 47.

68 **"jumping over [her] own shadow"** Han, *The Burnout Society,* 46.

68 **the capitalist "logic of increase"** Rosa, "De-synchronization, Dynamic Stabili-zation, Dispositional Squeeze," 34.

70 **the A–F grading system** J. Schneider and E. Hutt, "Making the Grade: A History of the A–F Marking Scheme," *Journal of Curriculum Studies* 46, no. 2 (2013): 15.

70 **early-twentieth-century "social efficiency" movement** Jonghun Kim, "School Accountability and Standard-Based Education Reform: The Recall of Social Efficiency Movement and Scientific Management," *International Journal of Educational Development* 60 (2018): 81.

70 **A socially efficient curriculum** Schneider and Hutt, "Making the Grade," 13–14.

70 **American schools were already using ranking** Franklin Bobbitt, *Some General Principles of Management Applied to the Problems of City-School Systems* (Chicago: University of Chicago Press, 1913), 15; Schneider and Hutt, "Making the Grade," 11–13; Kim, "School Accountability and Standard-Based Education Reform," 82–86.

71 **his efforts to make a "Beauty-Map"** Francis Galton, *Memories of My Life* (New York: E. P. Dutton, 1909), 315.

71 **creating an A–G scale** Francis Galton, *Hereditary Genius: An Inquiry into Its Laws and Consequences* (New York: D. Appleton, 1870), 338.

71 **"It would seem as though"** Galton, *Memories of My Life,* 312.

71 **Galton spends more time** Galton, *Memories of My Life,* 154–60.

71 **intelligence was inherently linked to speed** Galton, *Memories of My Life,* 248.

72 **"a new condition imposed upon men"** Galton, *Hereditary Genius,* 345.

72 **selecting against nomadism** Galton, *Hereditary Genius,* 347.

72 **"No man who only works"** Galton, *Heredtiary Genius,* 347–48.

72 **Not even Charles Darwin** Galton, *Memories of My Life,* 290–91.

73 **In 2021, California offered reparations** "California Launches Program to Compensate Survivors of State-Sponsored Sterilization," Office of Governor Gavin Newsom, December 31, 2021; Erin McCormick, "Survivors of California's Forced Sterilizations: 'It's Like My Life Wasn't Worth Anything,'" *The Guardian,* July 19, 2021, theguardian.com/us-news/2021/jul/19/california-forced-sterilization-prison-survivors-reparations.

73 **The magazine once offered $1,000** "$1,000.00 for the Most Beautiful Woman—One of Two Personal Beauty Prize Contests, the Other Being for the Most Handsome Man Based upon Perfection of Both Face and Figure—An Announcement," *Physical Culture* 45, no. 2 (February 1921): 54.

73 **A proponent of bodybuilding** Robert Ernst, *Weakness Is a Crime: The Life of Bernarr Macfadden* (Syracuse, N.Y.: Syracuse University Press, 1991), 18.

73 **Macfadden's articles** Bernarr Macfadden, "Vitalize with the Mono-Diet," *Physical Culture* 63, no. 6 (June 1930): 17; Bernarr Macfadden, "Make Your Vacation Pay Health Dividends," *Physical Culture* 59, no. 6 (June 1928): 27; Bernarr Macfadden, "Mountain Climbing in Your Own Home," *Physical Culture* 57, no. 5 (May 1927): 30; Bernarr Macfadden, "Are You Wasting Your Life?" *Physical Culture* 58, no. 3 (September 1927): 25.

74 **mental vitality was as important** Bernarr Macfadden, "Bernarr Macfadden's Viewpoint," *Physical Culture* 45, no. 2 (February 1921): 14.

74 *The genes themselves* **were productive** Amram Scheinfeld, "What You Can Do to Improve the Human Race," *Physical Culture* 78, no. 4 (October 1937): 20.

75 **"conspicuous busyness"** Michelle Shir-Wise, "Disciplined Freedom: The Productive Self and Conspicuous Busyness in 'Free' Time," *Time and Society* 28, no. 4 (2019): 1686.

76 **"The ones who happily claim"** Stefano Harney and Fred Moten, *The Undercommons: Fugitive Planning and Black Study* (New York: Minor Compositions, 2013), 140–41.

77 **as we work for policy changes** Burkeman, "Why Time Management Is Ruining Our Lives."

77 **"I'll have a double cheeseburger"** *Beavis and Butt-Head,* season 5, episode 15, "Tainted Meat," directed by Mike Judge, aired December 29, 1994, on MTV.

78 **In 2016, a young Chinese factory** David Bandurski, "The 'Lying Flat' Movement Standing in the Way of China's Innovation Drive," Brookings Tech Stream, July 8, 2021, brookings.edu/techstream/the-lying-flat-movement-standing-in-the -way-of-chinas-innovation-drive/.

78 **dismissed American "lie-flatters"** Allison Schrager, "'Lie Flat' If You Want, but Be Ready to Pay the Price," *Bloomberg,* September 13, 2021, bloomberg .com/opinion/articles/2021-09-13/-lie-flat-if-you-want-but-be-ready-to-pay-the -price?sref=2o0rZsF1.

79 **"It's just wild to me"** @slowdrawok, Twitter post, September 16, 2021, twitter .com/SlowdrawOK/status/1438568129320325122.

79 **"Billionaire: 'Quick, newspaper that I own'"** @w3dges, Twitter post, September 16, 2021; twitter.com/w3dges/status/1438556517297496069?s=20.

79 **"Why work hard?"** @JackJackington, Twitter post, September 16, 2021, twitter .com/JackJackington/status/1438615108402438146.

CHAPTER 3: CAN THERE BE LEISURE?

81 **"Work dominates everything around it"** Michael Dunlop Young and Tom Schuller, *Life After Work: The Arrival of the Ageless Society* (New York: HarperCollins, 1991), 93.

82 **This time is used** Michael Zhang, "Why Photographs of Watches and Clocks Show the Time 10:10," PetaPixel, June 27, 1013, petapixel.com/2013/06/27/why -photographs-of-watches-and-clocks-show-the-time-1010/.

82 **"All we have is now"** Lauren Bullen (@gypsea_lust), "All we have is now," Instagram, March 23, 2020, instagram.com/p/B-GJWpBJvX0/; Lauren Bullen (@gypsea_lust), "This too shall pass," Instagram, March 28, 2020, instagram .com/p/B-RgsTzJjFo/; Lauren Bullen (@gypsea_lust), "Currently stuck in paradise," Instagram, April 3, 2020, instagram.com/p/B-jH8NTp9hB/; Lauren Bullen (@gypsea_lust), "Face mask on, hair mask on; ready to soak," Instagram, April 4, 2020, instagram.com/p/B-lHUMLp1zM/. Because Instagram adjusts timestamps to your local time zone, viewers on Eastern Daylight Time will see the last two of these posts as having occurred on the same day, whereas they were posted on different days in Bali.

83 **slowness is "not necessarily the equivalent"** Filip Vostal, "Slowing Down Modernity: A Critique," *Time and Society* 28, no. 3 (2019): 1042.

83 **"paradoxically integral parts"** Vostal, "Slowing Down Modernity: a Critique," 1048.

83 **"Slow living is now 'for sale'"** Vostal, "Slowing Down Modernity: A Critique," 1046.

84 **"experience economy"** B. Joseph Pine II and James H. Gilmore, "Welcome to the Experience Economy," *Harvard Business Review,* July–August 1998.

85 **"acquisitive mood"** Susan Sontag, *On Photography* (New York: Farrar, Straus and Giroux, 2011), 4.

85 **in March 2022, Instagram announced** Aisha Malik, "Instagram Expands Its Product Tagging Feature to All US Users," *TechCrunch,* March 22, 2022, techcrunch.com/2022/03/22/instagram-product-tagging-feature/.

85 *Insta-bae* Isabella Steger, "The Japanese Words That Perfectly Sum Up How the Country Felt This Year," Quartz, December 1, 2017, qz.com/1144046/sontaku -japans-word-of-the-year-reflects-its-deep-political-unease/. The term is also some- times spelled "insuta-bae."

85 **two-fifths of American Millennials** Rachel Hosie, "'Instagrammability': Most Important Factor for Millennials on Choosing Holiday Destination," *The Independ- ent,* March 24, 2017, independent.co.uk/travel/instagrammability-holiday-factor -millenials-holiday-destination-choosing-travel-social-media-photos-a7648706 .html.

85 **"the outside world, the non-digital world"** *Inside,* directed by Bo Burnham (Netflix, 2021).

85 **poses enrobed amidst a lavender field** Lauren Bullen (@gypsealust), "Ful- filling my lavender field dreams," Instagram, July 3, 2021, instagram.com/p /CQ3k2zOBJea/.

86 **many uses of "social media envy"** B. Marder et al., "Vacation Posts on Face- book: A Model for Incidental Vicarious Travel Consumption," *Journal of Travel Research* 58, no. 6 (2019): 1027; H. Liu et al., "Social Media Envy: How Experi- ence Sharing on Social Networking Sites Drives Millennials' Aspirational Tour- ism Consumption," *Journal of Travel Research* 58, no. 3 (2019): 365. The latter paper suggests that "the low self-esteem segment is a potentially large yet under- explored market."

86 **rising from bed, cradling a coffee** Lauren Bullen (@gypsealust), "The one time we got up for sunrise," Instagram, September 15, 2021, instagram.com/p /CT1YY6Zr9Ij/.

86 **a series of photos** Anna Seregina, "Found a prison that has been converted to an influencer hotel," Twitter post, September 20, 2021, twitter.com/touching cheeses/status/1440052093788721155.

86 *The Shawshank Redemption* **is occasionally screened** Lucy Dodsworth, "A Night Behind Bars: The Malmaison Oxford Hotel Reviewed," On the Luce Travel Blog, March 3, 2021, ontheluce.com/reviewed-a-night-behind-bars-at-the-malmaison -oxford/.

86 **"Peace as a new luxury"** "Forestis Dolomites | Boutique Wellness Hotel in Brixen," forestis.it/en.

86 **As he watches the guests arrive** *The White Lotus,* season 1, episode 1, "Arrivals," directed by Mike White, premiered July 11, 2021, on HBO.

88 **somewhere vaguely European** Santana Row's preliminary website advertised a place "inspired by the grand boulevards of Europe and America," while its later website referred to pedestrian-friendly streets with a "European flair." "Santana Row [About]," archived February 5, 2004, web.archive.org/web/20040205185023 /www.santanarow.com/about.shtml; "Santana Row," archived August 3, 2005, web.archive.org/web/20050803081549/www.santanarow.com/.

89 **"The biggest insight we can glean"** Josh Allan Dykstra, "Why Millennials Don't Want to Buy Stuff," *Fast Company,* July 13, 2012.

90 **Nestlé selling us the public water** Tom Perkins, "The Fight to Stop Nestlé from Taking America's Water to Sell in Plastic Bottles," *The Guardian,* October 29, 2019, theguardian.com/environment/2019/oct/29/the-fight-over-water -how-nestle-dries-up-us-creeks-to-sell-water-in-plastic-bottles. Fulvia Serra, in "Reproducing the Struggle: A New Feminist Perspective on the Concept of Social Reproduction," *Viewpoint Magazine,* October 31, 2015, suggests something similar when she writes that the private enclosure of the public commons— a process that began in twelfth-century England—applies to the social and informational sphere as well. She writes that "intimacy, together with other social and intellectual practices that are necessary for the reproduction of our collectivity, is being appropriated today by the capitalist machine and, in the same movement, transferred from the collective sphere to that of the nuclear unit and from the sphere of reproduction to that of the market economy."

90 **Instagram has allowed ads** "Instagram Stories Ads—Now Available for All Businesses Globally," Instagram blog, March 1, 2017, business.instagram.com/blog /instagram-stories-available-globally.

90 **compensatory consumption** First coined in the 1960s, *compensatory consumption* has been described as "an umbrella term that captures consumer intentions and behavioural responses triggered by perceived deficits, needs and desires that cannot be fulfilled directly." Bernadett Koles, Victoria Wells, and Mark Tadajewski, "Compensatory Consumption and Consumer Compromises: A State-of-the-Art Review," *Journal of Marketing and Management* 34, nos. 1–2 (2018): 5.

90 **Asked about the term *Insta-bae*** Leo Lewis and Emma Jacobs, "How Business Is Capitalizing on the Millennial Instagram Obsession," *Financial Times,* July 12, 2018, ft.com/content/ad84c9d0-8492-11e8-96dd-fa565ec55929.

90 **internal paradoxes** Kathi Weeks, *The Problem with Work: Feminism, Marxism, Antiwork Politics, and Postwork Imaginaries* (Durham, N.C.: Duke University Press, 2011), 49.

90 **once assembly-line jobs** Jon D. Wisman and Matthew E. Davis, "Degraded Work, Declining Community, Rising Inequality, and the Transformation of the Protestant Ethic in America: 1870–1930," *The American Journal of Economics and Sociology* 72, no. 5 (November 2013): 1088. The authors quote Thorstein Veblen in his *Theory of the Leisure Class:* "The only practicable means of impressing one's pecuniary ability on . . . unsympathetic observers of one's everyday life is an unremitting demonstration of ability to pay."

91 **"It is only because possession involves"** Quoted in Weeks, *The Problem with Work,* 50.

91 **"leisure, without anybody planning it"** Tony Blackshaw, "The Man from Leisure: An Interview with Chris Rojek," *Cultural Sociology* 6, no. 3 (2012): 333.

91 **an "evidence-based" resort** John Chan, "Larry Ellison's $300m Hawaii Island

Will Transform Wellness," *Billionaire,* May 14, 2020, bllnr.com/travel/larry-ellison-s-us$300m-startup-will-transform-wellness; Avery Hartmans, "See Inside Larry Ellison's Hawaiian Island Wellness Retreat, a $1,200-Per-Night Luxury Spa Where Guests Track Their Health Data and Learn How to Live Longer Lives," *Business Insider,* February 21, 2021, businessinsider.com/larry-ellison-hawaii-wellness-spa-sensei-lanai-photos-2021-2.

91 **island he purchased nearly all of** Adam Nagourney, "Tiny Hawaiian Island Will See If New Owner Tilts at Windmills," *The New York Times,* August 22, 2012, nytimes.com/2012/08/23/us/lanai-a-hawaiian-island-faces-uncertain-future-with-new-owner.html.

91 **One app, Habitshare** Patrick Lucas Austin, "Need Some Help Reaching Your Goals? Try These 5 Habit-Tracking Apps," *Time,* July 8, 2019, time.com/5621109/best-habit-tracking-apps/.

91 **"Think of it like the algorithm"** Rachel Reichenbach, "Why Your Instagram Engagement Kinda Sucks Right Now," Rainylune, December 20, 2020, rainylune.com/blogs/blog/why-your-instagram-engagement-kinda-sucks-right-now.

92 **"no longer a photo sharing app"** Amelia Talt, "Why Instagram's Creatives Are Angry About Its Move to Video," *The Guardian,* August 8, 2021, theguardian.com/technology/2021/aug/08/instagram-artists-leaving-social-media-tiktok-shopping; Rebecca Jennings, "Nobody Wants More Crappy Videos on Instagram. Too Bad," *Vox,* March 29, 2022, vox.com/the-goods/23000352/instagram-algorithm-reels-video-following-favorites.

93 **Pieper's leisure** Josef Pieper, *Leisure, the Basis of Culture* (San Francisco: Ignatius, 2015), 46–47.

93 **as "an attitude of mind"** Pieper, *Leisure,* 46.

95 **"parks and libraries of the self"** Jenny Odell, *How to Do Nothing: Resisting the Attention Economy* (Brooklyn, N.Y.: Melville House, 2019), 13–14.

95 **Oakland's Morcom Rose Garden** "Morcom Amphitheatre of Roses—Oakland CA," The Living New Deal, livingnewdeal.org/projects/morcom-amphitheater-of-roses-oakland-ca/.

95 **the state's responsibility to provide leisure** Ida Craven, " 'Leisure,' According to the Encyclopedia of the Social Sciences," in *Mass Leisure,* eds. Eric Larrabee and Rolf Meyersohn (Glencoe, Ill.: Free Press, 1958), 8. Craven notes that concerns about the "leisure problem" arose with eight-hour work laws being discussed in 1916 and 1917.

95 **modernization would lead** John Maynard Keynes, "Economic Possibilities for our Grandchildren," *Essays in Persuasion* (London: Macmillan, 1933), 368.

95 **"blanket codes"** *Workers' Rights and Labor Compliance in Global Supply Chains,* eds. Doug Miller, Jennifer Bair, and Marsha Dickson (New York: Routledge, 2014), 9.

96 **"the birth of modern leisure"** Chris Rojek, *The Labour of Leisure: The Culture of Free Time* (London: SAGE, 2010), chap. 4.

96 **"what the American people do"** Statement quoted in National Recreation Association, *The Leisure Hours of 5,000 People: A Report of a Study of Leisure Time Activities and Desires* (New York: National Recreation Association, 1934). This report begins its section "Purpose, Method, and Scope of the Study" by stating, "Everyone knows that during the last few years there has been a rapid increase in the amount of time available to most people outside their working hours" (4).

96 **an important function of leisure** Gilbert Wrenn and D. L. Harvey, *Time on Their Hands: A Report on Leisure, Recreation, and Young People* (Washington, D.C.: American Council on Education, 1941), xx.

96 **A 1950 public education film** *A Chance to Play* (General Electric Company, in cooperation with the National Recreation Association, 1950), archive.org /details/Chanceto1950.

96 *Better Use of Leisure Time Better Use of Leisure Time,* directed by Ted Peshak (Coronet Instructional Films, 1950), archive.org/details/0034_Better_Use_of_Leisure _Time_10_22_15_00.

97 *until last year* Narrative interludes in this book are set in the summer of 2021; the change in park accessibility occurred in 2020.

97 *exclusively available to residents* Amanda Bartlett, "No Access: ACLU Sues Palo Alto over Decades-Long Ban of Non-residents from City Park," SFGate, September 16, 2020, sfgate.com/california-parks/article/ACLU-sues-Palo-Alto -Foothills-Park-nonresidents-15573266.php.

98 **asked people about enjoyable leisure experiences** George A. Lundberg, Mirra Komarovsky, and Mary Alice McInerny, *Leisure: A Suburban Study* (New York: Columbia University Press, 1934), 118.

99 **Varied and intoxicating, his walks** Garnette Cadogan, "Walking While Black," Literary Hub, July 8, 2016, lithub.com/walking-while-black/.

99 **"double-consciousness"** W.E.B. Du Bois, *The Souls of Black Folk* (New York: Cosimo Classics, 2007), 2.

100 **It's not until she visits home** Barbara May Cameron, "Gee, You Don't Seem Like an Indian from the Reservation," in *This Bridge Called My Back,* 3rd ed., eds. Gloria Anzaldúa and Cheríe Moraga (Berkeley, Calif.: Third Woman Press, 2002), 54.

100 **a Filipina American woman** Nicole Hong et al., "Brutal Attack on Filipino Woman Sparks Outrage: 'Everybody Is on Edge,'" *The New York Times,* March 30, 2021, nytimes.com/2021/03/30/nyregion/asian-attack-nyc.html.

101 **On a map from the 1930s** Denzel Tongue, "My Grandparents' Redlining Story Shows Why We Must Do Better," *Yes!* magazine, November 13, 2020, yesmagazine .org/opinion/2020/11/13/redlining-racial-inequity-covid.

101 **"even before the codification"** Victoria W. Wolcott, *Race, Riots, and Roller Coasters: The Struggle over Segregated Recreation in America* (Philadelphia: University of Pennsylvania Press, 2012), 16.

101 **Some amusement park owners** Wolcott, *Race, Riots, and Roller Coasters,* 16–18, 70–71, 118–19, 122.

101 **"We were allowed to swim"** Jackie Robinson, *I Never Had It Made: An Autobiography* (New York: HarperCollins, 2013), 7.

101 **Sammy Lee, a Korean American diver** The pool was Brookside Plunge, which between 1929 and 1945 was open to nonwhite people on Tuesdays between 2 and 5 P.M. It was eventually forced to reopen without restrictions after the NAACP obtained an injunction in 1947. Rick Thomas, "Throwback Thursday— Revisiting Our Racist Past," *South Pasadenan,* June 14, 2018, southpasadenan .com/throwback-thursday-revisiting-our-racist-past/.

102 **Wolcott writes that in 2005** Wolcott, *Race, Riots, and Roller Coasters,* 14, 232.

102 **"I can't really think"** Quoted in Jason Bittel, "People Called the Police on This Black Birdwatcher So Many Times That He Posted Custom Signs to Explain

His Hobby," *The Washington Post*, June 5, 2020, washingtonpost.com/science
/2020/06/05/people-called-police-this-black-birdwatcher-so-many-times-that
-he-posted-custom-signs-explain-his-hobby/.

102 **"ADVISORY! Have you seen this man?"** Walter Kitundu (@birdturntable), Twitter post, July 24, 2020, twitter.com/birdturntable/status/1286662401685893123.

102 **But when #blackbirdersweek content** Deborah Wang, "The Tale of the Black
Birders and Ruffled Feathers on Facebook," KUOW, July 5, 2020, kuow.org/stories
/black-birders-ruffle-feathers-on-facebook; "Seattle Audubon Statement on Facebook Group Censorship and 'No Politics' Policies," Seattle Audubon, July 1, 2020,
seattleaudubon.org/2020/07/01/seattle-audubon-statement-on-facebook-group
-censorship-and-no-politics-policies/.

102 **including the *Washington Post* article** Walter Kitundu, email to author, July 18,
2022.

103 **Madison Grant associated redwoods** Sam Hodder, "Reckoning with the League
Founders' Eugenics Past," Save the Redwoods League, September 15, 2020,
savetheredwoods.org/blog/reckoning-with-the-league-founders-eugenics-past/.

103 **the establishment of National Parks** Mark David Spence, *Dispossessing the
Wilderness: Indian Removal and the Making of the National Parks* (Oxford, UK:
Oxford University Press, 2000), 5. In his book *The Metaphysics of Modern Existence*
(Golden, Colo.: Fulcrum, 2012), Standing Rock Sioux writer Vine Deloria, Jr.,
gives a sense of the social function that the National Parks played after World
War II: "The interest in recreational activities in a natural surrounding was not a
profound philosophical or religious movement that recognized a value in natural entities themselves, but it did indicate that the aesthetic values of American
society could be expressed as a function of nature as well as an appreciation of
paintings, music, and other forms of art." He adds that the conservation movement saw nature "as a means of providing an emotional outlet for human frustrations," though, for him, the conservationist Aldo Leopold—with his emphasis
on a land ethic that would see nature as something more than property or an
amenity—proves an exception (181).

105 **the Amah Mutsun** Matt Dolkas, "A Tribal Band Reconnects with Ancestral
Lands," Peninsula Open Space Trust, March 3, 2020, openspacetrust.org/blog
/amah-mutsun/. The Amah Mutsun Tribal Band is chaired by Valentin Lopez,
whose teachings on fire I return to in chapter 5.

107 **"It isn't unusual for me"** Mark Hehir, email to author, March 5, 2021.

107 **activist hubs of all kinds** To give just one example from that same era in which
public leisure was in vogue, Black women workers in the tobacco industry created
spaces that combined spirituality, leisure, education, and activism. In their study
of women workers in the new industrial South from 1910 to 1940, M. Deborah
Bialeschki and Kathryn Lynn Walbert describe how, compared to white women
workers, Black women workers were more welcoming of the trade unionism
that was making its way through the southern tobacco industry, often stepping in
to union leadership roles themselves. At a time when most leisure facilities were
closed to nonwhite visitors and simply not built in Black neighborhoods, all-Black
tobacco union halls borrowed from and supplemented the role of Black churches
as spaces of communal affirmation, education, and activism. The women dressed
up for union events as if they were going to church, organized dances and shows,
and generally endeavored to provide the leisure they had been denied. M. Debo-

rah Bialeschki and Kathryn Lynn Walbert, "'You Have to Have Some Fun to Go Along with Your Work': The Interplay of Race, Class, Gender, and Leisure in the Industrial New South," *Journal of Leisure Research* 30, no. 1 (1998): 94–96.

107 **"Often the way people think"** Thora Siemsen, "On Working with Archives: An Interview with Saidiya Hartman," The Creative Independent, April 18, 2018, thecreativeindependent.com/people/saidiya-hartman-on-working-with -archives/.

107 **"Rest is not"** The Nap Ministry (@TheNapMinistry), Twitter post, October 10, 2020, twitter.com/TheNapMinistry/status/1314921775864651777. See also Tricia Hersey's book, forthcoming at the time of writing, *Rest Is Resistance: A Manifesto* (New York: Hachette, 2022).

108 **"Y'all be cranking"** The Nap Ministry (@TheNapMinistry), Twitter post, September 21, 2021. twitter.com/TheNapMinistry/status/1440296107527979028.

108 **For Hersey, rest is simultaneously** Tricia Hersey, "Our work has a framework: REST IS RESISTANCE!" January 11, 2021, thenapministry.wordpress .com/2021/01/11/our-work-is-has-a-framework/.

108 **"What do you say to people"** "Atlanta-Based Organization Advocates for Rest as a Form of Social Justice," *All Things Considered,* NPR, June 4, 2020, npr .org/2020/06/04/869952476/atlanta-based-organization-advocates-for-rest-as-a -form-of-social-justice.

109 **the divide between the world** Rojek, *Labor of Leisure,* chap. 4.

109 **he saw this trait** *Aristotle's Politics, Second Edition,* trans. Carnes Lord (Chicago: University of Chicago Press, 2013), book 3, chap. 14. "It is because barbarians are more slavish in their characters than Greeks (those in Asia being more so than those in Europe) that they put up with a master's rule without making any difficulties." In book 7, chap. 7, Aristotle attributes these differences to climate in Northern Europe and Asia when compared with Greece. Although these passages refer to people in remoter areas than those from which most enslaved people in Athens came, a footnote to Aristotle's initial discussion of natural slavery (book 1, chap. 6) in the Carnes Lord translation suggests "keep[ing] in mind the widespread sense of racial superiority among the polis-dwelling Greeks over the 'barbarians' of the Balkans and the Persian Empire, who made up a very large proportion of Greek slaves."

109 **if a polis were to have** *Aristotle's Politics,* book 1, chap. 4. "A possession . . . is an instrument for the purposes of life, and one's property is the aggregate of such instruments; and the slave is a possession of the animate sort. Every subordinate, moreover, is an instrument that wields many instruments, for if each of the instruments were able to perform its function on command or by anticipation, as they assert those of Daedalus did, or the tripods of Hephaestus . . . so that shuttles would weave themselves and picks play the lyre, master craftsmen would no longer have a need for subordinates, or masters for slaves."

Aristotle's Politics, book 7, chap. 9. "Since it was said earlier that happiness cannot be present apart from virtue, it is evident from these things that in the city that is most finely governed . . . the citizens should not live a worker's or a merchant's way of life, for this sort of way of life is ignoble and contrary to virtue. Nor, indeed, should those who are going to be citizens in such a regime be farmers; for there is a need for leisure both with a view to the creation of virtue and with a view to political activities."

109 **it was a good thing** Zeyad el Nabolsy, "Aristotle on Natural Slavery: An Analysis Using the Marxist Concept of Ideology," *Science and Society* 83, no. 2 (April 2019): 250.

109 **enslaved people, unable to deliberate** William Fortenbaugh, *Aristotle's Practical Side: On his Psychology, Ethics, Politics and Rhetoric* (Leiden, Netherlands: Brill, 2006), 249; Malcolm Heath, "Aristotle on Natural Slavery," *Phronesis* 53, no. 3 (2008): 266.

109 **This model of natural inferiority** Sylvia Wynter, "Unsettling the Coloniality of Being/Power/Truth/Freedom: Towards the Human, After Man, Its Overrepresentation—An Argument," *The New Centennial Review* 3, no. 3 (Fall 2003): 265–67, 296; Simone de Beauvoir, *The Second Sex* (New York: Vintage, 1989), xxii.

110 **"Thus by nature most things"** *Aristotle's Politics*, book 1, chap. 13; Frederick A. Ross, "Sermon Delivered in the General Assembly, New York, 1856," *Slavery Ordained of God* (Philadelphia: J. B. Lippincott, 1857), 47.

110 **a social hierarchy** Zeyad el Nabolsy has suggested that this hierarchy exhibits what he calls "proto-racialization." He recognizes that the modern concept of race came later, but finds that Aristotle's hierarchy is almost closer to modern racism than mere xenophobia, in the way that it conveniently "served to justify and stabilize" the institution of enslaving non-Greeks. El Nabolsy, "Aristotle on Natural Slavery: An Analysis Using the Marxist Concept of Ideology."

110 **During a surge in working-class self-education** David R. Roediger and Philip S. Foner, *Our Own Time: A History of American Labor and the Working Day* (New York: Verso, 1989), 21.

110 **"societies of hunter-gatherers"** Joan-Lluís Marfany, "The Invention of Leisure in Early Modern Europe (Debate)," *Past and Present* 156, no. 1 (August 1997): 190–91. "Draughts" is another name for checkers; Bagà is a medieval town in Catalonia, Spain.

112 **notable among labor leaders** Robert H. Zieger, *For Jobs and Freedom: Race and Labor in America since 1865* (Lexington: University Press of Kentucky, 2014), 25.

112 **"a blank—a negative"** Quoted in Roediger and Foner, *Our Own Time,* 99.

112 **"an indispensable *first* step"** Quoted in David Roediger, "Ira Steward and the Anti-Slavery Origins of American Eight-Hour Theory," *Labor History* 27, no. 3 (1986): 425.

112 **"make a coalition"** Roediger and Foner, *Our Own Time,* 95.

112 **"Fordist compromise"** Peter Frase, "Beyond the Welfare State," December 10, 2014, peterfrase.com/2014/12/beyond-the-welfare-state/.

112 **"Got a house, fridge, dishwasher, washer-dryer"** *Blue Collar,* directed by Paul Schrader (Universal Pictures, 1978).

113 **"The thing that is missed is"** Barbara Luck, "The Thing That Is Missed," *Processed World* 6 (November 1982): 49.

114 **aggressive policing** Niki Franco in discussion with the author, March 12, 2021.

CHAPTER 4: PUTTING TIME BACK IN ITS PLACE

116 **"But then, from the lower edge"** Helen Macdonald, "Eclipse," in *Vesper Flights* (New York: Grove, 2020), 80.

117 **"incase anyone is unsure"** James Holzhauer (@James_Holzhauer), Twitter post, March 17, 2020, twitter.com/James_Holzhauer/status/1239980923526889473;

jello (@JelloMariello), Twitter post, March 28, 2020, twitter.com/JelloMariello /status/1244120759162687490 [user no longer on Twitter]; Seinfeld Current Day (@Seinfeld2000), Twitter post, April 7, 2020, twitter.com/Seinfeld2000 /status/1247772104520421377; Mauroy (@_mxuroy), Twitter post, April 9, 2020, twitter.com/_mxuroy/status/1248228948686897152.

121 **Abstract time and abstract space** Henri Bergson, *Creative Evolution,* trans. Arthur Mitchell (Lanham, Md.: University Press of America, 1983), 156; Henri Bergson, *Matter and Memory,* trans. N. M. Paul and W. S. Palmer (Brooklyn, N.Y.: Zone, 1991), 210–11.

121 **"extraordinary . . . a kind of reaction"** Henri Bergson, *Time and Free Will: An Essay on the Immediate Data of Consciousness,* trans. F. L. Pogson (Mineola, N.Y.: Dover Publications, 2001), 97–101.

122 *élan vital* Bergson, *Creative Evolution,* 87–97. The authors translate *élan vital* as "life force." Bergson's use of the kaleidoscope analogy is from Bergson, *Matter and Memory,* 197.

122 **"the sundial directly models natural facts"** John Durham Peters, *The Marvelous Clouds: Toward a Philosophy of Elemental Media* (Chicago: University of Chicago Press, 2015), 220.

123 **"the clock were itself a materialization"** Carol J. Greenhouse, *A Moment's Notice: Time Politics Across Cultures* (Ithaca, N.Y.: Cornell University Press, 2018), 47.

123 **the language barrier** Kevin K. Birth, *Time Blind: Problems in Perceiving Other Temporalities* (Cham, Switzerland: Palgrave Macmillan, 2017), 21.

124 **"Explaining Aboriginal notions of time"** Tyson Yunkaporta, *Sand Talk: How Indigenous Thinking Can Save the World* (New York: HarperCollins, 2020), chap. 1.

124 **the assumption of uniform time** Birth, *Time Blind,* 31.

125 *Among these are chert* Andrew Alden, emails to author on April 6 and 9 and May 29, 2022. Most of the geological details about this beach are thanks to our correspondence and to his coverage for KQED. See Andrew Alden, "Geological Outings Around the Bay: Pebble Beach," KQED, March 3, 2011, kqed.org /quest/19198/geological-outings-around-the-bay-pebble-beach.

126 *saber-toothed cats and dire wolves* Daniel Potter, "The Bay Area During the Ice Age (Think Saber-Tooth Cats and Mammoths)," KQED, September 24, 2020, kqed.org/news/11839198/the-bay-area-during-the-ice-age-think-saber-tooth -cats-and-mammoths.

127 **"I see that the events"** Marcia Bjornerud, *Timefulness: How Thinking Like a Geologist Can Help Save the World* (Princeton, N.J.: Princeton University Press, 2018), prologue.

128 **Alfred North Whitehead eroded the concept** In *The Human Organization of Time,* Allen C. Bluedorn quotes Whitehead's opinion that "absolute time is just as much a metaphysical monstrosity as absolute space." Bluedorn, *The Human Organization of Time,* 28; see also W. Mays, "Whitehead and the Philosophy of Time," in *The Study of Time,* eds. J. T. Fraser, F. C. Haber, and G. H. Müller (Berlin: Springer, 1972), 358.

128 **"most of Western society remained Newtonian"** Vine Deloria, Jr., *The Metaphysics of Modern Existence* (Golden, Colo.: Fulcrum, 2012), 52.

128 **how relativity in physics resonates** Vine Deloria, Jr., "Relativity, Relatedness and Reality," *Winds of Change,* Autumn 1992.

129 **abstracting of time made it possible** Giordano Nanni, *The Colonisation of Time:*

Ritual, Routine and Resistance in the British Empire (Manchester, UK: Manchester University Press, 2012), 61.

129 **Until relatively recently** "Season (n.)," Online Etymology Dictionary, etymonline .com/word/season.

129 **"That silky oak tree"** Yunkaporta, *Sand Talk*, chap. 10.

130 **Tribes who lived** Deloria, "Relativity, Relatedness and Reality."

130 **each place exhibits a "personality"** Vine Deloria, Jr., and Daniel Wildcat, *Power and Place: Indian Education in America* (Golden, Colo.: American Indian Graduate Center and Fulcrum Resources, 2001), 23. "Power and place produce personality. This equation simply means that the universe is alive, but it also contains within it the very important suggestion that the universe is personal and, therefore, must be approached in a personal manner."

131 **"what would happen if human beings"** Daniel R. Wildcat, "Indigenizing the Future: Why We Must Think Spatially in the Twenty-first Century," *American Studies* 46, nos. 3/4 (Fall/Winter 2005): 430.

131 **coined the term *infraordinary*** Georges Perec, *Species of Spaces and Other Pieces*, trans. John Sturrock (New York: Penguin Classics, 2008), 210.

132 **novel without using the letter *e*** Georges Perec, *La Disparition* (Paris: Éditions Denoël, 1969).

132 **"A postal van"** Georges Perec, *An Attempt at Exhausting a Place in Paris*, trans. Marc Lowenthal (Cambridge, Mass.: Wakefield Press, 2010), 3.

133 **"time, unarrestable"** Marc Lowenthal, "Translator's Afterword," in Perec, *An Attempt at Exhausting a Place in Paris*, 49–50.

134 **it might "give us peace"** Quoted in Jacey Fortin, "The Birds Are Not on Lockdown, and More People Are Watching Them," *The New York Times*, May 29, 2020, nytimes.com/2020/05/29/science/bird-watching-coronavirus.html.

134 **a 37 percent increase in users** "Birdwatching Surges in Popularity During Covid-19 Pandemic," CBS Pittsburgh, March 3, 2021; Team eBird, "2020 Year in Review: eBird, Macaulay Library, BirdCast, Merlin, and Birds of the World," eBird blog, December 22, 2020, ebird.org/news/2020-year-in-review.

134 **Dollar sales of binoculars** Jacob Swanson, "Backyard Birds See a Popularity Surge During COVID-19 Pandemic," *The Herald-Independent and McFarland Thistle*, February 12, 2021, hngnews.com/mcfarland_thistle/article_be9d26f9-8d90 -58a8-b7c4-087c1579d473.html.

134 **largest-ever monthly increase in downloads** Marc Devokaitis, "Lots of People Are Discovering the Joy of Birding from Home During Lockdown," All About Birds, June 6, 2020, allaboutbirds.org/news/lots-of-people-are-discovering-the -joy-of-birding-from-home-during-lockdown/.

134 **Visits to Cornell's live bird cams** Gillian Flaccus, "Bird-watching Soars amid COVID-19 as Americans Head Outdoors," Associated Press News, May 2, 2020, apnews.com/article/us-news-ap-top-news-ca-state-wire-or-state-wire-virus -outbreak-94a1ea5938943d8a70fe794e9f629b13.

134 **eBird observation rates for suburban species** Team eBird, "Pandemic-Related Changes in Birding May Have Consequences for eBird Research," eBird blog, February 19, 2021, ebird.org/news/pandemic-related-changes-in-birding-may -have-consequences-for-ebird-research; Devokaitis, "Lots of People Are Discovering the Joy of Birding."

135 **"keep [my] eyes peeled for movement"** "Brown Creeper," All About Birds, all aboutbirds.org/guide/Brown_Creeper/.

135 **crows who memorize** Kat McGowan, "Meet the Bird Brainiacs: American Crow," *Audubon*, April 2016, audubon.org/magazine/march-april-2016/meet -bird-brainiacs-american-crow.

135 **the male black manakin** Jennifer Ackerman, *The Bird Way: A New Look at How Birds Talk, Work, Play, Parent, and Think* (New York: Penguin, 2021), 219.

135 **BirdNote's slowed-down recording** "What the Pacific Wren Hears," BirdNote, October 24, 2021, birdnote.org/listen/shows/what-pacific-wren-hears.

135 **predict hurricanes months in advance** Ackerman, *The Bird Way*, 322.

135 **if you showed an expert birder** Megan Prelinger, email to author, May 11, 2022.

136 **Gaviiformes, the order to which** Megan Prelinger, "Loons, Space, Time, and Aquatic Adaptability," in *These Birds of Temptation*, eds. Anna-Sophie Springer and Etienne Turpin (Berlin: K. Verlag, 2022), 258.

136 **new registrations for yard lists** A yard list is a list of species you have observed in your own yard.

136 **"patch lists"** "Patch and Yard Lists in eBird," eBird Help Center, support.ebird .org/en/support/solutions/articles/48001049078-patch-and-yard-lists-in-ebird.

136 **a sparrow-filled patch** J. Drew Lanham, "The United State of Birding," *Audubon*, December 19, 2017, audubon.org/news/the-united-state-birding.

137 **buckeyes normally live** Laura Lukes, "The Buckeye," The Real Dirt Blog (University of California Agriculture and Natural Resources), March 22, 2019, ucanr .edu/blogs/blogcore/postdetail.cfm?postnum=29729.

138 **unlike those other, animal-reliant species** Joe Eaton, "Fall of the Buckeye Ball," *Bay Nature*, October 1, 2008, baynature.org/article/fall-of-the-buckeye -ball/.

140 *Depending on how things go* Andrew Alden told me this would happen if, during the cold phase, the sea were to withdraw for long enough for the terrace, which is being tectonically uplifted, to remain out of reach of the waves when the sea returns during a warm period. (During cold periods, ocean levels drop as water is trapped in glaciers.) He added that "whether human activity has short-circuited the natural glacial cycle" was an open question. For more on how marine terraces are formed, see Doris Sloan, *Geology of the San Francisco Bay Region* (Berkeley: University of California Press, 2006), 18–19.

140 **Kimmerer is invited to consult** Robin Wall Kimmerer, "The Owner," *Gathering Moss: A Natural and Cultural History of Mosses* (Corvallis: Oregon State University Press, 2003), 125–40.

142 **Bergson's notion of intuition** Mark William Westmoreland, "Bergson, Colonialism, and Race," in *Beyond Bergson: Examining Race and Colonialism Through the Writings of Henri Bergson*, eds. Andrea J. Pitts and Mark William Westmoreland, (Albany: State University of New York Press, 2019), 174–78, 192.

142 **cellular decision-making** Barbara Ehrenreich, *Natural Causes: An Epidemic of Wellness, the Certainty of Dying, and Killing Ourselves to Live Longer* (New York: Twelve, 2018), 159.

143 **the "biological" perspective taught in school** Leticia Gallegos Cázares et al., "Models of Living and Non-Living Beings Among Indigenous Community

Children," *Review of Science, Mathematics, and ICT Education* 10, no. 2 (2016): 10, resmicte.library.upatras.gr/index.php/review/article/viewFile/2710/3052.

144 **some respondents grapple** "Are Rocks Dead or Alive?" Quora, quora.com/Are -rocks-dead-or-alive; some of the answers have since been taken down.

144 **the case that rocks can talk** George "Tink" Tinker, "The Stones Shall Cry Out: Consciousness, Rocks, and Indians," *Wicazo Sa Review* 19, no. 2 (January 2004): 106.

144 **"[It] was rooted"** Tinker, "The Stones Shall Cry Out," 106–7.

145 **as though they were inert** The word *inert* comes from the Latin word for "unskilled."

145 **how indigenous cultures have made use** Keith H. Basso, *Wisdom Sits in Places: Landscape and Language Among the Western Apache* (Albuquerque: University of New Mexico Press, 1996).

146 **he "worships" every bird he sees** J. Drew Lanham, interview with Krista Tippett, *On Being,* podcast audio, January 28, 2021, onbeing.org/programs/drew -lanham-pathfinding-through-the-improbable/.

146 **"lesser minds problem"** Juliana Schroeder, Adam Waytz, and Nicholas Epley, "The Lesser Minds Problem," in *Humanness and Dehumanization,* eds. P. G. Bain, Jeroen Vaes, and Jacques-Philippe Leyens (New York: Psychology Press, 2013), 49–67.

146 **"when [participants] are asked to engage"** Schroeder, Waytz, and Epley, "The Lesser Minds Problem," 61.

146 **compelling characters have fully formed hopes** Shonda Rhimes, "Creating Memorable Characters: Part 1," MasterClass, masterclass.com/classes/shonda -rhimes-teaches-writing-for-television.

147 **the filmmakers used lightweight cameras** *Winged Migration,* directed by Jacques Perrin, Jacques Cluzaud, and Michel Debats (BAC Films, 2001), tv.apple .com/us/movie/winged-migration/umc.cmc.6rcayre0fg8iooltodmagrshm.

147 **"in an ideal world"** Nicole R. Pallotta, "Winged Migration (2001) Sony Picture Classics," *Journal for Critical Animal Studies* 7, no. 2 (2009): 143–50; S. Plous, "Psychological Mechanisms in the Human Use of Animals," *Journal of Social Issues* 49, no. 1 (1993): 36.

147 **Their flight path tied these places** The birds' navigation reminds me of one of Bergson's observations: "Instead of saying that animals have a special sense of direction, we may as well say that men have a special faculty of perceiving or conceiving a space without quality." Bergson, *Time and Free Will,* 97.

148 ***Salt may be responsible*** In a 2013 study of the factors affecting tafoni formations, geomorphologist Thomas R. Paradise writes that "although influences of salinity, mineral solubility, lithology, and microclimatic influences are still considered essential," the formation of tafoni still "puzzles" those who study it. He recommends "examin[ing] the hierarchal associations between the many processes known to affect their incipience and progression," allowing that the mix of factors in each place might be different. Paradise also notes that such formations have been observed in rocks on Mars. Thomas R. Paradise, "Tafoni and Other Rock Basins," in *Treatise on Geomorphology, Volume 4,* ed. John F. Shroder (San Diego, Calif.: Academic Press, 2013), 125.

148 **a common origin with *experiment*** *Experience* stems from the Old French *esperience* ("experiment, proof, experience") and, further back, from the Latin *ex* ("out

of") + *peritus* ("experienced, tested"). "Experience (n.)," Online Etymology Dictionary, etymonline.com/word/experience.

148 **demonstrated their own form of experience** Mel Baggs, "In My Language," YouTube video, 8:36, January 14, 2007, youtube.com/watch?v=JnylM1hI2jc. I am indebted to Indira Allegra for sharing this piece with me.

149 **Ana, originally an animal trainer** Ted Chiang, "The Lifecycle of Software Objects," *Exhalation* (New York: Knopf Doubleday, 2019), 62–172. Special thanks to Joshua Batson for recommending and lending this book to me.

151 **"indigenous thinkers not only acknowledge"** Wildcat, "Indigenizing the Future," 422–23.

151 **running experiments** Francis Galton, *Memories of My Life* (New York: E. P. Dutton, 1909), 296.

152 **walking through a familiar town** Bergson, *Time and Free Will*, 129–30.

152 **learning and "creation events"** Yunkaporta, *Sand Talk*, chap. 8.

CHAPTER 5: A CHANGE OF SUBJECT

154 **"Alone, humanity has no future"** Achille Mbembe, "The Universal Right to Breathe," trans. Carolyn Shread, *Critical Inquiry* 47 (Winter 2021): S58–S62.

154 *footprint of an old house* Cornell Barnard, "Pacifica Home on Edge of Cliff Being Demolished," ABC 7 News, January 10, 2018, abc7news.com/pacifica-house-demolition-cliff-home-red-tagged-esplanade-avenue/2925264/.

155 **Solar panels, I read** Kevin Levey (@DrStorminSF), "Zero solar radiation getting the surface," Twitter post, September 9, 2020, twitter.com/DrStorminSF/status/1303763688310992896; Lori A. Carter (@loricarter), "Just checked my Tesla solar app," Twitter post, September 9, 2020, twitter.com/loriacarter/status/1303768636268527616.

157 **one that started much earlier** Elliott Almond, "Red Flag Warning in May? Fire Season Arrives Early in Northern California," *The Mercury News*, May 2, 2021, mercurynews.com/2021/05/02/red-flag-warning-in-may-fire-season-arrives-early-in-northern-california/.

158 **California and, indeed** Chad T. Hanson, *Smokescreen: Debunking Wildfire Myths to Save Our Forests and Our Climate* (Lexington: University Press of Kentucky, 2021), 55–56.

158 **Because this environment is so dry** John McPhee, *The Control of Nature* (New York: Farrar, Straus and Giroux, 1989), 208. McPhee notes that "in a sense, chaparral consumes fire no less than fire consumes chaparral."

158 **species like the lodgepole pine** National Park Service, "Wildland Fire in Lodgepole Pine," nps.gov/articles/wildland-fire-lodgepole-pine.htm; Hanson, *Smokescreen*, 28. A "snag" is a dead tree that is left upright to decompose naturally, providing an important nesting habitat for many species.

159 **the years after a burn** Berkeley Center for New Media, "A Conversation on Wildfire Ecologies [with Margo Robbins and Valentin Lopez]," April 21, 2021, bcnm.berkeley.edu/news-research/4485/video-now-online-margo-robbins-valentin-lopez.

160 **continually fired prairies** Stephen Pyne, *Fire in America: A Cultural History of Wildland and Rural Fire* (Seattle: University of Washington Press, 1997), 79–80. Before writing that "wherever the European went, forests followed," Pyne sug-

gests that "the general consequence of the Indian occupation of the New World was to replace forested land with grassland or savannah, or, where the forest persisted, to open it up and free it from underbrush."

160 **"park-like settings"** William G. Robbins and Donald W. Wolf, "Landscape and the Intermontane Northwest: An Environmental History," United States Department of Agriculture, Forest Service, Pacific Northwest Research Station, February 1994.

160 **"Our landscapes were"** Berkeley Center for New Media, "A Conversation on Wildfire Ecologies."

160 **the Spanish in the eighteenth century** Susie Cagle, " 'Fire Is Medicine': The Tribes Burning California Forests to Save Them," *The Guardian,* November 21, 2019, theguardian.com/us-news/2019/nov/21/wildfire-prescribed-burns -california-native-americans.

160 **California in the nineteenth** Kimberly Johnston-Dodds, "Early California Laws and Policies Related to California Indians," CRB-02-014, September 2002, courts .ca.gov/documents/IB.pdf.

160 **Section 10 of the 1850** "An Act for the Government and Protection of Indians," California State Legislature, §10 (1850), calindianhistory.org/wp-content/uploads /2015/09/04_22_1850_Law.pdf.

161 **some frontierspeople learned** Pyne, *Fire in America,* 100–101.

161 **the budding U.S. Forest Service** Martha Henderson et al., "Fire and Society: A Comparative Analysis of Wildfire in Greece and the United States," *Human Ecology Review* 12, no. 2 (2005): 175.

161 **complained of a New Jersey fire** Franklin Hough, *Report upon Forestry* (Washington, D.C.: Government Printing Office, 1878), 156.

161 **"the history of our race"** Char Miller, "Amateur Hour: Nathaniel H. Egleston and Professional Forestry in Post–Civil War America," *Forest History Today* (Spring/Fall 2005): 20–26. Miller's piece includes the full text of Egleston's "What We Owe to the Trees," originally published in *Harper's New Monthly Magazine* 46, no. 383 (April 1882): 675.

161 **promoting the idea that all fire** Rebecca Miller, "Prescribed Burns in California: A Historical Case Study of the Integration of Scientific Research and Policy," *Fire* 3, no. 3 (2020): 44, mdpi.com/2571-6255/3/3/44/htm#B25-fire-03-00044.

161 **"Paiute forestry"** Pyne, *Fire in America,* 102.

161 **A 1939 poster** James Montgomery Flagg, "Your Forests—Your Fault—Your Loss!" (Washington, D.C.: U.S. Government Printing Office, 1939), archive.org /details/CAT31359639/page/n1/mode/2up.

161 **"Forest Fires Aid the Enemy"** U.S. Department of Agriculture, Forest Service, "Forest Fires Aid the Enemy: Use the Ash Tray," U.S. Government Printing Office, 1943.

161 **One 1953 poster** "This Shameful Waste Weakens America!" (Washington, D.C.: U.S. Government Printing Office, 1953), commons.wikimedia.org/wiki/File: SmokeyBearShamefulWaste1953.jpg.

161 **California found itself at the forefront** "California and the Postwar Suburban Home," Calisphere, University of California Press, calisphere.org/exhibitions /40/california-and-the-postwar-suburban-home/#overview.

161 **a shoddy, cookie-cutter subdivision** The house, which my parents rented, was built in 1952 and was exemplary of the work of Cliff May, an architect who popu-

larized the prefab "miracle house." The houses in the subdivision were priced as low as $8,950 when they were built.

161 **Many of these suburbs** Miller, "Prescribed Burns in California."

162 **the Forest Service was created** Jan W. van Wagtendonk, "The History and Evolution of Wildland Fire Use," *Fire Ecology* 3 (2007): 3, fireecology.springeropen .com/articles/10.4996/fireecology.0302003; Bruce M. Kilgore, "Wildland Fire History—The History of National Park Service Fire Policy," *Interpretation*, Spring 1989, nps.gov/articles/the-history-of-national-park-service-fire-policy.htm; Andrew Avitt, "Tribal and Indigenous Fire Tradition," U.S. Forest Service, November 16, 2021, fs.usda.gov/features/tribal-and-indigenous-heritage; Hilary Beaumont, "New California Law Affirms Indigenous Right to Controlled Burns," Al Jazeera, December 3, 2021, aljazeera.com/news/2021/12/3/new-california-law -affirms-indigenous-right-to-controlled-burns.

162 **In 2021, after an initially** Lauren Sommer, "As California Megafires Burn, Forest Service Ditches 'Good Fire' Under Political Pressure," KQED, August 10, 2021, kqed.org/science/1976195/as-california-megafires-burn-forest-service-ditches -good-fire-under-political-pressure.

162 **"kicking the can"** Quoted in Sommer, "As California Megafires Burn, Forest Service Ditches 'Good Fire' Under Political Pressure."

162 **"If we don't resolve"** Quoted in Sophie Quinton, "To Control Forest Fires, Western States Light More of Their Own," Pew Stateline, May 16, 2019, pewtrusts .org/en/research-and-analysis/blogs/stateline/2019/05/16/to-control-forest -fires-western-states-light-more-of-their-own.

164 **"Continual repairs are required"** Doris Sloan, *Geology of the San Francisco Bay Region* (Berkeley: University of California Press, 2006), 260–61.

164 **In January 2021, a 150-foot chunk** Christopher Reynolds and Erin B. Logan, "23 Miles of Highway 1 near Big Sur Are Closed. Repairs Will Take Months," *Los Angeles Times*, February 1, 2021, latimes.com/travel/story/2021-02-01/23-miles -highway-1-near-big-sur-close-require-repairs.

164 **closed at least fifty-three times** Lisa M. Krieger, "Is Big Sur's Highway 1 Worth Saving?" *The Mercury News*, June 3, 2017. Kathleen Woods Novoa, a Big Sur resident and keeper of a local blog, told KQED in 2017, "The only question we ever have is where it's going to close, when and for how long. . . . We never question that it's going to close somewhere every single winter. And it does, just about."

164 **The San Gabriels** McPhee, *The Control of Nature*, 184.

164 **actually built the flat plain** In "Los Angeles Should Be Buried," Justin Nobel references McPhee, *The Control of Nature*, and adds that the city "was built not upon Hollywood or citrus groves or oil. It was built upon mud and sand and gravel, roughly 1 to 2 million years of debris flows layered upon one another to form a gigantic apron of sediment that rings the mountains and underlies the city. The problem is the mountains are still falling, and the city is still in the way." Justin Nobel, "Los Angeles Should Be Buried," *Nautilus*, June 14, 2019, nautil.us /los-angeles-should-be-buried-2-11054/.

164 **the story of a family** McPhee, *The Control of Nature*, 186.

165 **In the late 1980s** McPhee, *The Control of Nature*, 203.

165 **"The people who buy the houses"** Quoted in McPhee, *The Control of Nature*, 255.

165 **"To guide the flows"** McPhee, *The Control of Nature*, 189.

166 **"PROJECT AIMS TO HALT EROSION"** McPhee, *The Control of Nature*, 258.

166 **"The general public perceptions"** Henderson et al., "Fire and Society," 169–82.

166 **"From place to place"** Victor Steffensen, *Fire Country* (Richmond, Victoria: Hardie Grant Explore, 2020), 38–41.

167 **"The land is not really"** Paula Gunn Allen, "IYANI: It Goes This Way," in *The Remembered Earth: An Anthology of Contemporary Native American Literature,* ed. Geary Hobson (Albuquerque: University of New Mexico Press, 1989), 191.

167 **German scientific forestry** James C. Scott, *Seeing Like a State* (1998; repr. New Haven, Conn.: Yale University Press, 2020), 11–21.

168 **"flesh and things"** Zoe Todd, "Indigenizing the Anthropocene," in *Art in the Anthropocene: Encounters Among Aesthetics, Politics, Environments, and Epistemologies,* eds. Heather Davis and Etienne Turpin (London: Open Humanities Press, 2015), 246. Todd names other scholars pursuing related critiques: Juanita Sundberg, Sarah Hunt, Zakkiyah Iman Jackson, and Vanessa Watts.

169 **"Inherent to the Anthropocene discourse"** Daniel Hartley, "Anthropocene, Capitalocene, and the Problem of Culture," in *Anthropocene or Capitalocene? Nature, History, and the Crisis of Capitalism,* ed. Jason W. Moore (Oakland, Calif.: PM Press, 2016), 156.

169 **"Migration to cities"** Hartley, "Anthropocene, Capitalocene, and the Problem of Culture," 157.

170 **"Somebody, call the cops!"** *I Think You Should Leave,* season 1, episode 5, "I'm Wearing One of Their Belts Right Now," directed by Alice Mathias and Akiva Schaffer, released April 23, 2019, on Netflix.

170 **a religious conception of Man** Sylvia Wynter, "Unsettling the Coloniality of Being/Power/Truth/Freedom: Towards the Human, After Man, Its Overrepresentation—An Argument," *The New Centennial Review* 3, no. 3 (Fall 2003): 265–67.

171 **thinkers like Adam Smith contributed** serynada, "Real Human Being," *The New Inquiry,* March 12, 2015, thenewinquiry.com/real-human-being/. The view that serynada refers to could be described as social Darwinism, which, despite the name, is often used to refer to thinkers from before Darwin's time, such as Thomas Malthus, who wrote *An Essay on the Principle of Population* in 1798. Malthus, Adam Smith, and others would come to influence Darwin in turn, particularly in their emphasis on competition. In his study of Darwin's historical milieu, Loren Eiseley observed that "Darwin incorporated into the Origin of Species a powerful expression of the utilitarian philosophy of his time." Loren Eiseley, *Darwin's Century* (Garden City, N.Y.: Anchor Books, 1961), 348.

171 **"a nonreciprocal, dominance-based relationship"** Naomi Klein, *This Changes Everything: Capitalism vs. The Climate* (New York: Simon and Schuster, 2014), 169.

172 *Residents couldn't seem to agree* "Summary: City of Pacifica Beach Blvd. Infrastructure Resiliency Project Public Workshop," Beach Boulevard Infrastructure Resiliency Project, December 3, 2020. Archived: web.archive.org/web/20211111141854/cityofpacifica.org/civicax/filebank/blobdload.aspx?t=66412.89&BlobID=18000.

172 **"the controversy was at base"** Pyne, *Fire in America,* 101.

173 **In the 1960s** Kate Aronoff, *Overheated: How Capitalism Broke the Planet—and How We Fight Back* (New York: Bold Type, 2021), 133.

174 **scenario planning was much more** Aronoff, *Overheated,* 135.

174 **Shell needed a way** Aronoff, *Overheated,* 136.

174 **Shell "has a constitutive block"** Aronoff, *Overheated,* 140.

174 **Having shifted from funding** Aronoff, *Overheated,* 136–37; Rebecca Leber, "ExxonMobil Wants You to Feel Responsible for Climate Change so It Doesn't Have To," *Vox,* May 13, 2021, vox.com/22429551/climate-change-crisis-exxonmobil -harvard-study.

174 **"[A] 2008 ExxonMobil Corp advertorial"** Naomi Oreskes and Geoffrey Supran, "Rhetoric and Frame Analysis of ExxonMobil's Climate Change Communications," *One Earth* 4, no. 5 (May 2021): 706–8.

175 **It was BP that popularized** Rebecca Solnit, "Big Oil Coined 'Carbon Footprints' to Blame Us for Their Greed. Keep Them on the Hook," *The Guardian,* August 23, 2021, theguardian.com/commentisfree/2021/aug/23/big-oil-coined -carbon-footprints-to-blame-us-for-their-greed-keep-them-on-the-hook.

175 **Klein suggests that the well-off** Klein, *This Changes Everything,* 91.

175 **"if there is to be"** Aronoff, *Overheated,* 8; Douglas Rushkoff, *Survival of the Richest: Escape Fantasies of the Tech Billionaires* (New York: W. W. Norton, 2022), chap. 10.

175 **This rhetoric echoes Big Tobacco's effort** Oreskes and Supran, "Rhetoric and Frame Analysis of ExxonMobil's Climate Change Communications."

175 **Wells Fargo** *Banking on Climate Change: Fossil Fuel Finance Report 2020,* bankingon climatechaos.org/bankingonclimatechange2020/.

176 **"we should be clear"** Klein, *This Changes Everything,* 119.

176 **"In positing all of human existence"** Aronoff, *Overheated,* 141.

178 **Cancer Alley** Thom Davies, "Slow Violence and Toxic Geographies: 'Out of Sight' to Whom?" *Environment and Planning C: Politics and Space* 40, no. 2 (April 2019): 409–27.

179 **the Philippines has seen increased** William N. Holden and Shawn J. Marshall, "Climate Change and Typhoons in the Philippines: Extreme Weather Events in the Anthropocene," in *Integrating Disaster Science and Management: Global Case Studies in Mitigation and Recovery,* ed. Pijush Samui, Dookie Kim, and Chandan Ghosh (Amsterdam, Netherlands: Elsevier, 2018), 413. "The intensity of tropical cyclones making landfall in Southeast Asia increased by 12% from 1977 to 2014, with a doubling of the number of category 4 and 5 typhoons over this time." The authors note that with climate change, typhoons will also carry more moisture and track differently.

179 **In Sitio Nabong** Desmond Ng, "Why Manila Is at Risk of Becoming an Underwater City," Channel News Asia, March 14, 2020, channelnewsasia.com /cnainsider/why-manila-risks-becoming-underwater-city-climate-change-772141.

179 **"Lovely Weather Defined California"** Farhad Manjoo, "Lovely Weather Defined California. What Happens When It's Gone?" *The New York Times,* August 11, 2021, nytimes.com/2021/08/11/opinion/california-climate-change-fires-heat.html.

179 **California farmworker Martha Fuentes** Brian Osgood, "'What Choice Do We Have?' US Farm Workers Battle Deadly Heatwave," Al Jazeera, July 15, 2021, aljazeera.com/economy/2021/7/15/what-choice-do-we-have-us-farm-workers -battle-deadly-heat-wave.

179 **"configured in a future tense"** Kathryn Yusoff, *A Billion Black Anthropocenes or None* (Minneapolis: University of Minnesota Press, 2019), 53.

179 **"without invalidating"** Quoted in Nadine Anne Hura, "How to Centre In-

digenous People in Climate Conversations," *The Spinoff*, November 1, 2019, thespinoff.co.nz/atea/01-11-2019/how-to-centre-indigenous-people-in-climate-conversations.

179 **"post-apocalyptic"** Elissa Washuta, "Apocalypse Logic," *The Offing*, November 21, 2016, theoffingmag.com/insight/apocalypse-logic/.

181 *carrying lime, hay, and lumber* "Channel Islands National Marine Sanctuary Shipwreck Database" (archived), web.archive.org/web/20090120163755/channelislands.noaa.gov/shipwreck/dbase/mbnms/jamesrolph.html.

181 *protested work conditions* Imani Altemus-Williams and Marie Eriel Hobro, "Hawai'i Is Not the Multicultural Paradise Some Say It Is," *National Geographic*, May 17, 2021, nationalgeographic.com/culture/article/hawaii-not-multicultural-paradise-some-say-it-is.

181 **the native Hawaiian workers** Ronald T. Takaki, in *Pau Hana: Plantation Life and Labor in Hawaii, 1835–1920* (Honolulu: University of Hawai'i Press, 1984), 11–12.

181 *devastated by foreign diseases* Takaki, *Pau Hana*, 22.

181 *China, Japan, Norway* Takaki, *Pau Hana*, 21.

181 *arresting laborers* Takaki, *Pau Hana*, 129.

181 *To carry the sugar away* Noel J. Kent, *Hawaii: Islands Under the Influence* (Honolulu: University of Hawai'i Press), 80.

181 *"The land is chief"* Ho'okua'āina, "He ali'i ka 'āina, he kauwā ke kanaka," hookuaaina.org/he-alii-ka-aina-he-kauwa-ke-kanak. In the video that accompanies this explainer, Kānaka Maoli educator Danielle Espiritu discusses how this phrase relates to responsibility and mutual health: "In our communities that we live in now, do we know the health of our streams? Are they thriving? . . . Have they been channelized and cemented? Do we actually have an active relationship with our streams, or with our uplands, or with our valleys? Or are they covered with cement so that water cannot even go in the ground?"

181 *commercial interests on Hawaii* Deborah Woodcock, "To Restore the Watersheds: Early Twentieth-Century Tree Planting in Hawai'i," *Annals of the Association of American Geographers* 93, no. 3 (2003): 624–26; The Nature Conservancy, *The Last Stand: The Vanishing Hawaiian Forest*, nature.org/media/hawaii/the-last-stand-hawaiian-forest.pdf. "The rain follows the forest" is also a Hawaiian proverb.

182 *crashed in the exact spot* "Channel Islands National Marine Sanctuary Shipwreck Database"; Paul Slavin, "A Reminder of the Schooner of James Rolph," *Pacifica*, April 2016, 6.

182 **"an always developing, relentlessly intensifying"** McPhee, *The Control of Nature*, 208.

182 **"Cano was speculating"** McPhee, *The Control of Nature*, 24.

182 **insulating, time-outsmarting quality** Lewis Mumford, *Technics and Civilization* (New York: Harcourt, Brace and Company, 1934), 157.

182 **"a long cyclical change"** Mumford, *Technics and Civilization*, 59.

183 **Centuries after coal** Paul Meier, *The Changing Energy Mix: A Systematic Comparison of Renewable and Nonrenewable Energy* (New York: Oxford University Press, 2020), 102.

183 **legal acknowledgments of the nonhuman world** Elizabeth Kolbert, "A Lake in Florida Suing to Protect Itself," *The New Yorker*, April 11, 2022, newyorker.com

/magazine/2022/04/18/a-lake-in-florida-suing-to-protect-itself. See also Ashley Westerman, "Should Rivers Have Same Legal Rights as Humans? A Growing Number of Voices Say Yes," NPR, August 3, 2019, npr.org/2019/08/03/740604142/should-rivers-have-same-legal-rights-as-humans-a-growing-number-of-voices-say-yes.

183 **"Nature has no rights"** Vine Deloria, Jr., *The Metaphysics of Modern Existence* (Golden, Colo.: Fulcrum 2012), 180–81; Tiffany Challe, "The Rights of Nature—Can an Ecosystem Bear Legal Rights?" State of the Planet, Columbia Climate School, April 22, 2021, news.climate.columbia.edu/2021/04/22/rights-of-nature-lawsuits/; "Tribe Gives Personhood to Klamath River" (interview), *Weekend Edition*, NPR, September 29, 2019, npr.org/2019/09/29/765480451/tribe-gives-personhood-to-klamath-river.

184 **"We are unwell"** Nadine Anne Hura, "Those Riding Shotgun," *PEN Transmissions*, May 6, 2021, pentransmissions.com/2021/05/06/those-riding-shotgun/.

184 **talking about climate adaptation** Seth Heald, "Climate Silence, Moral Disengagement, and Self-Efficacy: How Albert Bandura's Theories Inform Our Climate-Change Predicament," *Science and Policy for Sustainable Development* 59, no. 6 (October 18, 2017): 4–15.

185 *Umwelt* **is a German word** Ed Yong, *An Immense World: How Animal Senses Reveal the Hidden Realms Around Us* (New York: Random House, 2022), 5–6.

186 **a "progressive state" of "improvement"** S. D. Smith, *Slavery, Family, and Gentry Capitalism in the British Atlantic: The World of the Lascelles, 1648–1834* (Cambridge, UK: Cambridge University Press, 2006), 315.

187 **the Greek word** *apokalypsis* Washuta, "Apocalypse Logic."

187 **"we need to lose the world"** Hélène Cixous, *Three Steps on the Ladder of Writing* (New York: Columbia University Press, 1990), 10.

187 **"a season from the planet"** Chen Chen, "When I Grow Up I Want to Be a List of Further Possibilities," from *When I Grow Up I Want to Be a List of Further Possibilities* (Rochester, N.Y.: BOA Editions, 2017).

188 **"employ[ed] the annoying term 'bodies'"** "SPN Bookworthy: 'How to Do Nothing: Resisting the Attention Economy' by Jenny Odell," *Silicon Prairie News*, August 19, 2021, siliconprairienews.com/2021/08/spn-bookworthy-how-to-do-nothing-resisting-the-attention-economy-by-jenny-odell/.

CHAPTER 6: UNCOMMON TIMES

189 **"We live according to the sun"** "Spain Considers Time Zone Change to Boost Productivity," BBC, September 27, 2013, bbc.com/news/world-europe-24294157. The woman, Lola Hidalgo Calle, was a reader responding to the prompt, "Should Spain change time zones?"

191 **"temporal commons"** Allen C. Bluedorn, *The Human Organization of Time: Temporal Realities and Experience* (Stanford, Calif.: Stanford University Press, 2002), 255.

191 **"The idea is not"** Bluedorn, *The Human Organization of Time*, 249.

191 **"quiet time did not just happen"** Bluedorn, *The Human Organization of Time*, 232–34.

191 **"Apparently key elements"** Bluedorn, *The Human Organization of Time*, 234.

192 **the German word** *zeitgeber* Bluedorn, *The Human Organization of Time*, 150.

192 **international trade agreements like NAFTA** Naomi Klein, *This Changes Everything: Capitalism vs. The Climate* (New York: Simon and Schuster, 2014), 71.

193 **"Yep. Thanks for taking my question"** "BP 3Q 2020 Results Webcast: Q&A Transcript," bp.com, 17, bp.com/content/dam/bp/business-sites/en/global /corporate/pdfs/investors/bp-third-quarter-2020-results-qa-transcript.pdf.

193 **"What interests [capital]"** Karl Marx, *Capital* (New York: Penguin, 1990), 1:376.

195 **a "correspondence"** William Grossin, "Evolution Technologique, Temps de Travail et Rémunérations [Technological Evolution, Working Time, and Remuneration]," *Information sur les Sciences Sociales* 29 (June 1990): 357, quoted and trans. in Gabriella Paolucci, "The Changing Dynamics of Working Time," *Time and Society* 5, no. 2 (1996): 150.

195 **"familects," dialects and shorthand** Kathryn Hymes, "Why We Speak More Weirdly at Home," *The Atlantic,* May 13, 2021, theatlantic.com/family/archive /2021/05/family-secret-language-familect/618871/. Thanks to Helen Shewolfe Tseng for mentioning this to me.

196 **Seventh-day Adventist** George R. Knight, *A Brief History of Seventh-day Adventists* (Hagerstown, Md.: Review and Herald Association, 1999), 14–19.

196 **Sabbath on Saturdays** Other Christian denominations observe a seventh-day Sabbath, but Seventh-day Adventism is one of the better-known examples.

196 **gained a toehold** Michael W. Campbell, "Adventist Growth and Change in Asia," *Adventist Review,* March 1, 2018, adventistreview.org/magazine-article/adventist -growth-and-change-in-asia/.

196 **"blasphemous interference"** Eviatar Zerubavel, "The Standardization of Time: A Sociohistorical Perspective," *American Journal of Sociology* 88, no. 1 (July 1982): 18.

196 **Twin Oaks, one of the 1960s communes** John R. Hall, *The Ways Out: Utopian Communal Groups in an Age of Babylon* (London: Routledge, 1978), 55.

196 **until 1911, the French stubbornly refused** Michael O'Malley, *Keeping Watch: A History of American Time* (New York: Viking, 1990), 109.

196 **Chairman Mao Zedong** Matt Schiavenza, "China Only Has One Time Zone— And That's a Problem," *The Atlantic,* November 5, 2013 theatlantic.com/china /archive/2013/11/china-only-has-one-time-zone-and-thats-a-problem/281136/.

196 **During World War II, Germany adopted** Ralf Bosen, "Opinion: Hitler Changed the Clocks—Let's Change Them Back," Deutsche Welle, October 27, 2018, dw.com /en/opinion-hitler-changed-the-clocks-lets-change-them-back/a-46060185.

197 **signal of solidarity with Hitler** Lauren Frayer, "Spain Has Been in the 'Wrong' Time Zone for 7 Decades," *Weekend Edition,* NPR, November 24, 2013, npr.org /sections/parallels/2013/11/30/244995264/spains-been-in-the-wrong-time -zone-for-seven-decades.

197 **For that reason alone** Feargus O'Sullivan, "Why Europe Couldn't Stop Daylight Saving Time," *Bloomberg CityLab,* March 10, 2021, bloomberg.com/news /articles/2021-03-11/will-daylight-saving-time-ever-end; Zoe Chevalier, "Debate over Daylight Saving Time Drags on in Europe," ABC News, November 6, 2021, abcnews.go.com/International/debate-daylight-saving-time-drags-europe /story?id=80925773.

197 **soon after the United States** Michael Downing, *Spring Forward: The Annual Madness of Daylight Saving* (Berkeley, Calif.: Counterpoint, 2005), 10–11.

197 **"In 1965, eighteen states"** Downing, *Spring Forward,* 13.

198 **"when you live in the desert"** Quoted in Scott Craven and Weldon B. John-

son, "Exchange: Arizonans Have No Use for Daylight Saving Time," *U.S. News & World Report,* March 13, 2021, usnews.com/news/best-states/arizona /articles/2021-03-13/exchange-arizonans-have-no-use-for-daylight-saving-time.

198 **Given the patchwork nature** See, for example, the section of Interstate 40 that passes through Chambers, Arizona, on Google Maps or on the Navajo Land Department's Navajo Nation Boundary Map: nnld.org/Home/Maps.

198 **In China, the one holdout** Barbara Demick, "Clocks Square Off in China's Far West," *Los Angeles Times,* March 31, 2009, latimes.com/archives/la-xpm-2009 -mar-31-fg-china-timezone31-story.html; Gardner Bovingdon, "The Not-So-Silent Majority: Uyghur Resistance to Han Rule in Xinjiang," *Modern China* 28, no. 10 (2002): 58.

199 **A sanitation worker in Ürümqi** Javier C. Hernández, "Rise at 11? China's Single Time Zone Means Keeping Odd Hours," *The New York Times,* June 16, 2016, nytimes.com/2016/06/17/world/asia/china-single-time-zone.html.

199 **Local TV networks** Gary Mao, personal communication, April 4, 2022.

199 **Uyghurs have been subjected** "Who Are the Uyghurs and Why Is China Being Accused of Genocide?" BBC News, June 21, 2021, bbc.com/news/world-asia -china-22278037.

199 **Downing jokes about the time** Downing, *Spring Forward,* xviii.

199 **A former Uyghur political prisoner** Maya Wang, "'Eradicating Ideological Viruses': China's Campaign of Repression Against Xinjiang's Muslims," Human Rights Watch, September 9, 2018, hrw.org/report/2018/09/09/eradicating -ideological-viruses/chinas-campaign-repression-against-xinjiangs.

199 **a Javanese proverb:** *Negara mawa tata* Scott, *Seeing Like a State,* 25.

200 **as the French language was imposed** Scott, *Seeing Like a State,* 72.

200 **"two systems of thought"** Quoted in Giordano Nanni, *The Colonisation of Time: Ritual, Routine and Resistance in the British Empire* (Manchester, UK: Manchester University Press, 2012), 179.

200 **One Xhosa group, after burning** Robert Godlonton, *A Narrative of the Irruption of the Kafir Hordes into the Eastern Province of the Cape of Good Hope, 1834–35, Compiled from Official Documents and Other Authentic Sources* (Grahamstown, South Africa: 1836), 140.

200 **a Xhosa group who refused** Nanni, *The Colonisation of Time,* 174–77.

201 **traditional dancing typically restricted** John W. Troutman, *Indian Blues: American Indians and the Politics of Music, 1879–1934* (Norman: University of Oklahoma Press, 2009), 51–53.

201 **Chinese citizens** See the section "Emergence of New Linguistic Forms" in *The Routledge Encyclopedia of the Chinese Language,* ed. Sin-Wai Chan (London: Routledge, 2016), 126–27.

201 **"grass-mud horse"** Xiao Qiang, "The Grass-Mud Horse, Online Censorship, and China's National Identity," Berkeley School of Information, December 5, 2012, ischool.berkeley.edu/news/2012/grass-mud-horse-online-censorship-and -chinas-national-identity.

202 **"public transcript"** James C. Scott, *Domination and the Arts of Resistance* (New Haven, Conn.: Yale University Press, 2008), 27.

202 **J. T. Roane uses the term** *plotting* J. T. Roane, "Plotting the Black Commons," *Souls* 20, no. 3 (2018): 242–44. Thanks to J. T. Roane for responding to my tweet asking about BIPOC writers on commons and commoning.

202 **enslaved people "used the plot"** Roane, "Plotting the Black Commons," 252.

202 **"By hiding in plain sight"** Roane, "Plotting the Black Commons," 244.

202 **"Lakota constructions of time"** Kathleen Pickering, "Decolonizing Time Regimes: Lakota Conceptions of Work, Economy, and Society," *American Anthropologist* 106, no. 1 (March 2004): 87.

203 **"time is money" implied laziness** Pickering, "Decolonizing Time Regimes," 93.

203 **"Time was never a specific minute"** Pickering, "Decolonizing Time Regimes," 92.

203 **"the settler who brings the center"** Fred Moten, "Come On, Get It!" *The New Inquiry,* February 19, 2018, thenewinquiry.com/come_on_get_it/.

203 **a local town council erected** Nanni, *The Colonisation of Time,* 226.

203 **"Filipino time"** Miguel A. Bernad, "Filipino Time," *Budhi: A Journal of Ideas and Culture* 3, nos. 2–3 (2002): 211–12. Bernad, who is not a fan of Filipino Time, nonetheless observes that "some Filipinos actually glory in their inability to be punctual, cheerfully citing 'Filipino time' as if it were something to brag about."

203 **Filipino time has outlived its welcome** Brian Tan, "Why Filipinos Follow Filipino Time," Medium, March 23, 2016, medium.com/@btantheman/why-filipinos-follow-filipino-time-d38e2c162927.

204 **she was "drawn to CPT"** *Colored People Time: Mundane Futures, Quotidian Pasts, Banal Presents* (curatorial text), Institute of Contemporary Art University of Pennsylvania, 2019, icaphila.org/wp-content/uploads/2018/10/Mundane-Futures-No-Bleed.pdf; John Hopkins, "A Look at Indian Time," *Indian Country Today,* September 13, 2018, indiancountrytoday.com/archive/a-look-at-indian-time.

205 **"Study is what you do"** Stefano Harney and Fred Moten, *The Undercommons: Fugitive Planning and Black Study* (New York: Minor Compositions, 2013), 110.

206 **"mental barrier"** Mehmet Bayram at "The Gig Economy, AI, Robotics, Workers, and Dystopia San Francisco," ILWU Local 24 Hall, San Francisco, July 10, 2019.

206 **used paper from local branches** Chris Carlsson (former editor of *Processed World*), emails to author, February 4 and July 2, 2022. Those interested can find the entire archive of *Processed World* on the Internet Archive at archive.org/details/processedworld.

207 **"the latest advances"** "BFB: Can Modern Technology Improve the Human Brain?" *Processed World* 6 (November 1982): 34.

207 **a doctored conference program** "Not Just Words, Disinformation," *Processed World* 5 (July 1982): 34.

207 **the certificate for "Team Spirit"** "Awarded to Gidgit Digit for Outstanding Service to the Bank—1982," *Processed Word* 5 (July 1982): 18. Elswhere in the issue, the name is spelled "Gidget Digit." Both spellings are pseudonyms for Stephanie Klein.

208 **Time is the one thing** "Labor Theory of Value?" *Processed World* 2 (July 1981): 34.

208 **Founders of the magazine** *Bad Attitude: The Processed World Anthology,* eds. Chris Carlsson and Mark Leger (New York: Verso, 1990), 7.

209 **"it is unlikely"** Gidgit Digit, "Sabotage! The Ultimate Video Game," *Processed World* 5 (July 1982): 25.

209 **a niche concern** One reader notably implored *Processed World* to pay "special attention . . . to not 'ghetto-izing' your concerns, to only addressing the young, the white, and the 'hip.'" *Processed World* 6 (November 1982): 6.

209 **"JUMPIN' JEHOSEPHAT!"** Letter to the editor, *Processed World* 2 (July 1981): 5.

210 **they'd received issues 4 and 5** Letter to the editor, *Processed World* 6 (November 1982): 10.

210 **"One day, at seven a.m."** Letter to the editor, *Processed World* 11 (August 1984): 6–7.

210 **food delivery gig workers** Arianna Tassinari and Vincenzo Maccarrone, "Riders on the Storm: Workplace Solidarity Among Gig Economy Couriers in Italy and the UK," *Work, Employment and Society* 34, no. 1 (February 2020): 45.

210 **"Don't take a poo!"** Emily Reid-Musson, Ellen MacEachen, and Emma Bartel, "'Don't Take a Poo!': Worker Misbehaviour in On-Demand Ride-Hail Carpooling," *New Technology, Work and Employment* 35, no. 2 (July 2020): 153, 156.

211 **"someone is always willing"** Russell Brandom, "The Human Cost of Instacart's Grocery Delivery," The Verge, May 26, 2020, theverge.com/21267669/instacart-shoppers-sick-extended-pay-quarantine-leave-coronavirus.

211 **"They'll just take the job somewhere else"** A. J. Wood, V. Lehdonvirta, and M. Graham, "Workers of the Internet Unite? Online Freelancer Organisation Among Remote Gig Economy Workers in Six Asian and African Countries," *New Technology, Work and Employment* 33 (2018): 105.

211 **Uber workers in different countries** Wilfred Chan, "Gig Workers of the World Are Uniting," *The Nation,* June 1, 2021, thenation.com/article/activism/global-gig-worker-organizing/.

212 **an actually creative activity** Oli Mould, *Against Creativity* (London: Verso, 2018), 50.

212 **"the universal market"** Harry Braverman, *Labor and Monopoly Capital: The Degradation of Work in the Twentieth Century* (New York: Monthly Review, 1998), 188–96.

212 **"the time they do spend"** Louron Pratt, "Awin to Introduce Four-Day Working Week," Employee Benefits, December 15, 2020, employeebenefits.co.uk/awin-four-day-working-week/. See also Perpetual Guardian, "We Are Taking the 4 Day Week Global," 4dayweek.com/; and Carolyn Fairbairn quoted in Larry Elliott, "John McDonnell Pledges Shorter Working Week and No Loss of Pay," *The Guardian,* September 23, 2019, theguardian.com/politics/2019/sep/23/john-mcdonnell-pledges-shorter-working-week-and-no-loss-of-pay.

212 **"humanizing" work** Braverman, *Labor and Monopoly Capital*, 26–27.

213 **"The individual 'freedoms' that are created"** Digit, "Sabotage! The Ultimate Video Game," 25.

213 **"mutant grammar" of refusal** Harney and Moten, *The Undercommons,* 52.

213 **"To refuse is to say no"** Carole McGranahan, "Theorizing Refusal: An Introduction," *Cultural Anthropology* 31, no. 3 (2016): 319.

214 **Marilyn Waring's 1988 book** The book was also published under the title *Counting for Nothing: What Men Value and What Women Are Worth.*

214 **"art of the dumb question"** *Who's Counting? Marilyn Waring on Sex, Lies and Global Economics,* directed by Terre Nash (National Film Board of Canada, 1995), available to stream for free at nfb.ca/film/whos_counting/.

216 **She and others in the movement** Camila Valle and Selma James, "Real Theory Is in What You Do and How You Do It," Verso Blog, January 11, 2021, versobooks.com/blogs/4962-real-theory-is-in-what-you-do-and-how-you-do-it.

216 **Guaranteed Adequate Income (GAI)** Johnnie Tillmon, "Welfare as a Women's Issue," in *Major Problems in American Urban History,* ed. Howard P. Chudacoff,

(Lexington, Mass.: D.C. Heath, 1994), 426–29. Tillmon was the chairperson of the National Welfare Rights Organization when she wrote this essay in 1972.

216 **women were slaves to wage slaves** Mariarosa Dalla Costa and Selma James, *The Power of Women and Subversion of the Community* (London: Falling Wall Press, 1972), 41.

216 **"where women are concerned"** Dalla Costa and James, *The Power of Women and Subversion of the Community*, 28.

216 **James had read the first volume** "Video: 'Sex, Race and Class'—Extended Interview with Selma James on Her Six Decades of Activism," Democracy Now!, April 18, 2012, democracynow.org/2012/4/18/video_sex_race_and_class _extended_interview_with_selma_james_on_her_six_decades_of_activism.

217 **"The ability to labor resides"** Dalla Costa and James, *The Power of Women and Subversion of the Community*, 11; emphasis in the original.

217 **a set of others** In her 1972 talk at the National Women's Liberation Movement Conference in Manchester, James's coinage of "wages for housework" appears under a demand for guaranteed income "for women and for men working or not working, married or not." Selma James, *Women, the Unions, and Work—Or . . . What Is Not to Be Done, and the Perspective of Winning* (London: Falling Wall Press, 1976), 67–69.

217 **By imputing value to women's work** Dalla Costa and James, *The Power of Women and Subversion of the Community*, 40.

217 **emphasizes its use of the *demand*** Kathi Weeks, *The Problem with Work: Feminism, Marxism, Antiwork Politics, and Postwork Imaginaries* (Durham, N.C.: Duke University Press, 2011), 131–36.

218 **Selma James does not support** Selma James, "I Founded the Wages for Housework Campaign in 1972—and Women Are Still Working for Free," *The Independent*, March 8, 2020, independent.co.uk/voices/international-womens -day-wages-housework-care-selma-james-a9385351.html.

218 **"To 'have time'"** Dalla Costa and James, *The Power of Women and Subversion of the Community*, 40.

218 **The demand for less work** Weeks, *The Problem with Work*, 145.

218 **In 1980, a year before** Chris Carlsson, "What Do San Franciscans Do All Day? Information Work," FoundSF, foundsf.org/index.php?title=WHAT_DO_SAN _FRANCISCANS_DO_ALL_DAY%3F_Information_Work; Chris Carlsson and Mark Leger, eds., *Bad Attitude: The Processed World Anthology*, 8.

219 **a melancholic collage** "Another Day at the Office: What Have We Lost?" *Processed World* 6 (November 1982): 32–33.

219 **"What does one do"** J. C., letter to the editor, *Processed World* 7 (Spring 1983): 10.

219 **"Dear PW"** J. Gulesian, letter to the editor, *Processed World* 5 (July 1982): 8.

219 **"making yourself worth more"** Walter E. Wallis, letter to the editor, *Processed World* 5 (July 1982): 5–8. Michaelson's response to this letter is the source of the expression "marching in lockstep toward the abyss," which I used in the introduction and in this chapter.

220 **Global Women's Strike** Some of the GWS's recent activities include supporting the proposed Care Income that is part of the Green New Deal for Europe, creating Election Action for Caregivers during the 2020 U.S. election, and working with the Poor People's Campaign.

220 **"I didn't feel, walking with them"** "Video: 'Sex, Race and Class.'"

223 **"innumerable sets of infrastructures"** Ivan Illich, *The Right to Useful Unemployment and Its Professional Enemies* (London: Marion Boyars, 1978), 26.

223 **"You just sign up"** A. J. Ravenelle, K. C. Kowalski, and E. Janko, "The Side Hustle Safety Net: Precarious Workers and Gig Work During Covid-19," *Sociological Perspectives* (June 2021): 10.

224 **"routine can demean"** Richard Sennett, *The Corrosion of Character: The Personal Consequences of Work in the New Capitalism* (New York: W. W. Norton, 1998), 43.

224 **"a palace that we build"** Abraham Joshua Heschel, *The Sabbath: Its Meaning for Modern Man* (New York: Farrar, Straus and Giroux, 2005), 14–15.

224 **"Tempo and intensity surround us"** Barbara Adam, *Timewatch* (Cambridge, UK: Polity, 1995), chap. 1.

CHAPTER 7: LIFE EXTENSION

227 **"Resonance, by contrast"** Hartmut Rosa, *Resonance: A Sociology of Our Relationship to the World*, trans. James C. Wagner (Cambridge, UK: Polity, 2019), chap. 6.

228 **A boy who is impatient** "The Magic Thread," in *Magic Fairy Stories from Many Lands*, ed. Susan Taylor (New York: Gallery, 1974), 123–38. Special thanks to my mom for digging up this book.

229 **he asks you to hold** Kevin Kruse, *15 Secrets Successful People Know About Time Management: The Productivity Habits of 7 Billionaires, 13 Olympic Athletes, 29 Straight-A Students, and 293 Entrepreneurs* (Philadelphia: Kruse Group, 2015), chap. 1.

229 **keeping a detailed log** Oliver Burkeman, "Why Time Management Is Ruining Our Lives," *The Guardian*, December 22, 2016, theguardian.com/technology /2016/dec/22/why-time-management-is-ruining-our-lives.

229 **questions about my lifestyle** "When Will I Die?—Calculator," apps.apple.com /us/app/when-will-i-die-calculator/id1236569653.

230 **Some automobile insurance companies use telematics** Alex Galley, "Your Auto Insurer Wants to Ride Shotgun With You. Are the Savings Worth It?" *Time*, April 26, 2021, time.com/nextadvisor/insurance/car/telematics-monitor-driving-insurance -discount/; "Beam Dental Blog: Group-Life-Benefits," blog.beam.dental/tag/group -life-benefits. "Beam Perks" are described at the bottom of the page in fine print.

231 **the product offered** Barbara Ehrenreich, *Natural Causes: An Epidemic of Wellness, The Certainty of Dying, and Killing Ourselves to Live Longer* (New York: Twelve, 2019), 111.

231 **"successful aging"** Ehrenreich, *Natural Causes*, 164–65.

233 **"crip time"** Alison Kafer, *Feminist, Queer, Crip* (Bloomington: Indiana University Press, 2013), 26–27. See also Ellen Samuels's "Six Ways of Looking at Crip Time," in *Disability Visibility* (New York: Vintage, 2020), 189–96.

233 **it can mean "bigger systemic fits"** Sara Hendren, *What Can a Body Do? How We Meet the Built Environment* (New York: Riverhead, 2020), 117.

233 **"questions about time"** Hendren, *What Can a Body Do?*, 173.

233 **"invitation to life on crip time"** Hendren, *What Can a Body Do?*, 181.

233 **"The economic tempo of the clock"** Hendren, *What Can a Body Do?*, 180.

233 **"a form of able-bodied productivity"** Hendren, *What Can a Body Do?*, 167.

233 **"economic productivity"** Hendren, *What Can a Body Do?*, 180.

234 **"the insistent, clock-driven measuring of childhood"** Hendren, *What Can a Body Do?*, 172.

234 **"How long does it take"** Hendren, *What Can a Body Do?*, 167.

235 **the idea of human enhancement** *Fixed: The Science/Fiction of Human Enhancement*, directed by Regan Brashear (New Day Films, 2014), fixed.vhx.tv/.

235 **with the Covid-19 pandemic** Ed Yong, "Our Pandemic Summer," *The Atlantic*, April 14, 2020, theatlantic.com/health/archive/2020/04/pandemic-summer-coronavirus-reopening-back-normal/609940/.

236 **standard schedules and expectations** The disabled organizer, writer, and social worker K. Agbebiyi has called for unlimited paid time off for disabled workers. See prismreports.org/2022/02/08/unlimited-paid-time-off-is-a-disability-justice-issue-that-needs-to-be-taken-seriously/.

236 **he developed a habit of swimming** Thanks to Steven for permission to share this story.

236 **"divisive tiredness"** Byung-Chul Han, *The Burnout Society* (Stanford, Calif.: Stanford University Press, 2015), 31–33.

238 **they would have been wintering** "Cedar Waxwing," All About Birds, allaboutbirds.org/guide/Cedar_Waxwing/overview.

239 **"People do not spring up"** Quoted in Hendren, *What Can a Body Do?*, 127.

239 **Desmond Tutu's description** Mia Birdsong, *How We Show Up* (New York: Hachette, 2020), chap. 1.

239 **a year that had "awakened us"** B. J. Miller, "What Is Death?" *The New York Times*, December 18, 2020, nytimes.com/2020/12/18/opinion/sunday/coronavirus-death.html.

240 *Crocker was certain* "The Chinese Workers' Strike," *American Experience*, PBS, pbs.org/wgbh/americanexperience/features/tcrr-chinese-workers-strike/.

240 *"strangers' plot"* Carolyn Jones, "Oakland Strangers' Plot Full of Mysteries," SFGate, January 24, 2011, sfgate.com/bayarea/article/Oakland-Strangers-Plot-full-of-mysteries-2478631.php. This article notes that twenty-two of the bodies were from an 1880 explosion at a dynamite plant (Giant Powder) in Berkeley. Marketing their dynamite as "Miner's Friend," Giant Powder also recommended it for blasting railroad tunnels. Andrew Mangravite, "Meeting the Miner's Friend," Science History Institute, May 1, 2013, sciencehistory.org/distillations/meeting-the-miners-friend.

241 **"a figurative silencing"** Sharon P. Holland, *Raising the Dead: Readings of Death and (Black) Subjectivity* (Durham, N.C.: Duke University Press, 2000), 18.

241 **"Reduced to the machinery"** Quoted in Holland, *Raising the Dead*, 28.

242 **The border between living and dead** Holland, *Raising the Dead*, 29.

242 **one out of twelve Black men** "Trends in U.S. Corrections," The Sentencing Project, May 2021, sentencingproject.org/wp-content/uploads/2021/07/Trends-in-US-Corrections.pdf.

242 **decline in prison educational programs** Angela Y. Davis, *Are Prisons Obsolete?* (New York: Seven Stories Press, 2011), 38–39.

242 **The ban was finally lifted** Lilah Burke, "After the Pell Ban," *Inside Higher Ed*, January 27, 2021, insidehighered.com/news/2021/01/27/pell-grants-restored-people-prison-eyes-turn-assuring-quality. In this article, Mary Gould, director of the Alliance for Higher Education in Prison, describes the challenge of ensuring the quality of post-secondary education in prison, and seeing that inmates are not simply exploited by institutions as a revenue stream. She also cautions against

treating education in prison as purely corrective, which, among other things, would limit opportunities only to those who are guaranteed release.

242 **the prison-industrial complex** Angela Y. Davis, *Are Prisons Obsolete?*, 84–104.

243 **"toxic-waste-dump prison"** Jonathan Simon, *Governing Through Crime: How the War on Crime Transformed American Democracy and Created a Culture of Fear* (Oxford, UK: Oxford University Press, 2007), 142.

243 **Rehabilitative programs do still exist** Nina Totenberg, "High Court Rules Calif. Must Cut Prison Population," *All Things Considered,* NPR, May 23, 2011, npr .org/2011/05/23/136579580/california-is-ordered-to-cut-its-prison-population; Jazmin Ulloa, "Despite an Emphasis on Inmate Rehab, California Recidivism Rate Is 'Stubbornly High,'" *Los Angeles Times,* January 31, 2019, latimes.com/politics /la-pol-ca-prison-rehabilitation-programs-audit-20190131-story.html.

243 **"Project Exile"** Simon, *Governing Through Crime,* 141–76.

243 **Over the last three decades** Ashley Nellis, "No End in Sight: America's Enduring Reliance on Life Imprisonment," The Sentencing Project, February 17, 2021, sentencingproject.org/publications/no-end-in-sight-americas-enduring-reliance -on-life-imprisonment/.

243 **the denial of educational programming** Ashley Nellis, "A New Lease on Life," The Sentencing Project, June 30, 2021, sentencingproject.org/publications/a -new-lease-on-life/.

244 "**If you've been in for ten**" Pendarvis Harshaw and Brandon Tauszik, "Lynn Acosta," in *Facing Life: Eight Stories of Life After Life in California's Prisons,* facing.life.

244 **"accelerated aging" in incarcerated populations** Maurice Chammah, "Do You Age Faster in Prison?" The Marshall Project, August 24, 2015, themarshallproject .org/2015/08/24/do-you-age-faster-in-prison.

244 **"What is Prison?"** Jackie Wang, *Carceral Capitalism* (South Pasadena, Calif.: Semiotext(e), 2018), 218.

244 **Sibil Fox Richardson, a formerly incarcerated** *Time,* directed by Garrett Bradley (Amazon Studios, 2020), amazon.com/Time-Fox-Rich/dp/B08J7DDGJY.

245 **the interior of the prison** Ismail Muhammad, "A Filmmaker Who Sees Prison Life with Love and Complexity," *The New York Times,* October 6, 2020, nytimes .com/2020/10/06/magazine/time-prison-documentary-garrett-bradley.html.

246 **"Never think or believe"** Joshua M. Price, *Prison and Social Death* (New Brunswick, N.J.: Rutgers University Press, 2015), chap. 8.

246 **exhaustive list of rights denied** Price, *Prison and Social Death,* chap. 8.

246 **In April 2022** Romina Ruiz-Goiriena, "Exclusive: HUD Unveils Plan to Help People with a Criminal Record Find a Place to Live," *USA Today,* April 12, 2022, usatoday.com/story/news/nation/2022/04/12/can-get-housing-felony-hud -says-yes/9510564002/; "Federal Financial Aid for College Students with Criminal Convictions: A Timeline," Center for American Progress, December 17, 2020, americanprogress.org/article/federal-financial-aid-college-students-criminal -convictions/; Claire Child and Stephanie Clark, "Op-ed: End the Lifetime Ban on SNAP for Felony Drug Convictions," Civil Eats, March 18, 2022, civileats.com/2022/03/18/op-ed-end-the-lifetime-ban-on-snap-for-felony-drug -convictions/.

246 *spirit murder* Patricia Williams, *The Alchemy of Race and Rights: Diary of a Law Professor* (Cambridge, Mass.: Harvard University Press, 1991), 73.

247 **"The hidden cost of spirit murder"** Price, *Prison and Social Death,* chap. 8.

247 **the closing of Rikers Island** Richard Schiffman, "The Secret Jailhouse Garden of Rikers Island," *The New York Times,* October 4, 2019, nytimes.com/2019/10/04/nyregion/garden-rikers-island.html.

248 **"It is a constitutive"** Viktor Frankl, "Self-Transcendence as a Human Phenomenon," *Journal of Humanistic Psychology* 6, no. 2 (1966): 97.

248 **"Does the report say anything"** Comment by "The Dodger" on Tom Jackman, "Study: 1 in 7 U.S. Prisoners Is Serving Life, and Two-thirds of Those Are People of Color," *The Washington Post,* March 2, 2021, washingtonpost.com/nation/2021/03/02/life-sentences-growing/.

249 **"an unmeasured expenditure of energy"** Ta-Nehisi Coates, *Between the World and Me* (New York: Spiegel and Grau, 2015), 91.

249 **Galanter and a colleague were visiting** Marc Galanter, *Cults: Faith, Healing, and Coercion* (Oxford, UK: Oxford University Press, 1999), 25–26.

250 **Seeing the human body** Michael Kranish and Marc Fisher, *Trump Revealed: The Definitive Biography of the 45th President* (New York: Scribner, 2017), 181.

250 **"My son doesn't need"** Hendren, *What Can a Body Do?,* 245.

251 **"I have hope for humankind"** Albert Woodfox, *Solitary: Unbroken by Four Decades in Solitary Confinement, My Story of Transformation and Hope* (New York: Grove Atlantic, 2019), 408.

251 **Woodfox cites** Woodfox, *Solitary,* 409.

251 **"Superiority? Inferiority?"** Quoted in Woodfox, *Solitary,* 407.

252 **A dehumanizing bias** Juliana Schroeder, Adam Waytz, and Nicholas Epley, "The Lesser Minds Problem," in *Humanness and Dehumanization,* eds. P. G. Bain, J. Vaes and J.-P. Leyens (New York: Psychology Press, 2014), 59–60.

253 **"a picture of the United Kingdom"** *Seven Up!* directed by Paul Almond (ITV [Granada Television], 1964); *7 Plus Seven,* directed by Michael Apted (ITV [Granada Television], 1970); *21 Up,* directed by Michael Apted (ITV [Granada Television], 1977); *28 Up,* directed by Michael Apted (ITV [Granada Television], 1984); *35 Up,* directed by Michael Apted (ITV [Granada Television], 1991); *42 Up,* directed by Michael Apted (BBC, 1998); *49 Up,* directed by Michael Apted (ITV, 2005); *56 Up,* directed by Michael Apted (ITV, 2012); *63 Up,* directed by Michael Apted (ITV, 2019).

253 **Apted begins to shift** Gideon Lewis-Kraus, "Does Who You Are at 7 Determine Who You Are at 63?" *The New York Times,* November 27, 2019, nytimes.com/2019/11/27/magazine/63-up-michael-apted.html.

254 **Apted himself passed away** Joe Coscarelli, "What Happens Now to Michael Apted's Lifelong Project 'Up'?" *The New York Times,* January 14, 2021, nytimes.com/2021/01/14/movies/michael-apted-up-series-future.html.

254 **"empathy machine"** Roger Ebert, "Ebert's Walk of Fame Remarks," June 24, 2005, rogerebert.com/roger-ebert/eberts-walk-of-fame-remarks.

256 **"deep-seated self"** Henri Bergson, *Time and Free Will: An Essay on the Immediate Data of Consciousness,* trans. F. L. Pogson (Mineola, N.Y.: Dover Publications, 2001), 125. Bergson writes of a "deep-seated self which ponders and decides, which heats and blazes up . . . a self whose states and changes permeate one another."

256 **covered in bark and sealed off** Seth Kugel, "In Indonesia, a Region Where Death Is a Lure," *The New York Times,* July 30, 2015, nytimes.com/2015/07/30/travel/frugal-traveler-indonesia-death-rituals.html.

256 **left at high elevations** *Vultures of Tibet,* directed by Russell O. Bush (New Day Films, 2013).

256 **noting from a physical standpoint** Miller, "What Is Death?"

256 **life is not yours alone** Yuri Kochiyama, *Passing It On: A Memoir* (Los Angeles: UCLA Asian American Studies Center, 2004), xx.

257 **"It is one thing to die"** Ehrenreich, *Natural Causes,* 208.

258 **Handke's description of touching** Han, *The Burnout Society,* 32.

258 **in a state of complete attention** Jiddu Krishnamurti, *Freedom from the Known* (New York: HarperOne, 2009), 102.

258 **"it is only a mind"** Krishnamurti, *Freedom from the Known,* 90.

258 **"no yesterday and no tomorrow"** Krishnamurti, *Freedom from the Known,* 87.

259 *a metamorphic rock that once bubbled* Doris Sloan, *Geology of the San Francisco Bay Area* (Berkeley: University of California Press, 2006), 55–56; Andrew Alden, "The Big Set of Knockers, Mountain View Cemetery," Oakland Geology, June 4, 2008, oaklandgeology.com/2008/06/04/the-big-set-of-knockers-mountain -view-cemetery/.

CONCLUSION: HALVING TIME

262 **"Scientists are saying"** *Southland Tales,* directed by Richard Kelly (Universal Pictures, 2007).

262 **"No one is responsible"** Michel Foucault, "Nietzsche, Genealogy, History," in *Language, Counter-Memory, Practice: Selected Essays and Interviews,* ed. D. F. Bouchard (Ithaca, N.Y.: Cornell University Press, 1977), 150.

262 **the California King Tides Project** "About the King Tides Project," California Coastal Commission, coastal.ca.gov/kingtides/learn.html.

263 **"You see this goblet?"** Quoted in Mark Epstein, *Thoughts Without a Thinker: Psychotherapy from a Buddhist Perspective* (New York: Basic Books, 2013), 80. I encountered this anecdote in a talk by Tara Brach. See Tara Brach, "Part 2: Impermanence—Awakening Through Insecurity," September 26, 2018, tarabrach .com/pt-2-impermanence-awakening-insecurity/.

264 **"It will no longer do"** Henri Bergson, *Time and Free Will: An Essay on the Immediate Data of Consciousness,* trans. F. L. Pogson (Mineola, N.Y.: Dover Publications, 2001), 198.

265 **bioregionalism, a sense of familiarity** Jenny Odell, *How to Do Nothing* (Brooklyn, N.Y.: Melville House, 2019), 149–51.

265 **"a steep night, a tangled week"** Unpublished work shared with permission, from an email with John Shoptaw, July 13, 2021.

265 **some speculated to have pushed** Catherine Brahic, "Sudden Growth Spurt Pushed the Andes Up Like a Popsicle," *New Scientist,* June 6, 2008, newscientist .com/article/dn14073-sudden-growth-spurt-pushed-the-andes-up-like-a -popsicle/.

265 **massive mudslide around 5,700 years ago** Joe D. Dragovich, Patrick T. Pringle, and Timothy J. Walsh, "Extent and Geometry of the Mid-Holocene Osceola Mudflow in the Puget Lowland—Implications for Holocene Sedimentation and Paleogeography," *Washington Geology* 22, no. 3 (September 1994): 3.

266 **The Nisqually oral tradition** Patrick Nunn, *The Edge of Memory: Ancient Stories, Oral Tradition and the Post-Glacial World* (New York: Bloomsbury, 2018), chap. 6;

Vine Deloria, Jr., *Red Earth, White Lies: Native Americans and the Myth of Scientific Fact* (Golden, Colo.: Fulcrum, 1997), chap. 8.

266 **the continent I am on** Sid Perkins, "Meet 'Amasia,' the Next Supercontinent," *Science,* February 8, 2012, science.org/content/article/meet-amasia-next -supercontinent.

266 **rupture can happen at ten times** "How Earthquakes Break the Speed Limit," UC Berkeley Seismology Lab, March 8, 2019, seismo.berkeley.edu/blog/2019/03/08 /how-earthquakes-break-the-speed-limit.html.

266 **Brood X** Michael D. Shear, "Cicadas Took On Biden's Press Plane. They Won," *The New York Times,* June 9, 2021, nytimes.com/2021/06/09/us/politics/cicadas -biden.html.

266 **An arborist in Royal Oak** Frank Witsil, "Trees Across Metro Detroit Suddenly Dropping Tons of Acorns: Here's Why," *Detroit Free Press,* September 30, 2021.

266 **masting in pecan trees** Robin Wall Kimmerer, *Braiding Sweetgrass: Indigenous Wisdom, Scientific Knowledge, and the Teachings of Plants* (Minneapolis, Minn.: Milkweed Editions, 2013), 19–20.

266 **a bristlecone pine tree** Alex Ross, "The Past and the Future of the Earth's Oldest Trees," *The New Yorker,* January 13, 2020, newyorker.com/magazine /2020/01/20/the-past-and-the-future-of-the-earths-oldest-trees.

266 **Neskowin Ghost Forest** Hugh Morris, "How a Powerful Earthquake Created Oregon's Eerie Seaside Ghost Forest," *The Telegraph,* July 31, 2018, telegraph .co.uk/travel/destinations/north-america/united-states/articles/neskowin -ghost-forest-oregon/; Justin Sharick, comment on "Neskowin Ghost Forest," Google Maps, goo.gl/maps/81EV3ng3H4Lqnwou9; Trypp Adams, comment on "Neskowin Ghost Forest," Google Maps, goo.gl/maps/wh2maRZ8SVCRYFpw5.

266 **a recognizable plant community** Carolyn J. Strange, "Serpentine Splendor," *Bay Nature,* April 1, 2004, baynature.org/article/serpentine-splendor/. Serpentine is also the state rock of California.

267 **skyscrapers tend to be built** *Rise of the Continents,* season 1, episode 3, "The Americas," directed by Arif Nurmohamed, aired April 3, 2013, on BBC Two; Helen Quinn, "How Ancient Collision Shaped the New York Skyline," BBC News, June 7, 2013, bbc.com/news/science-environment-22798563.

267 **An enormous fungal network** Anne Casselman, "Strange but True: The Largest Organism on Earth Is a Fungus," *Scientific American,* October 4, 2007, scientificamerican.com/article/strange-but-true-largest-organism-is-fungus/.

268 **Pando, a clonal colony** Christopher Ketcham, "The Life and Death of Pando," *Discover Magazine,* October 18, 2018, discovermagazine.com/planet-earth/the -life-and-death-of-pando.

268 **a summer fire in 1977** John McPhee, *The Control of Nature* (New York: Farrar, Straus and Giroux, 1989), 216–17.

269 **an event within an event** Stephanie Pappas, "The Longest Known Earthquake Lasted 32 Years," *Scientific American,* May 26, 2021, scientificamerican.com/article /the-longest-known-earthquake-lasted-32-years/.

269 **the process of personal deliberation** Bergson, *Time and Free Will,* 176.

269 **"For the individuality to be perfect"** Henri Bergson, *Creative Evolution,* trans. Arthur Mitchell (Lanham, Md.: University Press of America, 1983), 13.

270 **"it is hard to write"** Tyson Yunkaporta, *Sand Talk: How Indigenous Thinking Can Save the World* (New York: HarperCollins, 2020), chap. 3.

270 **Sarah (played by a young)** *Labyrinth,* directed by Jim Henson (Tri-Star Pictures, 1986). You can see this specific scene on YouTube: The Jim Henson Company, "Worm—Labyrinth—The Jim Henson Company," YouTube video, January 6, 2011, youtube.com/watch?v=l0K5T0AqVlY.

271 **a man who "tried going"** Yunkaporta, *Sand Talk,* chap. 1.

272 *kairos* **in modern Greek now means** Astra Taylor, "Out of Time," *Lapham's Quarterly,* September 16, 2019, laphamsquarterly.org/climate/out-time.

272 **The etymology of the word** *doubt* "Doubt (n.)," Online Etymology Dictionary, etymonline.com/word/doubt.

273 **Bergson was willing to grant** Bergson, *Time and Free Will,* 169.

273 **A similar practice is observed** Fei Lu, "Zuo Yuezi: Recovering from Gender Affirmation Surgery, Chinese American Style," *Atmos,* June 30, 2021, atmos.earth /zuo-yuezi-gender-affirmation-surgery-chinese-american/.

273 **Sofía had long thought about** Sofía Córdova, email to author, February 12, 2022.

276 **the seed of the "non-time"** Hannah Arendt, *Between Past and Future: Eight Exercises in Political Thought* (New York: Penguin, 2006), 13.

276 **European writers and intellectuals** Arendt, *Between Past and Future,* 3–4.

277 **"In the sociality of struggle"** Mariarosa Dalla Costa and Selma James, *The Power of Women and Subversion of the Community* (London: Falling Wall Press, 1972), 37.

277 **encountering freshness and agency** Arendt, *Between Past and Future,* 14.

278 **changes in culture and civilization** Vine Deloria, Jr., *The Metaphysics of Modern Existence* (Golden, Colo.: Fulcrum, 2012), 18.

279 **"earth tides"** "Earth tide," *Encyclopedia Britannica,* britannica.com/science /Earth-tide.

279 **while the moon pulls** Marina Koren, "The Moon Is Leaving Us," *The Atlantic,* September 30, 2021, theatlantic.com/science/archive/2021/09/moon-moving -away-earth/620254/.

279 **The road had originally gone** Bruce Haulman, "Raab's Lagoon: 16,000 Years Young," *Vashon-Maury Island Beachcomber,* December 26, 2018, vashon beachcomber.com/news/raabs-lagoon-16000-years-young/.

281 **The disease was even documented** "Raab's Lagoon Beach," Multi-Agency Rocky Intertidal Network (MARINe) Sea Star Site Observation History, marinedb.ucsc .edu/seastar/observations.html?site=Raabs%20Lagoon%20Beach. The wasting syndrome was observed affecting ochre sea stars here in 2014, 2016, and 2017.

282 **the director of the island's nature** Paul Rowley, "A Hope for Sea Stars, Healthy Oceans," *Vashon-Maury Island Beachcomber,* June 19, 2019, vashonbeachcomber .com/news/a-hope-for-sea-stars-healthy-oceans/.

282 **the resistance to the syndrome** "Starfish Make Comeback After Mysterious Melting Disease," EcoWatch, June 26, 2018, ecowatch.com/starfish-population -disease-climate-resilience-2581473075.html.

282 **more outbreaks had been observed** "SSWS Updates | MARINe" (September 28, 2021 update)," Multi-Agency Rocky Intertidal Network (MARINe), marine.ucsc.edu/data-products/sea-star-wasting/updates.html#SEPT28_2021.

282 **a narrator from the future** Ted Chiang, "What's Expected of Us," *Exhalation* (New York: Knopf Doubleday, 2019), 58–61.

283 **the zoologist Robert T. Paine** Ed Yong, "The Man Whose Dynasty Changed

Ecology," *Scientific American,* January 16, 2013, scientificamerican.com/article /the-man-whose-dynasty-changed-ecology/.

283 **For a long time, scientists** Aaron W. Hunter, "Rare Starfish Fossil Answers the Mystery of How They Evolved Arms," EcoWatch, January 21, 2021, ecowatch .com/starfish-evolution-mystery-2650057909.html.

284 **the compound eyes** Laura Geggel, "Starfish Can See You . . . with Their Arm-Eyes," Live Science, February 7, 2018, livescience.com/61682-starfish-eyes.html.

284 **the Coast Salish S'Homamish** Haulman, "Raab's Lagoon: 16,000 Years Young." The author uses an alternate version of the name, "sHebabS," which is also sometimes spelled "Sqababsh."

284 **it was a time when many** Emily Sohn, "How the COVID-19 Pandemic Might Age Us," *Nature,* January 19, 2022, nature.com/articles/d41586-022-00071-0.

Bibliography

Ackerman, Jennifer. *The Bird Way: A New Look at How Birds Talk, Work, Play, Parent, and Think*. New York: Penguin, 2021.

Adair, John. *Effective Time Management*. London: Pan, 1988.

Adam, Barbara. *Timewatch: The Social Analysis of Time*. Cambridge, UK: Polity, 1995.

Alexander, Roy, and Michael S. Dobson. *Real-World Time Management*. New York: American Management Association, 2009.

Allen, Paula Gunn. "IYANI: It Goes This Way." In *The Remembered Earth: An Anthology of Contemporary Native American Literature*. Edited by Geary Hobson. Albuquerque: University of New Mexico Press, 1989.

Arendt, Hannah. *Between Past and Future: Eight Exercises in Political Thought*. New York: Penguin, 2006.

———. *The Human Condition*. Chicago: University of Chicago Press, 1998.

Aronoff, Kate. *Overheated: How Capitalism Broke the Planet—and How We Fight Back*. New York: Bold Type, 2021.

Basso, Keith H. *Wisdom Sits in Places: Landscape and Language Among the Western Apache*. Albuquerque: University of New Mexico Press, 1996.

Beecher, Catharine. *Treatise on Domestic Economy*. Boston, Mass.: Thomas H. Webb, 1843.

Bennett, Arnold. *How to Live on Twenty-four Hours a Day*. Garden City, N.Y.: Doubleday, Doran and Company, 1933.

Bergson, Henri. *Creative Evolution*. Translated by Arthur Mitchell. Lanham, Md.: University Press of America, 1983.

———. *Matter and Memory*. Translated by N. M. Paul and W. S. Palmer. Brooklyn, N.Y.: Zone, 1991.

———. *Time and Free Will: An Essay on the Immediate Data of Consciousness*. Translated by F. L. Pogson. Mineola, N.Y.: Dover Publications, 2001.

Birdsong, Mia. *How We Show Up: Reclaiming Family, Friendship, and Community*. New York: Hachette, 2020.

Birth, Kevin K. *Time Blind: Problems in Perceiving Other Temporalities*. Cham, Switzerland: Palgrave Macmillan, 2017.

Bluedorn, Allen C. *The Human Organization of Time: Temporal Realities and Experience*. Stanford, Calif.: Stanford University Press, 2002.

Braverman, Harry. *Labor and Monopoly Capital: The Degradation of Work in the Twentieth Century*. New York: Monthly Review, 1998.

Brown, John. *A Memoir of Robert Blincoe, an Orphan Boy, Sent from the Workhouse of St. Pancras, London, at Seven Years of Age, to Endure the Horrors of a Cotton Mill, Through His Infancy and Youth, with a Minute Detail of His Sufferings, Being the First Memoir of the Kind Published*. Manchester, UK: J. Doherty, 1832.

Burkeman, Oliver. *4,000 Weeks: Time Management for Mortals*. New York: Farrar, Straus and Giroux, 2021.

Cameron, Barbara. "Gee, You Don't Seem Like an Indian from the Reservation." In *This Bridge Called My Back*. 3rd ed. Edited by Gloria Anzaldúa and Cherríe Moraga. Berkeley, Calif.: Third Woman Press, 2002.

Carlsson, Chris, and Mark Leger, eds. *Bad Attitude: The Processed World Anthology*. New York: Verso, 1990.

Chen, Chen. *When I Grow Up I Want to Be a List of Further Possibilities*. Rochester, N.Y.: BOA Editions, 2017.

Chiang, Ted. *Exhalation*. New York: Knopf Doubleday, 2019.

Cixous, Hélène. *Three Steps on the Ladder of Writing*. New York: Columbia University Press, 1990.

Coates, Ta-Nehisi. *Between the World and Me*. New York: Spiegel and Grau, 2015.

Craven, Ida. "'Leisure,' According to the Encyclopedia of the Social Sciences." In *Mass Leisure*. Edited by Eric Larrabee and Rolf Meyersohn. Glencoe, Ill.: Free Press, 1958.

Dalla Costa, Mariarosa, and Selma James. *The Power of Women and Subversion of the Community*. London: Falling Wall Press, 1972.

Davis, Angela Y. *Are Prisons Obsolete?* New York: Seven Stories Press, 2011.

———. *Women, Race, and Class*. New York: Vintage, 1983.

Deloria, Vine, Jr. *The Metaphysics of Modern Existence*. Golden, Colo.: Fulcrum, 2012.

———. *Red Earth, White Lies*. Golden, Colo.: Fulcrum, 1997.

Diaz, Natalie. "The First Water Is the Body." In *New Poets of Native Nations*. Edited by Heid E. Erdrich. Minneapolis, Minn.: Graywolf Press, 2018.

Downing, Michael. *Spring Forward: The Annual Madness of Daylight Saving*. Berkeley, Calif.: Counterpoint, 2005.

Dray, Philip. *There Is Power in a Union: The Epic Story of Labor in America*. New York: Anchor, 2011.

Du Bois, W.E.B. *The Souls of Black Folk*. New York: Cosimo Classics, 2007.

Dumas, John Lee. *The Freedom Journal*. Self-published, 2016.

Ehrenreich, Barbara. *Natural Causes: An Epidemic of Wellness, the Certainty of Dying, and Killing Ourselves to Live Longer*. New York: Twelve, 2018.

Eisenberger, Robert. *Blue Monday: The Loss of the Work Ethic in America*. New York: Paragon House, 1989.

Epstein, Mark. *Thoughts Without a Thinker: Psychotherapy from a Buddhist Perspective*. New York: Basic Books, 2013.

Ernst, Robert. *Weakness Is a Crime: The Life of Bernarr Macfadden*. Syracuse, N.Y.: Syracuse University Press, 1991.

Evans, Claire L. *Broad Band: The Untold Story of the Women Who Made the Internet*. New York: Portfolio/Penguin, 2018.

Fanon, Frantz. *Black Skin, White Masks*. New York: Grove Atlantic, 2007.

Fleming, Sandford. "Time-Reckoning for the Twentieth Century." In *The Smithsonian Report for 1886*. Washington, D.C.: Smithsonian Institution Press, 1889.

Fortenbaugh, William. *Aristotle's Practical Side: On His Psychology, Ethics, Politics, and Rhetoric*. Leiden, Netherlands: Brill, 2006.

Galanter, Marc. *Cults: Faith, Healing, and Coercion*. Oxford, UK: Oxford University Press, 1999.

Galton, Francis. *Hereditary Genius: An Inquiry into Its Laws and Consequences.* New York: D. Appleton and Company, 1870.

———. *Memories of My Life.* New York: E. P. Dutton and Company, 1909.

Glass, Fred. *From Mission to Microchip: A History of the California Labor Movement.* Berkeley: University of California Press, 2016.

Glickman, Lawrence B. *A Living Wage: American Workers and the Making of Consumer Society.* Ithaca, N.Y.: Cornell University Press, 2015.

Glime, J. M. *Bryophyte Ecology.* Houghton: Michigan Technological University, 2022. digitalcommons.mtu.edu/oabooks/4/.

Greenhouse, Carol J. *A Moment's Notice: Time Politics Across Cultures.* Ithaca, N.Y.: Cornell University Press, 2018.

Guendelsberger, Emily. *On the Clock: What Low-Wage Work Did to Me and How It Drives America Insane.* New York: Little, Brown, 2019.

Haber, Samuel. *Efficiency and Uplift: Scientific Management in the Progressive Era, 1890–1920.* Chicago: University of Chicago Press, 1964.

Han, Byung-Chul. *The Burnout Society.* Stanford, Calif.: Stanford University Press, 2015.

Hanson, Chad T. *Smokescreen: Debunking Wildfire Myths to Save Our Forests and Our Climate.* Lexington: University Press of Kentucky, 2021.

Hartley, Daniel. "Anthropocene, Capitalocene, and the Problem of Culture." In *Anthropocene or Capitalocene? Nature, History, and the Crisis of Capitalism.* Edited by Jason W. Moore. Oakland, Calif.: PM Press, 2016.

Hendren, Sara. *What Can a Body Do? How We Meet the Built Environment.* New York: Riverhead, 2020.

Hockney, David. *That's the Way I See It.* San Francisco: Chronicle Books, 1993.

Holden, William N., and Shawn J. Marshall. "Climate Change and Typhoons in the Philippines: Extreme Weather Events in the Anthropocene." In *Integrating Disaster Science and Management: Global Case Studies in Mitigation and Recovery.* Edited by Pijush Samui, Dookie Kim, and Chandan Ghosh. Amsterdam, Netherlands: Elsevier, 2018.

Holland, Sharon P. *Raising the Dead: Readings of Death and (Black) Subjectivity.* Durham, N.C.: Duke University Press, 2000.

Honoré, Carl. *In Praise of Slowness: Challenging the Cult of Speed.* New York: HarperOne, 2009. Ebook.

Hough, Franklin. *Report upon Forestry.* Washington, D.C.: Government Printing Office, 1878.

Illich, Ivan. *The Right to Useful Unemployment and Its Professional Enemies.* London: Marion Boyars, 1978.

James, Selma. *Women, the Unions and Work, or . . . What Is Not to Be Done, and the Perspective of Winning.* London: Falling Wall Press, 1976.

June 1868 Travelers Official Railway Guide of the United States and Canada. New York: National Railway Publication Company, 1968.

Kafer, Alison. *Feminist, Queer, Crip.* Bloomington: Indiana University Press, 2013.

Kent, Noel J. *Hawaii: Islands Under the Influence.* Honolulu: University of Hawai'i Press, 1983.

Keynes, John Maynard. "Economic Possibilities for Our Grandchildren." In *Essays in Persuasion.* New York: W. W. Norton, 2011.

Kimmerer, Robin Wall. *Braiding Sweetgrass: Indigenous Wisdom, Scientific Knowledge, and the Teachings of Plants.* Minneapolis, Minn.: Milkweed Editions, 2013.

———. *Gathering Moss: A Natural and Cultural History of Mosses.* Corvallis: Oregon State University Press, 2003.

Klein, Naomi. *This Changes Everything: Capitalism vs. the Climate.* New York: Simon and Schuster, 2014.

Knight, George R. *A Brief History of Seventh-day Adventists.* Hagerstown, Md.: Review and Herald Association, 1999.

Kochiyama, Yuri. *Passing It On: A Memoir.* Los Angeles: UCLA Asian American Studies Center, 2004.

Kranish, Michael, and Marc Fisher. *Trump Revealed: The Definitive Biography of the 45th President.* New York: Scribner, 2017.

Krishnamurti, Jiddu. *Freedom from the Known.* New York: HarperOne, 2009.

Kruse, Kevin. *15 Secrets Successful People Know About Time Management: The Productivity Habits of 7 Billionaires, 13 Olympic Athletes, 29 Straight-A Students, and 239 Entrepreneurs.* Philadelphia: Kruse Group, 2015.

Laird, Donald. *Increasing Personal Efficiency.* New York: Harper and Brothers, 1925.

Landes, David. *Revolution in Time: Clocks and the Making of the Modern World.* Cambridge, Mass.: Belknap Press / Harvard University Press, 2000.

Lundberg, George A., Mirra Komarovsky, and Mary Alice McInerny. *Leisure: A Suburban Study.* New York: Columbia University Press, 1934.

Macdonald, Helen. *Vesper Flights.* New York: Grove, 2020.

Macrì, Mario, Stefano Bellucci, Stefano Bianco, and Andrea Sansoni, eds. *Clocking and Scientific Research: The Opinion of the Scientific Community.* Istituto Nazionale di Fisica Nucleare (INFN), November 13, 1998. openaccessrepository.it/record /21217?ln=en.

Mancini, Marc. *Time Management (The Business Skills Express Series).* New York: Business One Irwin / Mirror, 1994.

Mangez, Eric, and Mathieu Hilgers, eds. *Bourdieu's Theory of Social Fields: Concepts and Applications.* New York: Routledge, 2015.

Marx, Karl. *Capital, Volume 1.* New York: Penguin, 1990.

Mays, Wolfe. "Whitehead and the Philosophy of Time." In *The Study of Time: Proceedings of the First Conference of the International Society for the Study of Time Oberwolfach (Black Forest)—West Germany.* Edited by J. T. Fraser, F. C. Haber, and G. H. Müller. Berlin: Springer, 1972.

McPhee, John. *The Control of Nature.* New York: Farrar, Strauss and Giroux, 1989.

Meisel, Ari. *The Art of Less Doing: One Entrepreneur's Formula for a Beautiful Life.* Austin, Tex.: Lioncrest, 2016.

Melossi, Dario, and Massimo Pavarini. *The Prison and the Factory: Origins of the Penitentiary System.* Translated by Glynis Cousin. Totowa, N.J.: Barnes and Noble Books, 1981.

Miller, Doug, Jennifer Bair, and Marsha Dickson, eds. *Rights and Labor Compliance in Global Supply Chains.* New York: Routledge, 2014.

Moten, Fred, and Stefano Harney. *The Undercommons: Fugitive Planning and Black Study.* New York: Minor Compositions, 2013.

Mould, Oli. *Against Creativity.* London: Verso, 2018.

Mueller, Gavin. *Breaking Things at Work: The Luddites Were Right About Why You Hate Your Job.* London: Verso, 2021.

Mumford, Lewis. *Technics and Civilization*. London: G. Routledge, 1934.

Nanni, Giordano. *The Colonisation of Time: Ritual, Routine, and Resistance in the British Empire*. Manchester, UK: Manchester University Press, 2012.

National Recreation Association. *The Leisure Hours of 5,000 People: A Report of a Study of Leisure Time Activities and Desires*. New York: National Recreation Association, 1934.

Nunn, Patrick. *The Edge of Memory: Ancient Stories, Oral Tradition, and the Post-Glacial World*. New York: Bloomsbury, 2018.

Odell, Jenny. *How to Do Nothing: Resisting the Attention Economy*. Brooklyn, N.Y.: Melville House, 2019.

O'Malley, Michael. *Keeping Watch: A History of American Time*. New York: Viking, 1990.

Perec, Georges. *An Attempt at Exhausting a Place in Paris*. Translated by Marc Lowenthal. Cambridge, Mass.: Wakefield Press, 2010.

———. *La Disparition*. Paris: Éditions Denoël, 1969.

———. *Species of Spaces and Other Pieces*. Translated by John Sturrock. New York: Penguin Classics, 2008.

Peters, John Durham. *The Marvelous Clouds: Toward a Philosophy of Elemental Media*. Chicago: University of Chicago Press, 2015.

Peters, Peter Frank. *Time, Innovation and Mobilities: Travel in Technological Cultures*. London: Routledge, 2005.

Pieper, Josef. *Leisure, the Basis of Culture*. San Francisco: Ignatius, 2015.

Price, Joshua M. *Prison and Social Death*. New Brunswick, N.J.: Rutgers University Press, 2015.

Pyne, Stephen. *Fire in America: A Cultural History of Wildland and Rural Fire*. Seattle: University of Washington Press, 1997.

Relax. The Jam Handy Organization and the Chevrolet Motor Division, 1937.

Roberts, Justin. *Slavery and the Enlightenment in the British Atlantic, 1750–1807*. Cambridge, UK: Cambridge University Press, 2013.

Robinson, Jackie. *I Never Had It Made: An Autobiography*. New York: HarperCollins, 2013.

Roediger, David R., and Philip S. Foner. *Our Own Time: A History of American Labor and the Working Day*. London: Verso, 1989.

Rojek, Chris. *The Labour of Leisure: The Culture of Free Time*. London: SAGE, 2010.

Rosa, Hartmut. "De-Synchronization, Dynamic Stabilization, Dispositional Squeeze." In *The Sociology of Speed: Digital, Organizational, and Social Temporalities*. Edited by Judy Wajcman and Nigel Dodd. Oxford, UK: Oxford University Press, 2016.

———. *Resonance: A Sociology of Our Relationship to the World*. Translated by James C. Wagner. Cambridge, UK: Polity, 2019.

Rosenthal, Caitlin. *Accounting for Slavery: Masters and Management*. Cambridge, Mass.: Harvard University Press, 2018. Ebook.

Ross, Frederick A. *Slavery Ordained of God*. Philadelphia: J. B. Lippincott, 1857.

Rushkoff, Douglas. *Survival of the Richest: Escape Fantasies of the Tech Billionaires*. New York: W. W. Norton, 2022.

Samuels, Ellen. "Six Ways of Looking at Crip Time." In *Disability Visibility: First-Person Stories from the Twenty-first Century*. Edited by Alice Wong. New York: Vintage, 2020.

Schroeder, Juliana, Adam Waytz, and Nicholas Epley. "The Lesser Minds Problem." In *Humanness and Dehumanization*. Edited by Paul G. Bain, Jeroen Vaes, and Jacques-Philippe Leyens. New York: Psychology Press, 2013.

Scott, James C. *Seeing Like a State*. New Haven, Conn.: Yale University Press, 2020.

Semple, Janet. *Bentham's Prison: A Study of the Panopticon Penitentiary*. Oxford, UK: Clarendon, 1993.

Sennett, Richard. *The Corrosion of Character: The Personal Consequences of Work in the New Capitalism*. New York: W. W. Norton, 1998.

Seymour, Richard. *The Twittering Machine*. London: Verso, 2020.

Simon, Jonathan. *Governing Through Crime: How the War on Crime Transformed American Democracy and Created a Culture of Fear*. Oxford, UK: Oxford University Press, 2007.

Sloan, Doris. *Geology of the San Francisco Bay Region*. Berkeley: University of California Press, 2006.

Smith, Mark M. *Mastered by the Clock: Time, Slavery, and Freedom in the American South*. Chapel Hill: University of North Carolina Press, 2000. Ebook.

Sontag, Susan. *On Photography*. New York: Farrar, Straus and Giroux, 2011.

Spence, Mark David. *Dispossessing the Wilderness: Indian Removal and the Making of the National Parks*. Oxford, UK: Oxford University Press, 2000.

Steffensen, Victor. *Fire Country*. Richmond, Victoria: Hardie Grant Explore, 2020.

Takaki, Ronald T. *Pau Hana: Plantation Life and Labor in Hawaii, 1835–1920*. Honolulu: University of Hawai'i Press, 1984.

Taylor, Charles. *Sources of the Self: The Making of the Modern Identity*. Cambridge, Mass.: Harvard University Press, 1989.

Taylor, Frederick Winslow. *Principles of Scientific Management*. Norwood, Mass.: Plimpton, 1911.

Troutman, John W. *Indian Blues: American Indians and the Politics of Music, 1879–1934*. Norman: University of Oklahoma Press, 2009.

Vitale, Alex. *The End of Policing*. London: Verso, 2017.

Wang, Jackie. *Carceral Capitalism*. South Pasadena, Calif.: Semiotext(e), 2018.

Weeks, Kathi. *The Problem with Work: Feminism, Marxism, Antiwork Politics, and Postwork Imaginaries*. Durham, N.C.: Duke University Press, 2011.

Williams, Patricia. *The Alchemy of Race and Rights: Diary of a Law Professor*. Cambridge, Mass.: Harvard University Press, 1991.

Wolcott, Victoria W. *Race, Riots, and Roller Coasters: The Struggle over Segregated Recreation in America*. Philadelphia: University of Pennsylvania Press, 2012.

Woodfox, Albert. *Solitary*. New York: Grove Atlantic, 2019.

Wrenn, Gilbert, and D. L. Harvey. *Time on Their Hands: A Report on Leisure, Recreation, and Young People*. Washington, D.C.: American Council on Education, 1941.

XPO Global Union Family. *XPO: Delivering Injustice*. February 2021. xpoexposed.org /the-report.

Yunkaporta, Tyson. *Sand Talk: How Indigenous Thinking Can Save the World*. New York: HarperCollins, 2020. Ebook.

Yusoff, Kathryn. *A Billion Black Anthropocenes or None*. Minneapolis, Minn.: University of Minnesota Press, 2019.

Zieger, Robert H. *For Jobs and Freedom: Race and Labor in America Since 1865*. Lexington, Ky.: University Press of Kentucky, 2014.

Index

Page numbers of illustrations appear in italics.

Credits and Permissions

PHOTO CREDIT INFORMATION

Page 194: ("Oh, You Unfit!") *Physical Culture,* October 1918

Page 207: (*Processed World* cover) Illustration by Tom Tomorrow; courtesy of Chris Carlsson

Page 209: (eight-hour graphic) Courtesy of Chris Carlsson

Page 230: (death calculator) Courtesy of DH3 Games

Page 245: (stills from *Time*) Courtesy of Garrett Bradley

Page 263: (wave in Pacifica) Courtesy of Alan Grinberg

Page 267: (schist outcrop) Courtesy of Caroline Eisenmann

Page 268: (outline of Pando) Lance Oditt, shared under a Creative Commons Attribution-Share Alike 4.0 International license. creativecommons.org/licenses /by/4.0/

Page 275: (stills from *Underwater Moonlight*) Courtesy of Sofía Córdova

All further images are courtesy of Jenny Odell.